GORILLA SOCIETY

GORILLA SOCIETY

Conflict, Compromise, and Cooperation Between the Sexes

ALEXANDER H. HARCOURT
and
KELLY J. STEWART

The University of Chicago Press
Chicago and London

Alexander Harcourt is professor of anthropology at the University of California at Davis and a faculty member of the University's Graduate Group in Ecology.
Department of Anthropology
University of California
1 Shields Avenue
Davis, CA 95616
USA
ahharcourt@ucdavis.edu

Kelly Stewart is a research associate in the Department of Anthropology at the University of California at Davis.
Department of Anthropology
University of California
1 Shields Avenue
Davis, CA 95616
USA
kjstewart@ucdavis.edu

The University of Chicago Press, Chicago 60637
The University of Chicago Press, Ltd., London

Printed in the United States of America

16 15 14 13 12 11 10 09 08 07 1 2 3 4 5

ISBN-13: 978-0-226-31602-4 (cloth)
ISBN-13: 978-0-226-31603-1 (paper)
ISBN-10: 0-226-31602-5 (cloth)
ISBN-10: 0-226-31603-3 (paper)

Library of Congress Cataloging-in-Publication Data

Harcourt, A. H. (Alexander H.)
Gorilla society : conflict, compromise, and cooperation between the sexes / Alexander H. Harcourt and Kelly J. Stewart.
p. cm.
Includes bibliographical references and index.
ISBN-13: 978-0-226-31602-4 (cloth : alk. paper)
ISBN-10: 0-226-31602-5 (cloth : alk. paper)
ISBN-13: 978-0-226-31603-1 (pbk. : alk. paper)
ISBN-10: 0-226-31603-3 (pbk. : alk. paper)
1. Gorilla—Behavior. 2. Gorilla—Ecology. 3. Social behavior in animals.
I. Stewart, Kelly J., 1951– II. Title.
QL737.P96H364 2007
599.884—dc22

2006037081

♾ The paper used in this publication meets the minimum requirements of the American National Standard for Information Sciences—Permanence of Paper for Printed Library Materials, ANSI Z39.48-1992.

To our parents and our teachers

CONTENTS

PART 2. GORILLAS, ECOLOGY, AND SOCIETY

ACKNOWLEDGMENTS

We are deeply grateful to our parents and our teachers for the interest in science and natural history that they fostered in us, and for the training they provided us that gives any worth to this book. Our parents are Ralph and Anne Harcourt and James and Gloria Stewart. We cannot thank them enough. Our teachers? They are too many to list, but Dian Fossey, at whose Karisoke Research Center (KRC) in Rwanda we started our work on gorillas, is obviously at the top of our list. Also at the top is Robert Hinde, with whom as Ph.D. students we started our professional careers in gorillaology, primate behavior, behavioral ecology, and socioecology. But all of our teachers are listed here in spirit, if not in name.

The KRC site used to lie (it was destroyed in the Rwandan chaos of the 1990s) in the Parc National des Volcans of Rwanda, and its study area included also the Parc National de Virunga-Sud of what was Zaire when we worked there, and is now the Democratic Republic of the Congo. We thank the governments, particularly the National Park Offices, of Rwanda and Zaire/Dem. Rep. Congo for their support and for authorization for our work, indeed for the support that continues to allow and encourage fieldwork in the area.

Many thanks to the KRC field assistants with whom we worked, particularly Alphonse Nemeye, now in prison convicted of a crime we are convinced he did not commit; and Emmanuel Rwelekana, killed during interrogation about a crime we know he did not commit; also André Vatiri, Kana Munyanganga, and Antoine Banyangandora, all killed during Rwanda's civil war in the 1990s. The full roster of all the Rwandan assistants who have worked at the KRC appears in the Acknowledgments of Martha Robbins et al., *Mountain Gorillas: Three decades of research at Karisoke.* We thank them all.

This book is far from only our own work, attempting as it does to describe and explain all gorilla society. We thank the many people on whose studies we have built, and whose ideas and data we have used, especially, of course, all the gorillaologists who have come before and after our time in the field. In a very strong sense, this book is written by all students of gorillas, perhaps especially those whose hard-earned data we have reworked. Martha Robbins and David Watts must be singled out as two of the main contributors to the gorilla literature.

Our thanks also to Peter Rodman, without whose invitations to lecture on gorilla society in his primate sociobiology course, we probably would never have written this book, and without whom we would not be at the comfortable University of California.

And finally, our gratitude to the friends and colleagues who were kind enough to take the time and energy to comment so fully on various chapters and parts of chapters: Tim Caro, Guy Cowlishaw, Tony Di Fiore, Diane Doran, Robin Dunbar, Colin Groves, Lynne Isbell, Peter Kappeler, Anne Pusey, Liz Rogers, Pascale Sicotte, Liesbeth Sterck, Juichi Yamagiwa, an anonymous reviewer, and especially Martha Robbins, who read the whole book. Their critiques considerably improved the result.

Alexander Harcourt, Kelly Stewart

Dept. Anthropology, Graduate Group in Ecology
University of California,
One Shields Ave., Davis, CA 95616, USA

ahharcourt@ucdavis.edu, kjstewart@ucdavis.edu

PART 1

INTRODUCTION

A gorilla group during its midday rest period. (Mountain gorillas.) © A. H. Harcourt

CHAPTER 1

Introduction

Summary

Socioecology is the study of how individuals' evolved survival, mating, and rearing strategies interact with the physical and social environments to produce the sort of society that we see. Primates are useful subjects of study in socioecology because they are so relatively well known, and their societies are so varied. Primates are also interesting because in an unusually large proportion of species for a mammal, the sexes live together even outside any breeding season. Gorillas are an interesting primate because being the largest bodied, they are an extreme, and extremes are always a good means of testing generalities.

The Book's Aim

Our main aim is to provide an easy introduction to modern thinking about socioecology in general, primate socioecology in particular, by presenting the equivalent of a worked example of how socioecologists attempt to explain the nature of an animal's society. We also hope, of course, to provide a cohesive explanation of gorilla society for those perhaps more interested in gorillas than in primate socioecology. And if we make readers sufficiently more interested in gorillas to do a bit more than they do at present to help conserve these animals, or any other of the world's threatened species, we will have achieved another aim.

1.1. What Is a Society? What Is Socioecology?

What is this book about? By "society" we mean attributes of animals such as the size and composition of the groups that animals live in, the degree of stability of groups, the movement of individuals in and out of groups, the competitive and cooperative nature of social relationships within groups, and the nature of interactions between groups. In other contexts, culture would be included, but this is the first and probably the last mention of the word in this book.

A useful description of "socioecology" as a scientific discipline is that it is an attempt to understand why species have the sort of society that they do, and why species differ in the nature of their society. Some of the major questions of primate socioecology, and hence of this book, have been nicely summarized elsewhere (e.g., Dunbar 1988; Isbell and Young 2002). With some paraphrasing, they include: Why do primates group? What determines the numbers of males and females in groups? What determines the nature of competition and cooperation among males and females within groups? What determines relationships between groups? Why is infanticide such a common male mating strategy in some species or populations, but not in others? What determines the main dispersing sex, and differences between species in the predominant dispersing sex?

Socioecology frames the questions and answers in terms of how individuals' evolved survival, mating, and rearing strategies interact with the physical and social environments to produce the sort of society that we see. Ecology is in the word *socioecology,* because of the well-verified assumption that the nature of the environment in which animals live interacts with the nature of the animals to influence the nature of the society. Take the question of why animals live in groups. Do they live in groups as a means of escaping predators? If so, how does the nature and distribution of food influence the inevitable competition that results from being close to other animals? How does it influence the opportunity for cooperation between individuals as they compete for food? How does whether the other group members are kin or not influence the outcome of the competition and cooperation, and hence the composition of the groups?

Answers to why and how questions can be given at many different levels (Alcock 2001; Hinde 1982; Tinbergen 1963). We could ask questions about the nature of society in relation to individuals' upbringing, their physiology, or their reactions to immediate events. If we did, we would not really be asking as a socioecologist. That is absolutely not to say that an understanding of physiology cannot enlighten socioecological understanding. It certainly can. Rather, it is not the normal level of questioning of socioecology. Socioecological explanations tend instead to be in terms of the influence of individuals' abilities to survive, mate, and rear offspring on the nature of society, and the influence of society on those abilities.

This level of explanation is often described as "evolutionary." That is very confusing, because the evolutionary history of a species is itself another valid level of explanation for the species' behavior or society. Another term used is the "ultimate" level, which is not self-explanatory at all. Indeed, it almost sounds teleological. The "functional" level gets closer to what is meant, because in biology a trait's function is often understood by the difference the trait makes to survival or mating or rearing. However, "function" also has mechanistic connotations or meanings. So, we will usually talk in terms of "payoffs" in this book: the payoff is the consequence that the trait has for an individual's ability to survive, or to mate, or to rear healthy offspring.

In talking about the nature of societies, some people distinguish between "social organization" and "social structure" (Kappeler and van Schaik 2002). They apply "organization" to characteristics of the population or species, for example, average size and composition of social groups. And by "structure" they mean the nature of the competitive and cooperative relationships between individuals within groups, for example, whether females are despotic or tolerant of one another. The conceptual distinction can be useful (even if the words *organization* and *structure* are effectively indistinguishable in plain English). However, we do not apply it here. Our *nature of society, structure of society, social system, society,* or whatever other synonymous terms we use, encompass both meanings, because a society is not described or understood unless both levels are described and understood (Hinde 1976).

The difference between socioecology and the perhaps more famous field of behavioral ecology is sometimes difficult to discern, because behavioral ecology also concerns itself with the survival, mating, and rearing strategies of individuals, and heavily concerns itself with understanding competition and cooperation. The difference is that whereas socioecology is the attempt to understand the nature of societies, behavioral ecology is the attempt to understand the nature of individuals. Behavioral ecologists ask why an individual behaves in the way it does, why it behaves in one way rather than another in a given environment, why it makes the apparent decisions that it does every instant of its life. Socioecologists use behavioral ecological understanding of the survival, mating, and rearing strategies of individuals to ask how those strategies interact to produce a species' society.

In socioecology, the aim is to build from individuals to society. In other words, given the nature of individuals, given the survival, mating, and rearing strategies of individuals (i.e., given what behavioral ecology tells us), how do the individuals and their strategies interact in different environments to produce different sorts of societies? In this book, therefore, we in effect accept behavioral ecology's understanding of why individuals behave as they do and build from there. So, we do not discuss sexual selection, for instance (hypotheses about competition for mates and choice of mates), because sexual selection theory more concerns understanding of individuals' behavior than understanding of the structure of society. While we mostly build from individual strategies to the nature of society, of course the nature and structure of a society, in turn, affect the survival, mating, and rearing strategies of individuals (Hinde 1976). This sort of two-way influence is very much part of our approach of interpreting society in terms of the interaction of strategies with the social environment.

And what of sociobiology, as in Edward Wilson's monumental classic *Sociobiology* (Wilson 1975)? Sociobiology as a concept usefully combines both socioecology and behavioral ecology. It asks questions both about the payoffs to individuals of the survival, mating, and rearing tactics that they use, and it asks questions about how the interaction of individuals' tactics and strategies affect the nature of the society in which the animals live, whether the animals be ants, apes, or Americans. "Sociobiology" almost immediately became a dirty word, because otherwise sensible biologists thought that it implied genetic determinism of humans' behavior (Allen et al. 1976). It meant nothing of the sort, of course (Wilson 1976). But the taint lives on, and it is now convenient to separate, as Wilson did, even if not so explicitly, the behavioral ecology of individuals and the socioecology of societies.

1.2. Why Primate Socioecology?

1.2.1. The tropics are little known

Primates live mostly in the tropics. A reason to study primate socioecology is simply that the tropics are far less well known than are temperate regions (Gaston and Blackburn 1999; Harcourt 1998a; Harcourt 2000; Janzen 1986). Study in the tropics can therefore potentially make more of a difference to knowledge and understanding of biology in general, socioecology in particular, than can yet more study of a temperate taxon.

With most of the world's biodiversity in the tropics, the ignorance of tropical biology cannot be too strongly stressed. For instance, three-quarters of studies on one of the main patterns in biogeography, the so-called distribution-abundance relationship, come from North America and Europe (Gaston and Blackburn 1999). Almost nothing is known about tropical terrestrial organisms' response to global warming, whereas shifts in time of first breeding of just a few days have been detected in temperate areas (Parmesan and Yohe 2003; Root et al. 2003; Walther et al. 2002). It is not that tropical organisms do not shift. They are simply not studied: in a review of 130 studies of terrestrial organisms' response to climate change, just 13 of those studies were from a tropical continent, and one, only one, was actually from within the tropics (Root et al. 2003). For lack of data, we do not even know for most tropical mammalian species whether they are rare or common (Yu and Dobson 2000). In the case of primates, the lack is especially obvious for the rarer species (fig. 1.1).

We will end this distressing list of tropical ignorance with an example of particular relevance to the gorilla. A tree species endemic to Gabon, *Cola lizae,* might require the gorilla to disperse its seeds (Tutin et al. 1991a). The

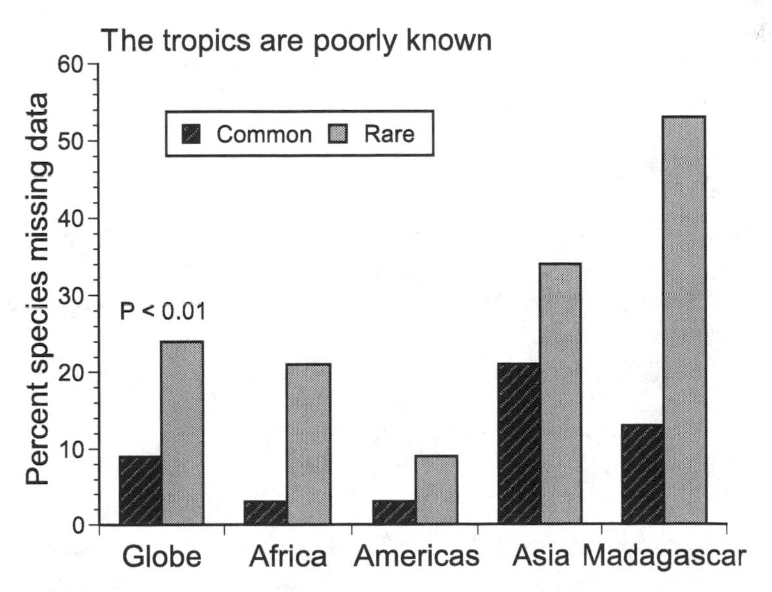

Fig. 1.1. Percent of primate species with geographic ranges of more than the median size (common) and less than the median size (rare) that lack published data on density, that is, whose numbers are unknown. *Details at end of chapter.* Source: Data from the literature, details in Coppeto and Harcourt (2005).

tree species was discovered by science only in the mid-1980s (Hallé 1987) and named after its discoverer, Liz Williamson, a Ph.D. student at the time, working on gorilla ecology. This new species, far from being a scattering of a few rare individuals in deepest, darkest African forest, happens to be the most common species in central Gabon (Tutin et al. 1991a).

The problem, of course, is that too few trained scientists study in the tropics. Norman Myers (1984, ch. 3, p. 65) pointed out this lack with a typically stunning statistic. While Colombia hosts perhaps 50,000 plant species and the United Kingdom about 1,400, Colombia employed in the 1980s no more than a few dozen trained botanists, whereas Britain employed more than one for every one of its 1,400 or so plant species!

1.2.2. Primates are relatively well known

There is, happily, a major exception to the ignorance of the tropics. Primates. Little as we know about some of them (fig. 1.1), primates are nevertheless an extraordinarily well-known tropical mammalian taxon (Campbell et al. 2007a; Nowak 1999; Rowe 1996), and an especially well-known tropical forest mammalian taxon. A main reason that primates are so well known is that they are, for the most part, very visible. Walk through a tropical forest—primates are mostly tropical forest animals—and it is highly likely that if you encounter any mammals, it will be a primate that you see or hear. Not only are many primates diurnal, but the diurnal primates are reasonably large, 3 kg and up. Also, they are arboreal, quite noisy at times, and many are gorgeously colored (see Rowe 1996).

We have had the good fortune to visit Ranomafana National Park in Madagascar. In one day we saw six primate species. The only other wild mammal that we saw was the ring-tailed mongoose—raiding the research center's kitchen. That's a slightly unfair contrast, because Madagascar is so lacking in non-primate mammals that over a third of its mammals are primates, compared to less than a quarter in the other three tropical continents (Emmons 1999). But you will find the same contrast in Yasuni National Park in Ecuador, Gunung Leuser National Park in Sumatra, and Bwindi Impenetrable National Park in Uganda.

The detail of primatology's data is remarkable (Rodman 1999; Janson and Terborgh 1986). Primatologists have data on the content of each mouthful of food (Altmann 1998; Clutton-Brock 1977), the duration of each bout of nursing (Stewart 1988), the frequency with which individual members of social groups threaten and attack other group members and the resultant food and energy intake (Janson 1985; Vogel 2005), and they have for a few species long-term demographic and life history data, sometimes over decades (e.g., Boesch and Boesch-Achermann 2000; Chapman, Struhsaker, and Lambert 2005; Goodall 1986; Gould, Sussman, and Sauther 2003; Robbins, Sicotte, and Stewart 2001; Rudran and Fernandez-Duque 2003; Takahata et al. 1999). If your interest is tropical mammalian biology, the *Primates* is your Order.

The other reason for this breadth and detail of information on primates, besides the fact that primates are easily studied, is that humans are primates, and we have intensively studied our closest relatives. Not only are humans primates, but three of the non-human primates, the two species of chimpanzees and the gorilla, are more closely related to humans than they are to their next closest primate relative, the orangutan (Goodman et al. 1998; Purvis 1995; Ruvolo et al. 1994; Wildman et al. 2003).

1.2.3. Primates are diverse

Primates are a relatively small Order—about 300 species. Nevertheless, the Order shows great diversity of body size, geography, habitat, and nature of society (Cowlishaw and Dunbar 2000, ch. 2–4; Kappeler and van Schaik 2002; Campbell et al. 2007; Smuts et al. 1987). For instance, the smallest primates, mouse lemurs, can sit comfortably in the palm of a human hand, as can the pygmy marmoset. The smaller mouse lemur species weigh less than 50 g (less than 2 oz) (Rowe 1996). Some insects weigh more than that. The larva of the goliath beetle *Goliathus goliatus* can weigh 70 g, and the adult male can weigh over 100 g (Campbell 2004). By contrast (to the mouse lemur), all adult great apes are larger than 30 kg, and the largest of the lot, the adult male gorilla, is around 160 kg, over 3,000 times larger than the mouse lemur.

Primates are terrestrial and arboreal. They live in desert and thick forest. They survive on lichen in the snow of the Himalayas and on lichen in the hot humid forests of central Africa. And in regard to primate societies, some are solitary, some are pair-living, some live in groups of scores, even hundreds. In some species, fathers do most of the carrying of infants, while in others, fathers never see their infants. And so on.

The Order Primates, therefore, provides variety that both stimulates the production of socioecological ideas and allows testing of them. It is worth adding that because primates have evolved more or less separately in four continents, we have the equivalent of four separate evolutionary "experiments." Explanations for the results of one experiment can be tested with the results of another—and the explanations are sometimes found wanting.

For instance, the males of polygynous lemurs should be larger than the females, and yet they are not (Kappeler and Ganzhorn 1993).

1.2.4. Primate males and females live together

With specific reference to socioecology, a main interest of primates is that so many of them live in (a) stable groups that (b) contain both sexes. Stable, bisexual grouping is quite rare for mammals. Most mammals do not live in stable groups, and if they do, the sexes usually live apart (Wilson 1975, ch. 23). In less than 15% of mammalian Orders and in only around 30% of the better known and more socially diverse mammalian Orders do we see stable groups of both sexes, whereas perhaps 75% of primate species so live (van Schaik 2000a; van Schaik and Kappeler 1997).

1.2.5. The study of primates

As we primatologists know, or reckon we know, so much about our Order, we also sometimes consider that non-primatologists pay too little attention to our work. Stephen Emlen and Lewis Oring's classic paper in *Science,* setting the "females to food, males to females" paradigm in vertebrate socioecology (details in ch. 2.1.2), does not once refer to a study of primates among the sixty or so references with named taxa (Emlen and Oring 1977). This omission is despite the fact that Wilson, someone who specializes on ants, had a long chapter on primate socioecology in his book written two years previously (Wilson 1975). Steve Emlen has continued to pretty much ignore primates. In another nice synthesis, this time on family structure, he makes almost no reference to primates (Emlen 1995), even though primatologists could probably validly claim that no other taxon-based group of scientists has as much information on family structure as do primatologists.

We must quickly add that primatologists, including us, are equally at fault. Primatologists surely pay too little attention to the rest of vertebrate socioecology and, most sadly, almost entirely ignore human socioecology, which, in turn, almost entirely ignores us (Harcourt 1998b; Hauser 1993). This book is not an exception, we are sorry to say. Our excuse is summarized by the Gary Larson cartoon of a boy in a classroom with his hand in the air saying to the teacher, please sir, my brain is full. The excuse is surely general: it is difficult enough to keep up with the literature on one's own taxon, let alone the literature on another taxon.

But why study non-human primates at all, given that we are members of an anthropology department, that is, a department devoted to the study of humans? The answer is Rudyard Kipling's "And what should they know of England who only England know?" (Kipling 1892). How can we know ourselves if we study only ourselves? We cannot. We cannot know the basic biological rules that apply to all beings if we study only humans (Foley 1987; Rodman 1999). And, of course, we cannot know what is uniquely human if we study only humans.

1.3. Why Gorilla Socioecology?

We are asking two questions when we ask "Why gorilla socioecology?" Why present an understanding of only one species (see ch. 3.1.2 for a discussion of the number of gorilla species)? And why is that species the gorilla?

Why only one species? Why is this not a book about primate socioecology in general? After all, we have been stressing the variety that we see in the Order Primates. What can one species tell us about such variety?

First, variety within species is as important for understanding socioecological principles as variety among species. The great British ornithologist, behavioral ecologist, and socioecologist David Lack suggested that comparison of closely related taxa was particularly valuable, because the more similar the compared taxa are in all other respects, the more chance the biologist has to detect the reason for any differences that exist.

Second, we are working on the principle that an example is worth a thousand words of general explanation. We hope that by concentrating on just one species, the concepts will be easier to grasp than if we surveyed the whole of primate socioecology.

Third, we hope that concentration on one species will help to highlight precisely where gaps still exist in socioecology. Our thinking here is that with just one species to be explained, it is far more difficult than with a multi-species comparison to pick and choose the best examples, and let slide those examples that don't quite fit. Everything that we say about gorilla males has to fit with what we say about the females. Everything we say about the females has to fit with what we say about the males. The whole story has to fit together. If it does not, the gaps will be easy to see.

Fourth, we hope that the deep look at one species will provide a certain sense of satisfaction among readers that by the end of the book they will have come to know one species of animal extremely well—and maybe even come to appreciate it enough that they will work for its conservation.

And finally, we are responding to the expressed wish of students that the story of primate socioecology as taught in our courses did not jump around the species so much, a wish that presumably had behind it one or more of the reasons that we have already listed for examination of a single species.

Why the gorilla? A reason to choose the gorilla to exemplify a field is that it is an extreme, even an exception—and extremes and exceptions are always good for testing the generality of hypotheses (Harcourt 2001). "It's the exception that tests the rule" is not an empty saying. Extremes call into question the generality of hypotheses, and hence test and stretch them. As an example, take the very simplest trait to measure, body size, a trait that affects or correlates with seemingly most aspects of an organism's life (Calder 1984; Clutton-Brock and Harvey 1983; Jungers 1985; LaBarbera 1989; McMahon and Bonner 1983; Owen-Smith 1988; Peters 1983). The gorilla is over twice as large as its next nearest-sized relatives (chimpanzees and orangutan), which themselves are over twice as large as the next largest primates (fig. 1.2). Indeed, the gorilla is a very large mammal, over 700 times the 0.1 kg mass of the median terrestrial mammalian species (Blackburn and Gaston 1998).

Even though the gorilla's society is exceptional in several ways, it is not unique. Plains zebras and some horse populations have in many respects the same social system as do gorillas (Rubenstein 1986; Watts 2000a). And

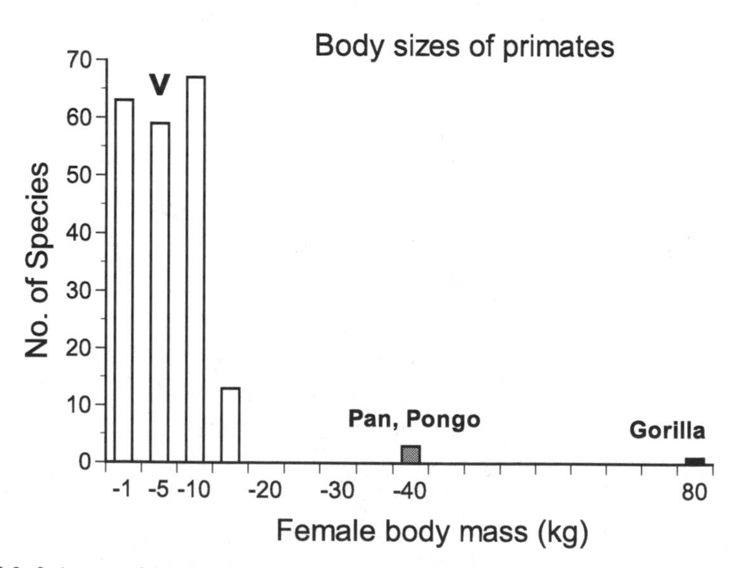

Fig. 1.2. Body sizes of female primate species. Median primate species' mass is 3.1 kg (**V**). *Pan* and *Pongo* females (chimpanzee and orangutan) are over ten times the median mass, and over twice the next heaviest primate species. *Gorilla* females, at 80 kg, are 25 times the median, and over twice *Pan* or *Pongo*'s mass.

so does the Thomas langur, an Asian arboreal colobine of only 7 kg or so (Steenbeek et al. 2000; Sterck and Steenbeek 1997), and maybe other arboreal Asian colobines too, all of roughly the same size (Sterck 1999). So also do the harem groups within the bands of the hamadryas baboon, a 10 kg desert-living primate (Kummer 1968; Sigg et al. 1982; Swedell 2002; Watts 2000a). If socioecologists can work out why such apparently disparate species should have roughly similar societies in many respects, we will have gone a long way toward understanding socioecological theory.

Structure of the Book

In animals in general, whether spiders or spider monkeys, females and males follow different main strategies to survive, mate, and rear their offspring (ch. 2). Think of the male spider that has evolved to be eaten by the female after mating, or even during mating (Andrade 1996). Glance at adult male and female gorillas (male 160 kg, silver backed; female 75 kg, black all over), and you can see that they approach life in different ways. That idea that the sexes follow different strategies is also very much part of common perceptions of humans.

The different survival, mating, and rearing strategies of the sexes underpin much current thinking about socioecology, as we will explain. So, we have ordered the book around those separate strategies. Within the chapters, we occasionally expand on certain topics in text boxes.

We open with a brief account of some of the main relevant themes in current socioecological theory, that is, the framework on which our analysis of gorilla society is based, and which we hope our analysis will illuminate (ch. 2).

Then we provide background information on the gorilla. We present current information on its general biology (ch. 3), its use of the habitat (its "ecology") (ch. 4.1), and the nature of its society (ch. 4.2). Here we essentially update the gorillaologist's bible, George Schaller's (1963) *The Mountain Gorilla,* in our detailed description of what the gorilla does, with some consideration of why it does it. But the bulk of the "why" as it relates to the nature of the gorilla's society is what the rest of the book is about.

We start with female strategies and their influence on the structure of the gorilla's society (ch. 5), concentrating on the roles of competition and cooperation in relation to grouping. We move on to briefly consider the strong influence of the gorilla male on competition and cooperation among females (ch. 6) and then discuss how the females might benefit from association with a male for protection from predation and infanticide (ch. 7).

We close the discussion of female strategies by considering how males might influence females' decisions to emigrate, given the free exercise of female choice of mates (ch. 8).

With male strategies, we start with the influence of the environment and females on the male strategies (ch. 10) and then discuss male mating strategies and how they influence the nature of gorilla society (ch. 11).

The order of these chapters has no connection with the relative importance of females, males, or the interaction of their strategies. It will be seen that the strategies of the two sexes and their influence on one another are so intertwined that we could have presented the argument in almost any order.

The remaining chapters are summaries of our arguments concerning the socioecology of the two sexes' strategies (ch. 9, 12); suggestions for future research in gorilla and primate socioecology (ch. 13); and finally, the future of gorillas as a species, the why, what, how much, where, and how of their conservation, and the relevance of socioecology to conservation (ch. 14).

No species or society can be understood without comparison to other species and societies. That, after all, is why primatologists are so often housed in anthropology departments. We therefore end each chapter (ch. 3–8, 10–11) with a comparison of gorillas to the other great apes, the two chimpanzee species (bonobo and chimpanzee, *Pan*) and the orangutan, *Pongo.* We do so in part because they are close relatives of the gorilla, but largely because, despite being relatives, their use of the environment and hence their societies are so different that the contrasts with the gorilla help test and confirm the socioecological explanations we give for the gorilla (see Wrangham 1979).

Anyone reading the book from beginning to end will experience a fair bit of repetition. That is deliberate. We assume that most will not read from start to finish, but will instead dip in. That being the case, and with each sex's strategy so dependent on the other's, we wanted to ensure easy comprehension by easy accessibility to summaries of the rest of our arguments.

To help readability, all statistics and other data-heavy substantiation of arguments are put separately at the end of each chapter (and so are also easily skipped). All new statistical tests are two-tailed unless otherwise stated and performed using Siegel, Statview SE+, or JMP 5.0 (Abacus Concepts 1990–91; SAS Institute Inc. 2002; Siegel 1956).

Figure Details

Fig. 1.1. Tropical species understudied. N = Globe: 212 species for which geographic range size known, 10 common/25 rare species for which density unknown; Africa: 58, 1/6; Americas: 65, 1/3; Asia: 57, 6/10; Madagascar: 30, 2/8. Globe, contingency test, $G = 7.9, p < 0.01$.

Fig. 1.2. The main divisions of the Order Primates, the infra-orders, differ greatly in body mass. Strepsirrhines (lemurs, bushbabies): median = 0.75 kg (range, 0.03–7.2 kg); Platyrrhines (New World monkeys): 0.9 kg (0.12–9.3); Catarrhines (Old World monkeys and apes): 6.2 kg (1.1–14.5, excluding great apes); *Pan, Pongo* females: 35 kg; *Gorilla* females: 80 kg. (Smith and Jungers [1997] is an excellent compendium of primate body masses.)

A large-bodied folivore feeding on abundant, evenly distributed vegetation. (Western gorilla.)

CHAPTER 2

Primate Socioecology: A Brief Introduction

Summary

For the sake of brevity, we present this summary of our understanding of the current state of vertebrate socioecology as if little is left for debate. Of course, much is debatable and debated.

1. Primates have a great variety of societies. Socioecological theory ties that variety to the contrasting survival, mating, and rearing strategies of the two sexes. It is access to resources, not mates, that usually separates the successful female mammal from the unsuccessful one. Females are therefore distributed according to the distribution of their main food, which differs according to the body size of the species. By contrast, it is access to mates that separates the successful male mammal from the unsuccessful one. Males are therefore distributed according to the distribution of females.
2. Different distributions of food and different foods correlate with different forms and intensities of competition, different costs and benefits of grouping, and different payoffs of using cooperation with others in competition. Rich, clumped food promotes contest (fighting) competition. The competition causes individuals to disperse if food sources are small, but they can benefit from grouping and cooperating in contest if sources are large enough for more than one. Poor food (often densely distributed, e.g., leaves) produces scramble

competition (get to the food first and eat fastest). Under this scenario, grouping is possible, but cooperation in competition is not especially useful.

3. Body size affects antipredation strategies. Small animals, the distribution of whose food forces them to be solitary or live in small groups, hide as an antipredator strategy. Larger-bodied animals that can forage in groups benefit from safety in numbers. Consequently, body size influences the nature of societies.
4. In mammalian species in which females are in more or less large groups (often, large-bodied species feeding on poor resources), males compete intensively for the rich, clumped resource of females, and evolve extra large body size. The species is then sexually dimorphic. Females of such species can benefit from associating with the larger, protective male more than they could in less dimorphic species.
5. Primates are an unusual mammal in the proportion of species that live in more or less stable heterosexual groups. Whether males associate with females depends in part on the payoffs to males from staying with only a few females compared to searching for many estrous females, and the payoffs to the males from defense of females and their offspring.
6. Infanticide by non-father males is seen in primate societies in which usually only one male mates with a number of females, and lactation is of far longer duration than gestation. The infanticide effectively increases the number of available estrous females, that is, mates, given that death of an infant causes earlier return to estrus.
7. Females have some counterstrategies to infanticide. The main one relevant to understanding the nature of societies is association with a protective male.
8. If animals remain for their lives in a group, inbreeding and genetic homogeneity of offspring are threats. Individuals that leave (for whatever benefit) could then be at an advantage, despite the demonstrable costs of dispersal to new sites and groups. In mammals, generally, more males than females leave the group or site of their birth, and often go farther. Why the sex difference? In polygynous societies (males mate with several females), the benefits of emigration for the successful male can be greater than those for the successful female. Additionally, if avoiding mating with kin is how individuals that stay prevent inbreeding, then males (that benefit from mating

many females) suffer a greater cost than do females (that need only one male). However, if resources are easy to find, benefits to females from staying are not obvious. In that case, if a related male stays, the costs of inbreeding can force females out.

9. The unusual proportion of primate species in heterosexual groups is explained in part by the fact that primate males can help by carrying offspring, and thus can contribute more than can males of many other mammals to survival of their offspring. Nevertheless, in many species the males associate with females but do not actively care for the offspring. The other part of the explanation for males associating with females is that the slow maturation of primates provides a potential for greater benefits to both sexes from protection of offspring by males. Where help from other animals in caring for offspring is highly beneficial, females sometimes actively prevent others from successfully breeding, a severe form of contest competition for resources in groups.

2.1. Socioecology

2.1.1. Introduction

We touched on the great variety of societies of primates in chapter 1.2.3. Primates live alone. They live as closely bonded monogamous pairs in which the father carries the offspring most of the time. They live as single-male harems. They live as large multi-male, multi-female groups. In some groups obvious contrasts in competitive ability among females (dominance hierarchies) exist, and strong cooperation among kin; sometimes no obvious contrasts in competitive ability exist, and cooperation is infrequent; and so on, and so on (Dunbar 1988; Kappeler and van Schaik 2002; Campbell et al. 2007; Smuts et al. 1987). The society of gorillas is only one among many.

For over forty years, the variety of societies in primates and other taxa has been categorized and analyzed in many ways as a means of understanding the ecological determinants of both any one society and of the diversity of societies (Clutton-Brock 1989b; Crook 1965; Crook and Gartlan 1966; Davies 1991; Eisenberg, Muckenhirn, and Rudran 1972; Isbell 2004; Isbell and Young 2002; Jarman 1974; Nunn and van Schaik 2000; Sterck, Watts, and van Schaik 1997; van Hooff and van Schaik 1992; Wilson 1975, Part III; Wrangham 1980). In this chapter we provide an introduction to what we see as the basics of socioecological theory that has resulted from the past four decades of analysis.

Two interlocking frameworks of understanding underlie our approach. (A framework is, roughly, a way of thinking about the world—as we use the word. All sorts of words are used to describe how scientists think about the world. We try to sort them out in box 2.1, which indicates our definitions of several of the words used to describe various stages in the development and testing of scientific ideas.)

First, the nature of a society is determined by interaction of the constituent individuals' evolved strategies to survive to adulthood, then to mate, and finally to rear their offspring to adulthood (Crook, Ellis, and Goss-Custard 1976; Dunbar 1988). It is determined by many other facets of individuals' propensities and behaviors too, but it is this strategies level on which we concentrate (ch. 1.1; see also ch. 13.3.1).

Second, the sexes use different strategies at each step (Bradbury and Vehrencamp 1977; Clutton-Brock 1989b; Davies 1991; Emlen and Oring 1977). As we suggested in the final section of chapter 1, the female spider on her web eating the male as it mates her (Andrade 1996) epitomizes the difference. This understanding that the sexes use different strategies is the framework around which the book is most obviously built. In brief, the successful female is distinguished from the unsuccessful female mainly by a difference between them in access to resources, particularly food. The distribution of food in the environment thus determines the distribution of females. By contrast, the successful male is distinguished from the unsuccessful male mainly by a difference between them in access to mates. The distribution of females then determines the distribution of males.

We start the chapter, then, with a discussion of how the nature of the animal determines the nature of its food, which determines the nature of competition and cooperation over the food, which in turn determines the nature of the society. The nature of the animal also determines the nature of its predators and the means that the animal might use to escape being preyed upon. Those means of escape might prevent grouping or induce grouping. Either way, predation strongly influences the nature of the prey's society. The way that females are distributed over the environment (determined by food and predators) then influences the strategies that males' use to gain access to fertile females, which strategies determine the competitive and cooperative relationships between and within the sexes. Those relationships are, of course, fundamental to the nature of a society. And finally, the sexes' strategies for successful rearing of offspring interact to influence the nature of society. With rearing too, the distribution of food, through its influence on the distribution of females, influences the sexes' interactions with one another and their offspring, and hence influences the nature of society.

Box 2.1. Inspired guess or substantiated hypothesis? Some definitions

Model, hypothesis, generalization, framework. Many words are used in scientific writing that can sound grander and more specific than they actually are. There often is not much difference in meaning, even in scientific writing, between a hypothesis and a bright idea, between a model and a generalization. Indeed, dictionaries often do not distinguish between these words, giving them all pretty much the same meaning.

We think, though, that it is useful to distinguish, if only to demystify scientific writing, and help readers know what they are getting. The most important distinctions to make are, we suggest, statements of fact from explanations, and quantitative tests of explanations from verbal presentations of them. However the words are used, they are not science unless what they say can be shown to be wrong. (Creation science, or its current manifestation, intelligent design, cannot be shown to be wrong—"God did it" cannot be disproved. To argue its teaching in science courses is, therefore, misplaced.)

Generalization—the phenomenon (fact) will be roughly as we have seen before, if the context is roughly the same.

Guess or *Idea*—an explanation for a phenomenon that has little to no substantiating data.

Hypothesis—the scientific word for a guess or idea or explanation, but, crucially, a hypothesis must be carefully worded to ensure that it can be disproved.

Model—a quantitative expression of an explanation, whether the explanation be termed a guess, bright idea, or hypothesis.

Prediction—a forecast of what will be seen in a *new* situation if a hypothesis is correct, not simply a generalization that the future will be like the past.

Theory—a well-constructed set of interconnected explanations (hypotheses), especially ones that have been strongly substantiated by making predictions that have been tested by new observations or models.

Framework—fundamental theory, but so general, so all-encompassing, so accepted, that it is more useful to think of it as a background way of thinking on which to hang more recent development of hypotheses. *Paradigm* might have the same sense.

Schema/scheme/schematic—a flow diagram, an organizational chart, of how a system works.

2.1.2. Females to food, males to females

Male and female animals follow very different main strategies to survive, mate, and reproduce. The difference in the strategies can be understood by considering the difference between the sexes in the resource that makes the most difference to the reproductive success of each sex.

The resource that makes the most difference to a female vertebrate, the resource that most improves her ability to survive, mate, and, especially, produce offspring and then rear them, is food (Bradbury and Vehrencamp 1977; Emlen and Oring 1977). The female that produces many surviving healthy offspring is the female that has fed well, that has outcompeted other females for food. She is rarely the female that has mated with more males, especially in species that produce only one offspring at a time, such as the larger primates. Most males produce enough sperm in one ejaculate to fertilize a female. Once a female, especially a female mammal, has mated with one male, she probably has enough sperm to fertilize all her eggs. Once a female mammal is pregnant, she cannot get more pregnant. Instead, the best way for a female to increase her reproductive success (the number of surviving offspring that she produces) is usually to obtain more resources for herself and her offspring.

By contrast, the successful male is usually the one that gets access to more or better mates (Bradbury and Vehrencamp 1977; Emlen and Oring 1977; Trivers 1972). Especially for mammals, in which more often than not the female can raise the offspring herself, it is the number of females mated, not the amount of food eaten, that distinguishes the successful from the unsuccessful male (Clutton-Brock 1988, Part 4). In quite a strong sense, the distinction forces a comparison of the nature, outcome, and relative strength of ecological influences on the structure of society with the nature, outcome, and relative strength of the influence of sexual selection on the structure of society (Janson 2000; Nunn and van Schaik 2000).

Of course, the distinction between the influence of females and ecology on the one hand and males and sexual selection on the other hand is a gross simplification. Females can benefit from multiple sires for their offspring (Jennions and Petrie 2000; Madsen et al. 1992), and a starving male is going to be a poor competitor for mates. Nevertheless, for nearly thirty years, the distinction has been an extraordinarily useful framework around which to hang the interesting societal complications that arise. It explains a lot of what we see. We can see rodents, canids, felids, ungulates, and primates all fitting the framework (Clutton-Brock 1989b; Clutton-Brock, Guinness, and Albon 1982; Davies 1991; Dunbar 1984).

In sum, females' survival, mating, and rearing strategies are different from males'. Females compete by surviving and rearing offspring. Males compete by mating. Females go to where the food is. Males then map themselves onto the females. To quite a large extent, discussions of the influence of survival and rearing strategies on the nature of society are discussions about the influence of females on society; discussions about the influence of mating strategies on society are discussions about the influence of males on society.

2.2. Food and Society

2.2.1. Body size, food, and society

Small-bodied animals feed on different foods than do large-bodied animals. As food determines the nature of society, small-bodied animals have different sorts of society than do large-bodied animals. John Crook (1965) first explained how in the early 1960s with work on weaverbirds. Happily for primate socioecology, Crook then moved to primates (Crook 1972; Crook et al. 1976; Crook and Gartlan 1966). Small fruit- and insect-eating primates range alone, foraging in trees at night, with fathers sometimes helping to rear offspring, for example, by carrying them. By contrast, larger-bodied omnivorous primates forage in large groups during the day in trees or on the ground, with offspring raised with little to no direct help from the sometimes relatively large males, except perhaps via protection against danger (Crook 1972; Crook et al. 1976; Crook and Gartlan 1966).

In sum, the body size of an animal, whether weaverbird or monkey, is associated with its diet and habitat, and therefore its society. These factors of body size, diet, and group size were nicely integrated in Peter Jarman's (1974) explanation for the variety of societies seen among ungulates. Small-bodied animals (with relatively high metabolic rates) require a highly nutritious diet, other things being equal. Nutritious foods are usually small and sparsely distributed in the environment. Small-bodied animals therefore have to be sparsely distributed, that is, live solitarily or in pairs or small groups. Conversely, large-bodied animals can cope with a less nutritious diet. Poor-quality food quite often occurs in abundance in large patches (grass in savanna). Therefore, large-bodied animals can occur in large groups (see chapter frontispiece).

Tim Clutton-Brock and Paul Harvey (1977a) considerably advanced the field of socioecology by applying quantitative analyses to the previously more or less qualitative distinctions between societies. They statistically

Box 2.2. Accounting for taxonomic relatedness: Phylogenetic correction

A major means of testing socioecological hypotheses is to compare species to see if the environment correlates in some way with some aspect of the nature of their society. For instance, one might argue that if five baboon species were terrestrial and lived in large groups, while five tamarin species were arboreal and lived in small groups, the hypothesis is statistically supported that terrestrial species live in large groups to escape the extra predation on the ground ($p < 0.05$, binomial test). One would be wrong.

Statistical tests require independent data. Species are independent data only if creationists are correct, and even then, only if it was by pure accident that God made all baboons more similar to one another than they are to tamarins. In fact, given evolution, given that we can lump species into deeper taxonomic categories, that is, given that we know that some species are more similar to one another than they are to others, there is no valid way to claim without proof that all species are independent data in a statistical analysis. Once you know the main substrate and the group size of one baboon species, you pretty much know them for all baboon species. Substrate and group size of five species of baboon are not five data points in a statistical test; they are one. To count them as five data points is to commit the statistical sin of data inflation, double-counting, pseudoreplication (Harvey and Pagel 1991; Nunn and Barton 2001; Purvis and Webster 1999).

Various methods exist to cope with the fact of non-independence of species through common descent. A simple one is to compare genera, not species (Harcourt et al. 1981c). However, of course, some genera are more related than are others. A common method now is comparative analysis by independent contrasts (CAIC) (Harvey and Pagel 1991; Nunn and Barton 2001; Purvis and Webster 1999). Not only is CAIC an excellent method, but the program to implement it

confirmed, for instance, correlations between body size, diet, group size, and nature of society. A main subsequent advance has been to take explicit and quantitative account of the fact that the more closely related species are, the more similar they are (obviously), and therefore the less independent they are as data points in a statistical test (box 2.2). Justin Brashares, Garland, and Arcese's (2000) testing of Jarman's arguments for antelope and Charlie Nunn and Carel van Schaik's (2000) testing of the variety of hypotheses to explain variation in some of the main facets of primate societies illustrate use of this phylogenetic correction, which sometimes supported the previous analyses and sometimes negated them.

comes with a very comprehensible manual and is free on the Web (Purvis 2004; Purvis and Rambaut 1995).

In simplified brief, CAIC needs an evolutionary tree of the taxonomic group that one is interested in. It then examines the data, chooses the shallowest taxa that differ on the traits of interest (e.g., 2 baboon species, 2 tamarin species), and asks in which direction and by how much they differ on one trait (substrate use) and then the same for the hypothesized associated trait (group size). The independent contrast here is the comparison between the baboons with that between the tamarins—the difference between the baboons is independent of the difference between the tamarins, and vice versa. CAIC then moves to the next deeper evolutionary node (cercopithecines vs. callitrichines), and asks the same questions. Each deeper comparison averages (in the broad sense) the trait values from the shallower taxonomic level (average baboon vs. average tamarin). And so it continues (Old World monkeys vs. New World Monkeys, and so on). At each level, only one contrast between equivalent taxa is being made (i.e., no data inflation). The result is a list of contrasts in the independent trait, which can then be compared with the list in the dependent trait, to see whether, for instance, most contrasts in substrate use are associated with the same direction of contrast in group size.

You will see criticism of various sorts of this program and others that do the same job (e.g., Martin, Genoud, and Hemelrijk 2005; Westoby, Leishman, and Lord 1995). The criticisms are largely directed at exactly how the programs account for taxonomic similarity and difference, and the inferences one can make from the phylogenetically corrected results, rather than at the idea that species are not independent data—because species are fundamentally not independent, unless one can show otherwise.

Note that in Jarman's (1974) formulation, food does not force large-bodied animals to be in large groups; it allows them to be in large groups if there are other reasons to group. This distinction between what the environment (physical or social) forces animals to do (in the sense that the animals benefit if they do it), and what it allows them to do is important. For instance, a main benefit of grouping is escape from being eaten through safety in numbers (sec. 2.3.3), but whether or not animals do form groups is heavily influenced by what the distribution of their food allows.

Among primates, the extremes of the Jarman contrast are the small-bodied insectivores, such as bushbabies, and the large-bodied folivores, such as the gorilla (Clutton-Brock and Harvey 1977a; Crook and Gartlan 1966; Eisenberg et al. 1972). Insects are nutritious, but they are small, and ephemeral, and so reliance on them means a more solitary life. Foliage is everywhere (forests are green), and therefore folivores can be group-living, and usually are. Of course, exceptions occur, such as some small-bodied folivorous lemurs. Indeed, lemurs provide many interesting exceptions to general rules (Kappeler and Heymann 1996).

2.2.2. Food and competition

> The prodigious waste of human life occasioned by this perpetual struggle for room and food. . . . **THOMAS MALTHUS (1798, Ch. 3, p. 48)**

2.2.2.1. The nature of the food affects the nature of competition

There is more to the story than animals being as thinly or as densely distributed as their food. Organisms compete for resources, as first Thomas Malthus and then Charles Darwin told us well over a century ago (Darwin 1859; Malthus 1798). Nevertheless, Lynne Isbell has suggested that some primate species, folivores in particular, are unlimited by food supply (Isbell 2004; Isbell and Young 2002).

Occasionally, food might not be in such short supply that there is competition over it. For instance, after an epidemic, or in a site or period of severe predation, there should be plenty of food for everyone: hence rapid rebounds of populations after crashes in their numbers, especially if the crash was not caused by lack of food (Gilg, Hanski, and Sittler 2003). The Virunga gorillas might be a case in point. Since protection was improved in the region, first in the late 1960s in the study area of Dian Fossey's Karisoke Research Center, and subsequently from the late 1970s throughout the protected area of the Virunga Volcano region, the population has increased rapidly (Harcourt 1986; Harcourt et al. 1983; Kalpers et al. 2003), with birth rates as high as in zoo populations (ch. 3.2.7).

However, almost any study that is long-term in relation to the generation length of the organism studied has found that as density increases, reproduction drops and mortality rises, even in insects (Hassell, Latto, and May 1989; Sibly et al. 2005). A main explanation for such density-dependent regulation is competition for resources.

From the standpoint of basic biology, we thus disagree with Isbell. Moreover, some primate data negate Isbell's idea: biomass of the folivorous subfamily of colobines in various forests in Africa and Asia correlates remarkably closely with quality of the forest, as measured by abundance of legumes or by average quality of mature leaves in the forest (Chapman et al. 2002; Davies 1994). The most reasonable explanation for that correlation is limitation by food supply. In other words, even folivores are competing over food.

The quality and distribution of resources influences the nature of competition over them. Animals will contest resources, that is, fight over them, if the resources are worth defending, and defensible; otherwise the animals will compete by scrambling for access, that is, getting to the resources first and consuming them fastest (Davies and Houston 1981, 1984; Gill and Wolf 1975). Large amounts of high-quality resources are worth defending or taking, for example, a large patch of flowers for a sunbird or a fruiting tree for a primate group. Defensible resources are those that are clumped, as opposed to spread out, and those over which there are not so many competitors as to make defense too costly. (We need to warn that some population ecologists use the words *contest* and *scramble* to describe not the process of competition, but instead the outcome. For them, contest and scramble competition distinguish between uneven and even distribution of possession of resources, whatever the competitive behavior, fights or race, that led to that possession.)

Isbell and co-workers have argued that the important issue is not whether resources are defensible, but whether they are takeable (Isbell, Pruetz, and Young 1998). A very small resource is defensible, but not takeable, she argues—a vervet can put a small fruit in its mouth, and it would be very difficult for even a dominant group member to take that fruit away. Thus, animals do not compete over defensible resources, but over takeable ones. However, if the resource is small, it is probably not worth aggressively taking. We are then back to the original formulation of the argument, in which animals do not contest resources that are not worth fighting over but instead scramble for them.

While food certainly influences the nature of the competition, no linear relation from contested high-quality, clumped resource to scrambled low-quality, dispersed resource exists, because number of competitors and the animals own ability to defend come into the picture too. For instance, rich resources can lead to so many competitors that animals have to give up on defense, and scramble instead (Gill and Wolf 1975). And independently of the resources, if the animal cannot quickly travel across its home range, it cannot defend it (Mitani and Rodman 1979).

The interaction of influences on whether resources are going to be contested or not can be illustrated by considering whether frugivores or folivores might be more likely to be territorial, that is, to aggressively defend their home range. While fruit is perhaps worth defending, it is thinly and widely distributed, which means that frugivores have larger home ranges than do folivores (Clutton-Brock and Harvey 1977a; Janson and Goldsmith 1995)—and large areas are less easily defended than are small ones. At the same time, while folivores might have smaller, more defensible ranges, they tend to travel less far per day than do frugivores (Clutton-Brock and Harvey 1977a; Janson and Goldsmith 1995), and so are less able to defend those ranges. Consequently, in primates, diet does not clearly predict territoriality (Grant, Chapman, and Richardson 1992).

None of these explanations contradicts the idea that at a fundamental level, the nature of food influences not only how animals are distributed, but the nature of competition over the food. Food thus determines the nature of society.

2.2.2.2. The nature of competition affects the nature of the society

An egalitarian society is a different sort of society than is a stratified society, among animals as among humans. With scramble competition over widely distributed resources, differences in ability between individuals do not become as obvious as with contest competition over clumped resources. Animals that win in contest competition, the dominants, monopolize the rich, clumped foods over which animals are prepared to fight; the subordinate animals lose access (Harcourt 1987a; Koenig et al. 1998; Sugiyama and Ohsawa 1982; Whitten 1983). With severe enough contest competition, dominants can outreproduce subordinates (Harcourt 1987a), even in chimpanzee society in which females often range alone (Pusey, Williams, and Goodall 1997).

It is not only the human observer who can recognize hierarchical societies. Animals in them can recognize their own position relative to others, and the more clever among them, corvids (crow family) and primates, recognize relative rank among others (Bergman et al. 2003; Cheney and Seyfarth 1990, ch. 3, 9; Harcourt 1992; Paz-y-Mino et al. 2004).

Observations of dominance hierarchies and of daily distance traveled increasing with group size have been used as measures of the existence of competition in groups, so obvious does it seem that the associations with group size are the result of competition (Isbell 1991, 2004; Isbell and Young 2002). However, the next step in the argument should not automatically be taken. It is not necessarily the case that if daily distance traveled does not increase with increasing group size, or if no dominance hierarchies exist, animals are not competing, nor that food does not limit female reproductive success.

First, as we said at the start of this section, competition is likely to be a main limiting factor for all species, even insects. And second, lack of detection of competition does not necessarily mean that animals are not competing (Janson and Goldsmith 1995; Koenig 2002). Folivores, which do not increase daily distance traveled with group size as obviously as do frugivores, could face constraints on travel not faced by frugivores, such as greater gut fill and hence less ability to increase intake by moving more (Janson and Goldsmith 1995). They could also adopt other means than increased daily distance traveled to cope with the extra competition, such as greater group spread, or feeding on lower-quality vegetation (Janson and Goldsmith 1995; Koenig 2002).

2.2.3. Food, competition, and grouping

2.2.3.1. Grouping increases competition

The interest in grouping among vertebrate socioecologists arises in part because of the strongly supported hypothesis that as far as competition for resources is concerned, being with others must be costly. Primate socioecologists are particularly interested because such a large proportion of primate species live in more or less stable groups (van Schaik and Kappeler 1997).

There is almost no way to escape the logic that being with others, proximity to others, increases the frequency and intensity of competition for resources, both contest and scramble (Alexander 1974; Dunbar 1988, ch. 7; Krause and Ruxton 2002, ch. 3). The closer an animal is to others, and the more other animals present, the more likely it is that two animals will try to obtain the same resource, that is, will compete. Thus, in a variety of species, clusters of feeding animals are smaller, the smaller the patch of food (Cashdan 1992; Goss-Custard and Sutherland 1997; Krebs and Davies 1993, ch. 5). The process can easily be tested by throwing to ducks different amounts of bread in different places: more ducks will gather where more bread is thrown. Similarly, in primate societies in which the individuals range in parties of continually changing size—fission-fusion societies—party size matches the size of the food patch, for example, the size of a fruiting tree (Isabirye-Basuta 1988; McFarland Symington 1988; Utami et al. 1997; White 1989; Wrangham 1977).

In general, though, group-living primates live in stable groups that do not change size from moment to moment as food patches change in size. Instead, the cost of competition that arises from grouping with others results in larger groups traveling farther per day, lower feeding rates per animal in larger groups, subordinate animals doing worse in larger groups, increased aggression in larger groups, and so on (Chapman 1990; Chapman and Chapman 2000; Clutton-Brock and Harvey 1977a; Isabirye-Basuta 1988; Isbell 1991; Janson 1988; Janson and Goldsmith 1995; McFarland Symington 1988; Williams, Liu, and Pusey 2002; Wrangham 1977).

In sum, grouping must usually increase competition, grouping is costly as far as feeding is concerned, and competition sets an upper limit to group size (Barton 2000b; Dunbar 1988, ch. 7; Janson and Goldsmith 1995; Mangel 1990; though see Sussman and Garber 2006; Terborgh and Janson 1986; van Schaik 1983).

2.2.3.2. Food and the benefits of cooperation in competition as a cause of grouping

> It is said that God is always for the big battalions. **VOLTAIRE (1770)**
>
> In battles . . . ants from the colony with the greater local density of workers were more likely . . . to retain control of the contested area. The decision to attack or withdraw depended upon the relative number of nestmates and intruders. **ELDRIDGE ADAMS (1990, p. 321)**

Grouping to cooperate in defense. If the crucial resource is worth defending and defensible, and comes in packets large enough for more than one animal, then two cooperating animals can probably defeat a single animal in competition for it. (By cooperate, we mean simply, act together—more details are in box 5.1). If so, individuals will benefit from being in a group with others, and therefore they will form cooperative groups, or remain in them (Bygott, Bertram, and Hanby 1979; Davies and Houston 1981; Dugatkin 1997; Hannon et al. 1985; Harcourt 1992; Koenig 1981; Packer and Pusey 1982; van Schaik 1983, 1989; Wrangham 1980).

Nick Davies and Alasdair Houston (1981) provide still one of the loveliest examples of how the quality and distribution of resources, along with the intensity of contest competition over those resources, could cause animals to benefit from grouping with others and cooperating in competition. When there is enough food in a territory for more than one animal, and when competition for the food is intense, then the territory owner will be forced to allow in a second individual, from whose presence it benefits if the newcomer also defends the territory. So precisely did Davies and Houston calculate the relative amounts of food and intrusion necessary for cooperative defense that their predictions about whether they would see the cooperation or not were correct on fully 85% of forty days of observation of pied wagtails (birds) on an Oxfordshire riverbank.

Many primate species have been so well studied that not only species, but populations within species, have been contrasted for frequencies and intensity of competition. The influence of competition on cooperation and grouping can then be tested. Thus Nunn and van Schaik (2000) found that females of species in which competition between individuals or groups was classified as more intense or frequent were more likely to remain in the group of their birth and cooperate with kin (stats. 2.1). Within species, Robert Barton and colleagues found the same in two populations of baboons, and Sue Boinski in two populations of squirrel monkey (Barton, Byrne, and Whiten 1996; Boinski 1999).

Group-living animals can benefit from cooperation with others in competition within their own group, or with other groups (Isbell 1991; Isbell and Van Vuren 1996; Isbell and Young 2002; Smuts 1987; Sterck et al. 1997; van Hooff and van Schaik 1992; van Schaik 1989; Walters and Seyfarth 1987). The distinction between within-group and between-group competition and the conditions under which one is a more important influence than the other is a strong component of attempts to explain primate societies (good reviews in Isbell and Young 2002; Koenig 2002; van Schaik 1989). We are not convinced that an obvious dichotomy exists (see also Koenig 2002; Ménard 2004). Instead, it seems to us that whether within-group competition is more important than between-group competition will vary with the size of the patch, the nature of the patch, and the nature of movement of individuals between patches (Isbell 2004; see also Matsumura 1999).

When cooperation is particularly advantageous, cooperation might be the main benefit to forming or remaining in a group. For instance, it appears that lone male lions are highly unlikely to gain a group on their own, and even unrelated males will cooperate to gain access to a pride of females (Bertram 1975; Bygott et al. 1979; Packer et al. 1991; Packer and Pusey 1982). The same occurs in populations of red howler monkeys (Pope 2000). Once animals remain in a group for whatever reason, but perhaps especially if they remain for benefits from cooperation, groups are likely to be composed of kin (Wilson and Hölldobler 2005).

Cooperation as a consequence, not a cause, of grouping. Many animals live in a group, or remain in a group, without apparent cooperation, including many primate species (Isbell 2004; Isbell and Young 2002; Sterck et al. 1997). Therefore, cooperation alone does not explain grouping. Instead, as in some birds in which offspring remain with their parents to help them raise subsequent broods, the fact that offspring are present to help does not necessarily arise as a result of the inclusive fitness benefits to the offspring of remaining to help. Instead, their grouping with the parents often appears to occur as a result of the offspring not being able to find good territories to breed on their own. They would do better if they could breed on their own. However, failing lone breeding, the next best option is to remain where it is safe—after all, they have survived there. And if they are going to remain, then under the right conditions, it can pay to help feed or protect subsequent siblings, or defend the breeding territory, that is, to cooperate (Brown 1987; Emlen 1991, 1995; Komdeur 1992; Stacey and Koenig 1990; Vehrencamp 1983).

Similarly, callitrichine (marmosets and tamarins) males have been described as being more likely to stay and help in their natal group when surrounding territories are taken, implying that they would breed better on their own (Goldizen 1987; Pope 2000). At the same time, they have also been described as remaining in their natal group to wait to breed (waiting for dominant group members to die) because lone breeding is disadvantageous (Goldizen 1987; Pope 2000). Staying must often be the best option for any animal, simply because leaving is dangerous. If emigration is dangerous, as it must normally be (predators' hideouts and escape routes unknown, lack of knowledge of resources, aggression to newcomers), the default should be to remain. Whether the residents then actively cooperate with one another in competition within the group or with other groups will depend on the nature of competition, in other words, on the nature of the habitat and how the animals use it (Dunbar 1988, ch. 7; Isbell 2004; Isbell and Van Vuren 1996; Isbell and Young 2002; Sterck et al. 1997; van Schaik 1983; Wrangham 1980).

Cooperation as a benefit (main or subsidiary) to grouping, particularly to remaining in the natal group, will be especially likely in situations where its benefits to the cooperating individuals are not just a temporary rise in competitive ability at the time of the contest, but a permanent rise in competitive ability. That can happen in several situations. One is when groups compete, and the larger group wins, as the opening quotes of this section indicate. Similarly, cooperation can be beneficial if individuals in larger groups are better protected from predation, and cooperation maintains or increases the size of the group (Clutton-Brock 2005). Cooperation can also cause grouping when help is frequent enough that helped animals and their opponents learn that an individual's competitive ability is effectively that of its most powerful main helper, usually its mother (Cheney 1977; Lee and Johnson 1992). One then gets the classic baboon and macaque "inheritance of maternal rank" society, whereby daughters have effectively the same competitive ability as their mother. They outrank all that their mother can defeat, and are defeated by all that defeat their mother. A society results in which whole families rank in relation to one another (Chapais 1995; Datta 1992; Kawai 1958/1965; Kawamura 1958/1965; Walters and Seyfarth 1987).

Kin and cooperation. The more beneficial cooperation is, the more animals will work, it seems, to maintain friendly partnerships with potentially cooperative individuals (Aureli and de Waal 2000). Kin appear to work the hardest, that is, they are actively the friendliest to one another (Emlen 1995, 1997; Kapsalis 2004; Silk 2002a). Why? Should not kin help one another without having to work to maintain a cooperative partnership, because of the inclusive fitness benefits of help among them?

However, in group-living species, animals will often have alternative partners, some of which might be more powerful partners than any kin (Chapais 2005; de Waal 1989b; Noë 1992, 1994; Seyfarth 1977). Indeed, females will compete over partnerships with particularly useful non-kin group members (Chapais 2005; Cheney and Seyfarth 1990, ch. 2). Partnerships with kin therefore need to be maintained. Additionally, the returns from helping kin are surer than with non-kin. Not only can each obtain inclusive fitness benefits from cooperation, but precisely because of the inclusive fitness benefit, and probably also because of greater familiarity, kin are more likely to agree to be cooperative partners; in essence, they are more trustworthy (Harcourt 1991b). With benefits more likely from kin, it is more worth working to maintain a partnership with them than with non-kin.

Finally, a variety of foraging benefits of grouping have been suggested. We are not convinced that these are relevant to understanding primate grouping (box 2.3).

2.3. Predation and Society

If grouping inevitably induces competition, why do animals group? A main answer is escape from predation: in brief, the more animals in the group, the safer any one animal is. Grouping and group size are thus a function of the conflict between the costs of competition (minimizing group size) and the benefits of safety (maximizing group size). Grouping and group size cannot be understood unless both are accounted for (Barton 2000b; Dunbar 1988, ch. 7; Janson and Goldsmith 1995; Mangel 1990; Terborgh and Janson 1986; van Schaik 1983). Very nice experimental studies using fish and birds show that when competition is likely, groups are small; when predation is likely, groups are large (Elgar 1986; Hoare et al. 2004). Indeed, sometimes the effect of grouping on competition, for example, reduction in vigilance, will affect susceptibility to predation, and vice versa (Isbell 1994).

2.3.1. Primates are preyed upon and have evolved antipredator strategies

Observers of habituated primates rarely see predation, in part because the observers scare away the predators, which have not become habituated (Isbell and Young 1993). Nevertheless, plenty of evidence exists that primates are

Box 2.3. Food, grouping, and foraging benefits: Unlikely hypotheses for primates

Several foraging benefits of grouping have been suggested, that is, benefits associated with finding food (as opposed to gaining it once found). The hypotheses all originated from ornithologists, and we suspect that all are more applicable to birds than to terrestrial mammals, including primates, depending as they do on ease of travel between continually reforming groups. We suggest that the hypotheses are especially unimportant for folivores: foliage is everywhere, and no animal needs help in finding it. We briefly mention them, though, because others disagree with us (Isbell 2004; Rodman 1988a).

Henry Horn showed that if food is in quite large patches but ephemeral, animals would exploit it most efficiently by *foraging from a central point* in their range (Horn 1968). Grouping then results from gathering at the central point and, given some common knowledge of where the resources are (see also discussion that follows), joint travel to the resources. Use of central place foraging depends on long-distance travel being easy and fast. It is for birds. It is not for primates. In contrast to Edward Wilson, who suggested that the Horn principle might explain much grouping in primates, with predation having a subsidiary role (Wilson 1975, ch. 26), we suggest that in primates, central place foraging is very rare. It occurs only where night-resting sites that are safe from predators are scarce, as is the case for some baboon populations (Cowlishaw 1997; Kummer 1968).

Grouping *prevents wasted travel to already depleted sources* (Cody 1974). The idea here is that animals not in a group would not know where other animals had been, and could waste time traveling to a source, only to find that others had recently depleted the source. Although this hypothesis has been suggested for gorillas (Fossey and Harcourt 1977), for primates in general (Isbell 2004), and for associations between groups of different species (Enstam and Isbell 2007), we cannot see how the costs of permanent competition from group members in permanent close proximity can usually be less than the costs occasionally incurred by ignorant travel to an already visited site. Stuart Altmann (1974) suggested that the costs of competition could be minimized if groups spread at right angles to the line of travel. However, animals get left behind, and so in ignorance cross others' feeding paths; and when the group turns, the outside individuals have to travel fast and far to keep in line, or cross everyone else's feeding paths. Furthermore, if groups' home ranges overlap, as they do in the case of many primates, including gorillas (ch. 4.1.5), grouping has not solved the problem of unknown, unscheduled visits by others.

Grouping allows animals to *use others' knowledge to find food* (Krause and Ruxton 2002, ch. 2). The argument works well for ephemeral foods that appear in patches large enough for more than one animal to feed, and for species that can see and easily move long distances, such as birds. Alternatively, where one species might know its range better than another (e.g., a species with a small home range, or a group in its core range), the large-range or peripheral-range group or species could potentially use the small- or core-range species to lead it to food (Norconk 1990). Data to substantiate the suggestion are largely lacking, and we maintain that the benefit of the occasional discovery by another animal of an unknown food source is unlikely to outweigh the cost of a continually present competitor. Indeed, two studies, one of capuchin/squirrel monkey association and the other of two guenon species, substantiate our assertion: the ignorant species were probably outcompeted by the knowledgeable species (Norconk 1990), making it likely that escape from predation is the cause of the associations (Terborgh 1983, ch. 8).

A specific example of this foraging benefit of grouping applies to situations when animals gather from a wide foraging region in nightly sleeping sites. The site can then act as an *information center,* with unsuccessful individuals being able to use others' strength of directed movement the next day as an indication of a good food source (Krause and Ruxton 2002, ch. 2). However, as we have already suggested, if anything like central place foraging occurs in primates, we suspect that it is a result of scarcity of night-resting sites that are safe from predators, not because of the advantages of parasitized information.

Finally, animals often call in the presence of good sources of food, some with seemingly evolved *food calls* given largely in this one context. This observation seems to negate our claim of no food-finding benefits to being with others, especially as group members sometimes apparently punish others for not calling in the presence of useful resources (Hauser 1992). Given that continual close proximity outweighs the benefit of an occasional discovery of food (mentioned earlier), two other hypotheses for food calls seem more likely to us, although both need experimental testing (Krause and Ruxton 2002, ch. 4). The calls keep group members in the vicinity of the calling animal, with consequent antipredator benefits to the caller of safety in numbers (Elgar 1986; Harcourt, Stewart, and Hauser 1993). The calls advertise prowess—calling animals have been clever enough to find food and are well-fed and healthy enough to be able to announce the find and share it—with mating and other competitive advantages (Clark and Wrangham 1994; Krause and Ruxton 2002, ch. 4). The same happens in humans (Bliege Bird and Bird 1997).

preyed upon, some of them severely (Anderson 1986; Cheney and Wrangham 1987; Isbell 1994; Karpanty 2006; Miller and Treves 2006; Shultz et al. 2004). For instance, primates can account for over a third of prey remains in some raptor nests (Stanford 2002); up to 15% of some primate populations and over a third of mortality can be due to predation (Cheney and Wrangham 1987; Goodman, O'Connor, and Langrand 1993; Shultz et al. 2004); and population sizes of some primate species appear to be lowered by predation (Stanford 2002; Terborgh et al. 2001). Even the great apes, including the largest of them, the gorilla, are killed by predators (Anderson 1986; Rijsken 1978, ch. 8; Schaller 1963, ch. 7).

Not surprisingly, therefore, many primate species have evolved or developed what seem to be obvious antipredator tactics (Miller and Treves 2006). Some of these are relevant whatever sort of society the animals live in. For instance, primates rest and sleep in apparently safe refuges during the day and especially at night (up trees, in cliffs) (Anderson 1998; Boinski, Treves, and Chapman 2000; Cowlishaw 1997; Dunbar 1988, ch. 7; Sterck 2002; Uhde and Sommer 2002; Wright 1998); and some species have different alarm calls for different predators (as do several non-primate species), playback of which calls (both primate and non-primate) produces appropriate responses from listeners (Cheney and Seyfarth 1990, ch. 5, 9; Evans, Evans, and Marler 1993; Fichtel and Kappeler 2002; Templeton, Greene, and Davis 2005).

The question here is whether primates group as an antipredation strategy, as do so many other animals (Hoare et al. 2004; Krause and Ruxton 2002, ch. 8, 9).

2.3.2. Body size and predation

If species that live in large groups tend to be the larger-bodied species (sec. 2.2.1), then any comparison across species of predation in relation to group size needs to consider first the effects of body size on predation. Certainly, the larger the species, the fewer predators can kill it (Sinclair, Mduma, and Brashares 2003; Struhsaker 1967). However, are findings that larger-bodied primate species suffer less predation than do smaller-bodied better explained by contrasts in group size or in body size?

The apparently simple question of whether species that live in large groups are better protected from predation than those that live in small groups is even more complex than simply accounting for the relationship between group size and body size (Miller and Treves 2006). Small-bodied species use different means of escape from predation than do large-bodied species—hiding solitarily instead of grouping (Brashares et al. 2000; Clutton-Brock and Harvey 1977a; Crook and Gartlan 1966; Jarman 1974; Terborgh and Janson 1986). Also, large species, because of the relative safety of their large size, might live in riskier situations than do small species. For instance, in one West African forest, leopards prey more on larger-bodied primate species (Zuberbühler and Jenny 2002; see also Shultz et al. 2004).

In an attempt to minimize some of these confounding factors, we separated the larger-bodied, diurnal taxa, which might group as an antipredator tactic, from the small-bodied, nocturnal taxa, which are more likely to hide solitarily as an antipredator tactic.

Among the diurnal primates, we found no relation between body mass and group size (stats. 2.2). This finding does not negate suggestions that body size influences group size (sec. 2.2.1), because we have omitted the extreme small-bodied animals, and the hypotheses stated that large-bodied animals can group, not that they always do. If body mass does not correlate with group size among the diurnal primates, analysis of body mass in relation to predation on them should be unconfounded by group size. Even so, and in contrast to some previous studies (Cheney and Wrangham 1987; Isbell 1994), we found no relation between body mass and predation rate among diurnal taxa once we accounted for taxonomic relatedness (stats. 2.3). However, as said, the relationship might be confounded by the fact that the large-bodied taxa can afford riskier habitats, precisely because they are large.

The advantage, then, of analysis among the small-bodied, largely solitary nocturnal taxa is that neither grouping nor risk is a confounding factor. It turns out that among them, body mass correlated perfectly with predation rate across the three genera for which data are available (*Aotus* owl monkey, *Galago* bushbaby, *Microcebus* mouse lemur), with the largest suffering the least predation. However, still the story is not clean, for not only is the sample size very small, but each genus lives on a different continent, and the order of predation rate is the same as the order among the continents of number of carnivore species per primate species (Emmons 1999).

In sum, we have to conclude that in contrast to some previous studies, we cannot find for primates a statistically demonstrated relationship between body mass and either predation rate or predation risk.

2.3.3. Predation and grouping

2.3.3.1. Grouping within species protects from predation

An individual of any species—beetle, bird, or baboon—is less likely to be caught by an attacking predator if it is in a group than if it is alone. There is almost no way of escaping the mathematical logic of safety in numbers

(Alexander 1974; Dunbar 1988, ch. 7). And many nice experiments empirically substantiate the logic (Bertram 1978; Krause and Ruxton 2002; Pulliam and Caraco 1984).

But what if larger groups attract more predators? They certainly sometimes do so (Krause and Ruxton 2002, ch. 3). For instance, in Gombe Stream National Park, Tanzania, a doubling of the group size of chimpanzee prey, the red colobus, doubled the probability that an encounter between prey and predator would turn into an attack (Stanford 1998a, ch. 6). However, across a variety of predator-prey systems, it seems that the safety afforded by numbers increases faster than do the attacks (Krause and Ruxton 2002, ch. 2), with the result that individuals in larger groups are usually safer (Anderson 1986; Bertram 1978; Caro 2005, ch. 8; Dunbar 1988, ch. 7; Krause and Ruxton 2002; Pulliam and Caraco 1984; Shultz et al. 2004).

Other antipredator benefits of grouping can include greater likelihood of detection of the predator, with more animals vigilant, even if individuals are less vigilant (Bertram 1978; Caro 2005, ch. 4; Krause and Ruxton 2002, ch. 2; Pulliam and Caraco 1984). More animals also means more frightening mobbing or communal defense against the predator, and greater confusion of it as several animals mill in different directions (Bertram 1978; Caro 2005, ch. 11; Krause and Ruxton 2002; Pulliam and Caraco 1984). Thus, larger groups of red colobus contain more males, which can more successfully fight off chimpanzees (stats. 2.4) (Stanford 1998a, ch. 6; 2002). Indeed, individuals of some species will in the presence of predators associate with other species that are larger or more aggressive (Krause and Ruxton 2002, ch. 6; Noë and Bshary 1997). Finally, whatever the species involved, but perhaps especially in multi-species groups, several different sorts of antipredator benefits can accrue to grouped individuals, and these benefits can then have added benefits of their own, such as greater foraging efficiency (Wolters and Zuberbühler 2003).

2.3.3.2. Grouping evolved for protection from predation?

Even if animals in larger groups are safer than are those in smaller groups, the relative safety does not mean that the animals first grouped to avoid predation. Escape from predation might be a subsidiary benefit, not the main benefit (see Hinde 1956; 1982, ch. 4, on the general topic of evolved functions of behaviors). Maybe the main advantage of being in a larger group is better competitive ability (sec. 2.2.3.2). So, we ask here about the evidence that primates indeed group to avoid predation, that is, whether escaping predation is the main benefit of grouping, and therefore the evolutionary cause of grouping.

If we want to ask whether grouping is a response to risk of predation, we cannot use rate of predation as a measure of risk, because then the adaptation is confounded with the factor to which it is adapted (Boinski et al. 2000; Cheney and Wrangham 1987; Hill and Dunbar 1998; Janson 1998a). As Charlie Janson (1998a) so clearly put it, risk and rate measure very different things. Risk is a measure of an individual's chance of being caught; rate is a measure of the number of individuals caught in a population; it is more or less meaningless to talk of an individual's rate of being caught.

Various, mostly indirect, indices of predation risk have been used. Janson and Michelle Goldsmith (1995) found that adding assumed measures of predation risk (body mass, terrestriality—small-bodied animals on the ground would be most at risk) to a quantitative model of the association of diet with group size of diurnal species explained about twice as much (two-thirds) of the variance in group size across populations as was explained with diet alone, and indeed made diet no longer a mathematically significant explanation of variation in group size. They did not, however, correct for taxonomic relatedness.

Nunn and van Schaik (2000) did so correct (as well as correct for other factors, for example, they used only diurnal species, which are more likely to use grouping as protection than are nocturnal species). And they found that primates are indeed in larger groups, including females, under conditions of assumed greater predation risk (stats. 2.5). Similarly, Russell Hill and Phyllis Lee (1998) found larger groups of cercopithecoids in riskier habitats, using qualitative observations of density of predators as their measure of risk (stats. 2.6).

One way to control for several confounding factors, including taxonomic dependence, is to compare populations within species (Boinski et al. 2000). When we used the Hill/Lee data, we found no significant within-species association of risk with group size. Connie Anderson (1986) also found little within-genus variation of group size with predation rate (not risk), and nor did we when reanalyzing her within-species data. One problem, of course, is that other environmental factors than predation risk will differ between populations. As Tom Struhsaker (2000) has pointed out, these other factors are a problem with the argument that the hypothesis of grouping in response to predation is supported by the finding of smaller social groups in island populations, which experience less predation (van Schaik and van Noordwijk 1985).

Another problem is assessment of risk. Are terrestrial primates really at greater risk than are arboreal ones, or open-country primates more exposed to predators than are forest-dwelling primates? We have found little to no evidence to support these ideas. For instance, arboreal primates are sometimes preyed upon quite heavily (Olupot and Waser 2001; Shultz et al. 2004). The best measure of risk is experimental manipulation of presence or absence of a predator in the vicinity of an observed group. Such experiments indicate that animals, including primates, clump more tightly, or clump temporarily into larger groups, in the presence of a predator, or simulated predator (Boinski et al. 2000; Hoare et al. 2004; Krause and Ruxton 2002, ch. 8, 9; Noë and Bshary 1997; van Schaik and Mitrasetia 1990).

In sum, the larger the group, the safer an individual is, whether primate or other vertebrate; group size is then limited by the increasing costs of competition from increasing numbers of group members (Alexander 1974; Barton 2000b; Dunbar 1996; Janson and Goldsmith 1995; Mangel 1990).

2.3.3.3. Grouping with a male for protection from predation

Males in large-bodied, sexually dimorphic species of group-living animals are often at the forefront in attacking predators (Cowlishaw 1994; Stanford 2002). So formidable are baboon males, for example, that they have killed leopards (Cowlishaw 1994). Similarly, pacific bird species, such as godwits, pigeons, or geese, can benefit from nesting near aggressive species, such as skuas or falcons, although there is some trade-off between protection afforded from other predators and aggression or predation from the protective species (Krause and Ruxton 2002, ch. 6; Quinn and Kokorev 2002). Thus an obvious hypothesis for why females might associate with males in species in which the male is substantially larger than is the female is that the females benefit from the males' protection.

In primates, higher predation risk has been found to correlate with higher number of males in groups, as if females are readier to associate with more males, the greater the danger (Kappeler and van Schaik 2002; Nunn and van Schaik 2000). However, as the authors point out, more males could be attracted to larger groups of females (for which effect see sec. 2.4.2.5), and larger groups of females could form where risk of predation is high.

Hill and Lee (1998) specifically tested for this possibility of male numbers being determined by female numbers, and found more males per female as predation risk increased. Nunn and van Schaik did the same, but found no effect (Nunn and van Schaik 2000). When we used Hill and Lee's data, but corrected for taxonomic relatedness, which they argued was not necessary because they analyzed within one primate family, we also found no effect. Also, Struhsaker (2000) did not find number of males per female to vary with risk (measured as density of main predator, the crowned hawk eagle) across populations of red colobus. However, extra males might be useless as defenders against raptors, which are too quick for defense to be useful (Peter Kappeler, pers. comm.).

Where number of males per female does not vary with risk, the inference has been made that predation does not affect association between the sexes (van Schaik and Kappeler 1997). However, given that grouping with any other animal, let alone a powerful, mobbing male (in sexually dimorphic species), fully halves the chances of being the victim when a predator attacks, maybe a threshold effect operates? If there is any risk, females benefit from grouping with one male, but after the initial large benefit, having others in the group is not so beneficial. Instead, the number of males in primate groups is explained by competition among the males in relation to the number of females (sec. 2.4.2.5).

2.4. Mating and Society

> [T]he greater size and strength of man, in comparison with woman, together with his broader shoulders, more developed muscles, rugged outline of body . . . have been preserved or even augmented . . . by the strongest and boldest men having succeeded best . . . in securing wives, and thus having left a large number of offspring. It is not probable that the greater strength of man was primarily acquired through the inherited effects of his having worked harder than woman for his own subsistence and that of his family; for the women . . . are compelled to work at least as hard as the men. **CHARLES DARWIN (1871)**

Mates are the major resource for males, the resource that most distinguishes the successful from the unsuccessful male (sec. 2.1.2). Therefore a discussion of mating and society is mostly a discussion of male-male mating competition and society.

That does not mean that mates are not important for females too, or that females do not compete for males. Of course, the concentration on males in this section on the relationship between mating strategies and the nature of society does not mean that females do not influence male mating competition (Harcourt 1997). For over 125 years, choice by females of suitable mates and its influence on competition between males has been a major component of sexual selection theory (Darwin 1871). Go to a cock-of-the-rock display ground in Ecuador—scarlet, black, white birds, but mainly scarlet, leaping and shrieking at dawn in the treetops—and while the males pretty much ignore the arrival of another male at the lek (the gathering of males), the arrival of a female drives the males to frenzy.

But there is little to no way of getting around the fact that competition for mates is most intense in the sex with the higher reproductive rate (Clutton-Brock and Parker 1992; Clutton-Brock and Vincent 1991). And in mammals, because pregnancy severely slows females' reproductive rate (a normal human female cannot reproduce more than once a year), mating competition is most intense among males.

Several excellent reviews, books, and edited books on sexual selection in general and in primates mean that our discussion here can be shorter than it might have otherwise been (Clutton-Brock 1989b; Davies 1991; Kappeler and van Schaik 2004; Setchell and Kappeler 2003; van Schaik and Janson 2000). If we make an unsubstantiated generalization about mating competition, take one or more of these references as the source (unless our generalization is in error, in which case we accept blame).

2.4.1. Mating competition among males and sexual dimorphism

Sexual dimorphism (i.e., differences between the sexes in their morphology, in what they look like) in body size, weaponry, and other traits is significant to understanding the nature of society when it influences which sex can dominate the other. For instance, the smaller body size and weaker explosive strength or aggressive power of females in the human species has surely led to the female sex's subordinance in many (most? all?) human societies. As Thomas Malthus put it over two hundred years ago,

> *. . . the women are represented as much more completely in a state of slavery to the men . . . the misery that checks the population falls chiefly, as it must always do, upon that part whose condition is lowest in the scale of society . . . the women . . . condemned as they are to the inconveniences and hardships of . . . the constant and unremitting drudgery of preparing everything for the reception of their tyrannic overlords.*
>
> (Malthus 1798, ch. 3, p. 41–42)

Competition between males for access to females has long been associated with dimorphism (two body forms, often two body sizes) of many sorts between the sexes (Darwin 1871). Much subsequent work has strongly substantiated that the dimorphism of the sort we are talking about here (body size, weaponry, ornaments) arises largely through mating competition (selection in relation to sex), even if some details have to be sorted out, and some other evolutionary mechanisms than sexual selection are a possible cause of some sexual dimorphism (Hedrick and Temeles 1989; Plavcan 2004). In short, the larger male with the better weapons wins contests, and females choose large males, either as good protectors and providers, or as suppliers of good genes (Clutton-Brock 1989b; Davies 1991; Kappeler and van Schaik 2004; Setchell and Kappeler 2003; van Schaik and Janson 2000). Experiments in birds and fish in which the secondary traits are changed (males' tails elongated or shortened, for instance) illustrate the influence of mating competition, including by female choice, on these traits: females prefer the males with artificially enlarged tails (Andersson 1982).

Nevertheless, in primates, there is little good evidence that variation within species in male secondary sexual traits in fact correlates with variation in mating success (Plavcan 1999, 2004; Setchell and Kappeler 2003), however much variation between species substantiates the connection between the two (Alexander et al. 1979; Clutton-Brock and Harvey 1978; Mitani, Gros-Louis, and Richards 1996b).

2.4.1.1. Sexual dimorphism and body size

Body size, as said before, influences (or correlates with) almost all aspects of animals' (and plants') physiology and behavior, including the nature of their societies. Across large taxonomic groupings of animals, larger-bodied species are more sexually dimorphic (Clutton-Brock, Harvey, and Rudder 1977; Smith and Cheverud 2002), including in taxa in which the female is the larger sex (Abouheif and Fairbairn 1997). However, in primates, the relation is due to a subset of the largest species. Apart from them, essentially no relationship exists between body mass and dimorphism (stats. 2.7) (Lindenfors and Tullberg 1998; Smith and Cheverud 2002). Indeed the small-bodied prosimians/strepsirrhines (largely nocturnal species, except for the larger-bodied diurnal lemurs of Madagascar), and the relatively small-bodied New World primates are noticeably monomorphic, despite great variation in mating system, that is, in intensity of intermale competition.

Prosimian/strepsirrhine monomorphy has been suggested to be due to a very short breeding season producing scramble competition for mates, not contest competition (van Schaik and Kappeler 1994). Certainly, the largest of the Old World primates are non-seasonal, but the nearly monomorphic New World monkeys are not obviously more seasonal than are the dimorphic Old World ones (Harcourt, Purvis, and Liles 1995; Mitani, Gros-Louis, and Manson 1996a). One answer to the greater sexual dimorphism of the larger Old World primates, and presumably therefore to societies of females dominated by males, might lie in the higher proportion of Old World simians that are terrestrial, where there might be less constraint on body size and more fighting advantage to it (stats. 2.8). However, the orangutan is an exception, as it so often is.

2.4.1.2. Sexual dimorphism and group size

Richer resources can induce more competition over them, up to a point (sec. 2.2.2). If there is any connection between competition and dimorphism, in other words if dimorphism can tell us anything about the nature of a society, we should see greater sexual dimorphism in species with larger groups of defensible females. We do, in both primates and ungulates (Clutton-Brock et al. 1977; Pérez-Barbería, Gordon, and Page 2002).

The relationship in primates, though, is not close. A better indication of intensity of competition than either female group size, or the number of females per male in a group, is what is called the operational sex ratio (OSR), which most closely matches level of dimorphism if it is measured as the number of males (competing) per estrous female per day (Mitani et al. 1996a; Plavcan 1999, 2004). The gorilla, because of its slow birthrate and very short estrous periods, has by far the highest OSR and is highly dimorphic; the other highly dimorphic ape, the orangutan, has the next highest OSR and is far above the next primate (Mitani et al. 1996a).

Lack of dimorphism in body size does not mean lack of competition between males. Fighting is not the only way for competing males to win access to females. Plenty of other ways to defeat a rival exist than to fight, or at least than to fight alone (Setchell and Kappeler 2003; Soltis 2004). Males can form coalitions to get access to females (Noë 1992); they can inseminate large amounts of sperm (male access to the female is, after all, only the sperm's means of access to an egg) (box 11.1) (Harcourt 1997; Harcourt et al. 1981c; Harcourt et al. 1995); and males can be friendly to females (Seyfarth 1978a; Smuts 1985, ch. 8).

2.4.2. Male-male mating competition and access to females: Male-female association

Males being friendly to females as a means of gaining mating access raises the topic of association between the sexes. Primates are, as we have said (ch. 1.2.4), an unusual mammal in the high proportion of taxa in which the sexes associate more or less permanently, perhaps 75% of species compared to 30% in other mammals (stats. 2.9). How does this association reflect or influence mating competition?

2.4.2.1. Association for male care

Although in this brief introduction to primate socioecology we are treating mating and rearing strategies separately, there is no point in a male mating with a female (or vice versa) if the offspring do not survive. When females cannot raise offspring on their own, and the main helper is the father, the mating system is equivalent to the rearing system. The male competes with other males both by associating with his mate, and by helping his mate to rear the resultant offspring, for example, by feeding them or carrying them (Clutton-Brock 1989b; Davies 1991; Emlen and Oring 1977). We will consider this sort of society further in section 2.5, Rearing and Society.

2.4.2.2. Association without active care?

If females can raise offspring without male help, why in so many primate species and some other mammalian species do males associate with females outside the mating season? Time spent with a female that is not going to be fertile again until she has produced the current set of offspring, and maybe even not until she has weaned the current set, is wasted mating time.

A male could find it advantageous to associate permanently with a female or group of females if females are so difficult to locate that a male needs to remain with them in order to ensure that he is with a female when she is in estrus (Carlson and Isbell 2001; Dunbar 1988, ch. 13; 2000; Wrangham 1979). The classic cases here are deep-sea organisms that find it so difficult to find each other that as soon as a male finds a female, he attaches to her, even gets inside her and stays there, effectively nothing more than a sperm sac (e.g., Rouse, Goffredi, and Vrijenhoek 2004). The female benefits too, as she has access to a ready supply of sperm when she needs it.

Some evidence that males might associate with females as a means of simply finding them at estrus is that in many species in which the sexes range apart, males have evolved means to stimulate estrus in the females, such as penile spines (Dixson 1987; Dixson 1998, ch. 9; Harcourt and Gardiner 1994; Stockley 2002). An extension of the argument that males stay because they cannot find females is that they stay to be the first with the female when she comes into estrus, or to be the favored mate of the female, because of familiarity.

If male-female association is more prevalent among primates, are female primates more difficult to find than are females of other species? A potential measure of relative difficulty of detection might be their density. John Robinson and Kent Redford (1989) provided data on density of mammal species in South American forests. Primates are not at unusually low density. Indeed, all orders but rodents appear to have extraordinarily similar densities for their body mass, even carnivores. Furthermore, Robin Dunbar and Carel van Schaik calculated from size of territories, densities of each sex, and distances traveled per day, that in several primate species, notably gibbons, males would more or less easily get access to over four times as many females by ceasing to associate with just one female, and instead enlarging their territory (Dunbar 1988, ch. 12; van Schaik and Dunbar 1990). So why do gibbon, and other, males associate with females?

A male can help his offspring in ways other than carrying them. He can protect them and their mother from predators, for instance. Or he can reduce feeding competition faced by the offspring and their mother by keeping other males out (Schülke 2005). But if the benefit of protection is the reason for males to associate with females, why should primate males apparently accrue the benefits more than do other mammals? Part of the answer could be that the period of susceptible immaturity is longer in primates (Read and Harvey 1989), and therefore the male has more chance of making a difference, and therefore gains more benefit from making a difference.

2.4.2.3. Association to mate-guard

Preventing other males from mating is one of the more common reasons for males to associate with females, including in insects (Thornhill and Alcock 1983). Males can mate-guard by defending the land on which the female(s) lives at the time, or by defending the female(s) directly. Whether the society is a "resource defense" or "female defense" one as far as its mating systems are concerned will depend on the distribution of females relative to the male's ability to defend them.

If the area over which females roam is small in relation to the male's ability to patrol it, the male can defend the land, that is, be territorial as a mating strategy (Mitani and Rodman 1979). However, if the area is indefensible, then the male must defend the females (Clutton-Brock 1989b; Davies 1991; Emlen and Oring 1977).

Area defense or resource defense is more prevalent in birds than in mammals (Davies 1991; Greenwood 1980). One reason for the difference is surely that birds can more easily traverse their range than can mammals (Wrangham 1979). Both sorts of defense exist in mammals and primates (Clutton-Brock 1989b; Davies 1991; Dunbar 1988; Smuts et al. 1987). Where males defend a territory (resource defense), females might or might not be active partners in defense also. An example is the gibbon, the females of which take an active vocal part in defense of the territory and have as long canines as males do (Leighton 1987; Plavcan 1999). By contrast, female black-and-white territorial colobus tend to leave most defense up to the male (Fashing 2001).

But if the defense of the land or females, and the consequent association of the male with females, is only to prevent access to estrous females, why does the male stay when females are not in estrus, or not receptive? Especially, why does he stay in species, such as gibbons, where females are at sufficient density for the male to defend access to more than one female (Dunbar 1988, ch. 12; van Schaik and Dunbar 1990)?

2.4.2.4. Association to guard against infanticide

Richard Wrangham's (1979) hypothesis that male and female primates associate with one another to benefit from the male's ability to protect against infanticide by non-father males is gaining ground (van Schaik and Janson 2000), despite Luis Ebensperger's (1998) suggestion that coalitions in general among primates, but including male-female coalitions, are so often unsuccessful that protection against infanticide cannot explain male-female association. We go into more detail about infanticide in section 2.4.3. Here we consider just the hypothesis that it is the basis to the unusual prevalence of male-female association in primates.

A main reason for a male to remain with non-fertilizable females is that a rival male does not necessarily need to wait for a female to come into estrus. He can induce estrus—by killing the female's infant (sec. 2.4.3). Once estrus can be induced, then preventing other males having access to estrous females becomes a full-time business for a male. Even seasonal breeders are not freed from the necessity to guard, for loss of this year's offspring can hasten return to estrus the next year (sec. 2.4.3).

A male remaining to protect his offspring against infanticidal males is usually interpreted as a rearing strategy—protection of offspring from harm. However, infanticide itself is clearly a mating strategy. Prevention of it is therefore a mating strategy, specifically a mate-guarding strategy, and we think is usefully considered as such, as well as a rearing strategy.

Primatologists have argued more strongly than have students of other taxa that females associate with males to benefit from the males' protection against infanticide. Why might primates benefit more from anti-infanticide mate-guarding and offspring protection than do other taxa?

Infanticide is most advantageous to the killing male when the period of lactation is a lot longer than the period of gestation, and females are not fertile during lactation (van Schaik 2000c; van Schaik, van Noordwijk, and Nunn 1999). This situation is likely to arise in taxa in which the young mature slowly. While in general, gestation length correlates with duration of maturation (time to weaning) (Read and Harvey 1989), some taxa have longer times to weaning for the duration of gestation than do others. Primates are one of these (Read and Harvey 1989; van Schaik 2000a). Terrestrial carnivores are another (Read and Harvey 1989; van Schaik 2000a). Infanticide is common in terrestrial carnivores also (Packer and Pusey 1984; van Schaik 2000a, c). Indeed lions are a classic case of adaptive infanticide as a male reproductive strategy (Bertram 1975). Thus, Craig Packer and his colleagues argue that a main reason why female lions live in prides is for defense against infanticidal males (Packer, Scheel, and Pusey 1990).

Another argument for the high proportion of primate species with infanticide, and hence with long-term male-female association, concerns seasonality of breeding. Infanticide should be most advantageous in non-seasonal breeders, because return to estrus is quickest with them, and the infanticidal male is therefore most likely not only to be near the female when she comes into estrus again, but to be alive (van Schaik 2000c). We do not know whether a greater proportion of primate species than species of other mammalian orders are non-seasonal, but even the often highly seasonal lemurs commit infanticide, and protection against infanticide has been offered as an explanation for male-female association in this taxon too (Kappeler and Ganzhorn 1993).

2.4.2.5. More than one male associating

Mammalian and primate species and populations differ in the usual number of males per breeding group, or the proportion of breeding groups with one male or more than one male (Clutton-Brock 1989b; Davies 1991; Kappeler 2000). The explanations for the variation and for why sometimes more than one male associates with a group of females are the same as those that apply to competition for any resource (sec. 2.2.2).

More males occur in groups with more females, especially more estrous females (Andelman 1986; Cords 2000; Dunbar 1988, ch. 7; Harcourt et al. 1995; Mitani et al. 1996a; Nunn 1999b; Struhsaker 2000). While one male might be able to defend a small group of females (a relatively poor resource), if the group becomes large, not only does it become less defensible (spread over a greater area), but it attracts more contenders, which are probably readier to fight harder for the richer resource (Andelman 1986). Monopoly becomes impossible.

Additionally, a co-defender might be positively useful, as in the case of the wagtails (Davies and Houston 1981), and as we have already discussed when talking about cooperation and grouping (sec. 2.2.2; 2.2.3). Thus, larger coalitions of male lions are able to retain tenure of prides for longer than are smaller coalitions, because the larger coalitions win contests (Bygott et al. 1979; McComb, Packer, and Pusey 1994; Packer et al. 1988). The same is true in red howler monkeys, and with them too, as with wagtails, coalitions are more likely the more competitors (Pope 2000).

Patrolling chimpanzee parties are a very nice primate example of the Voltaire principle of larger groups winning contests (sec. 2.2.3.2). Larger patrolling parties of chimpanzee males defeat smaller ones; smaller patrolling parties sometimes retreat without a fight; and with chimpanzees as with lions (McComb et al. 1994), judgment of relative size of party is at least partly mediated through sound—either larger parties make more sound from more different places, or the animals can, in effect, count (Wilson, Hauser, and Wrangham 2001; Wilson and Wrangham 2003). Within groups also, the same Voltaire principle operates of larger groups defeating smaller: coalitions of middle-ranking baboons can defeat a high-ranking male in contests over estrous females (Noë 1992).

2.4.2.6. Breaking the association—emigration

Animals that have bred unsuccessfully, or can breed even more successfully elsewhere, sometimes emigrate (Pusey 1987). For mammals, some of the best evidence for such emigration to improve breeding opportunities comes still from Packer's study of dispersal by successfully breeding baboon males, that is, males with above average mating activity in their current group, which nevertheless moved to groups with a greater number of estrous females than their present group (Packer 1979). Similarly, capuchin males that emigrate subsequent to their first emigration usually move to groups with relatively more females (Jack and Fedigan 2004).

Turning to the emigration of animals from their natal group, if animals remained for their lives in the group in which they were born, inbreeding would be a potential problem. Plenty of evidence exists that both in captivity and in the wild, breeding with close relatives results in greater proportions of poor-quality offspring that have, for example, lower survival and reproductive rates (Frankham, Ballou, and Briscoe 2002; Keller and Waller 2002; Ralls and Ballou 1983; Ralls, Harvey, and Soulé 1986; Shields 1987; Shields 1993; Thornhill 1993). An obvious way to avoid inbreeding is to leave the natal group (Johnson and Gaines 1990; Pusey 1987), and in most species, one or both sexes leave, with the result that inbreeding appears to be rare in wild populations (Ralls et al. 1986).

Either sex could avoid inbreeding by emigration. In many species, both sexes do leave, including many primate species. But in many mammalian species and societies, including primates, a greater proportion of one sex than the other leaves or disperses farther—and that sex is more often the male than the female (Dobson 1982; Greenwood 1980; Harcourt 1978a; Johnson 1986; Pusey and Wolf 1996; Pusey 1987, 1992; Pusey and Packer 1987). Why the sex bias?

In addition to sex bias, the other question is whether the emigration occurs because of costs of inbreeding. Or is the resultant absence of inbreeding a fortuitous consequence of emigration for another reason, that is, another benefit (Johnson and Gaines 1990). Thus, one argument is that young animals emigrate to decrease competition by, for example, finding emptier habitat (Dobson 1982; Pope 2000; Shields 1987). There is also the suggestion that females emigrate, or are forced to emigrate, in species in which larger groups attract more males, and hence suffer more takeovers, and hence higher rates of infanticide (Crockett and Janson 2000; Steenbeek and van Schaik 2001; Sterck and Korstjens 2000).

The starting point to any consideration of dispersal is, as said (sec. 2.2.3.2), that dispersal is initially costly to both sexes—unknown food sites, unknown predator sites, unknown escape routes from predators, loss of opportunity to help and be helped by kin, and increased aggression from strangers (Harcourt 1978a; Isbell and Van Vuren 1996; Johnson and Gaines 1990; Steenbeek 1996; Stenseth 1983). Frugivores should benefit more than do folivores from an intimate knowledge of their range (Barton 2000a; Clutton-Brock and Harvey 1980; Garber 2000; Janson 1998b; Menzel 1997; Milton 1988), and hence suffer a greater cost of emigration. Nevertheless, even foliage differs in quality from place to place and tree to tree (Moore and Foley 2005). Remaining in the natal home range would seem, therefore, to be the default, as Isbell and Truman Young have argued for primates (Isbell 2004; Isbell and Young 2002). But it should be the default for both sexes, other things being equal. Of course, other things are not equal.

If access to resources other than mates determines female more than male reproductive success (sec. 2.1.2), might females suffer more than males from leaving known sites (Pusey 1987)? Additionally, if breeding females disperse, then any female traveling with a fetus or infant will impose costs not only on herself but on her offspring. That is not the case for males. This second argument is similar to the argument that males can more afford to leave in dimorphic species because their breeding is delayed, that is, at the same age, a male will be moving without a dependent offspring (Johnson 1986).

Even if dispersal is somehow equally costly to both sexes, the benefit of moving could be greater for a successful polygynous mammalian male than for a successful female: the increase in reproductive output gained by the male is potentially greater than that gained by the female (Clutton-Brock 1988). Therefore, a successful male is more likely to achieve a gain sufficient to outweigh the costs of inbreeding and dispersing than is a female (Clutton-Brock and Harvey 1976). This argument would apply also if competition, not avoidance of inbreeding, were the main cause of emigration by animals born into a group or area (Dobson 1982; Moore and Ali 1984). Not only might males of polygynous species be more likely to emigrate, but they might have to go farther before they found a site or group in which their relative competitive ability was greater than in their place of origin.

Given that animals do disperse, then as long as immigration occurs, an animal that remains in the group or area of its birth (natal residence) can find non-kin in a group and avoid inbreeding. By the previous arguments, because females are less likely to disperse than are males, and males more likely to disperse than are females, females that remain are more likely to meet non-kin mates than are males that remain in a group. The sex difference in initial payoffs of natal residence and dispersal are reinforced.

However, given that the costs of emigration are high, it could be better for both sexes to stay and inbreed than to emigrate and probably die (Bengtsson 1978; Johnson and Gaines 1990; Lehmann and Perrin 2003). After all, costs of inbreeding are probabilistic: some offspring are fine (Elgar and Clode 2001; Frankham et al. 2002; Ralls and Ballou 1983). If so, then to understand differences between the sexes in propensity to disperse, we need to turn from comparing the sexes to examining the payoffs within each sex to the disperser relative to the stayer, because, after all, animals are largely competing with their own sex, not the other sex.

As long as animals can recognize kin, even if indirectly by past association, inbreeding is not an inevitable consequence of staying. A means of avoiding inbreeding exists—do not mate with kin. Plenty of evidence exists to show that animals, including humans, can effectively avoid mating with kin by avoiding mating with animals that they knew well as immatures, the classic Westermarck effect (Bischof 1975; Manson 2007; Muniz et al. 2006; Paul and Kuester 2004; Pusey and Wolf 1996; Rendall 2004; Shepher 1971; Thornhill 1993; Wolf 1970).

The problem is that staying in a group with sexually unavailable kin reduces the number of mates relative to what would be available in the average new group (Isbell 2004; Shields 1987). That reduction should be a greater cost to males: a female needs only one mate, whereas males succeed in relation to their number of mates (other things being equal, of course). If so, a female staying in her natal group suffers little reduction in reproductive output, whereas a staying male could suffer more than a leaving male—in proportion to the fraction of females in the natal group that are kin.

What evidence is there for mammals that inbreeding affects dispersal, and affects sex bias in dispersal? First, it is almost always the case that either both sexes or one goes from the immediate natal social group or area (Greenwood 1980; Pusey 1987; Stenseth 1983; Thornhill 1993), or if both stay (e.g., pilot whale), males mate outside the natal group (Amos, Schlotterer, and Tautz 1993). Second, across primate species and genera, a greater proportion of females emigrate from groups into which a smaller proportion of males immigrate (Pusey 1988). And across mammalian species, if the tenure of breeding males is longer than the time females take to mature, females emigrate (Clutton-Brock 1989a).

But what determines the duration of the males' tenure? Might it be the case that if it is very beneficial for females to stay, a male that ousted the resident male could be guaranteed the whole package of females, whereas if residence was not obviously beneficial to females, the incoming male might fight furiously, only to find that all females leave? If so, females' readiness to stay and leave determines male tenure (Harcourt 1978a; Stewart and Harcourt 1987), which readiness is determined by, for example, benefits of cooperation in competition (sec. 2.2.3).

In other words, the nature of food, and hence of competition between females over it, determines the duration of male residence, and hence female residence. Food affects the distribution of females, which affects the distribution of males, which affects the distribution of females.

2.4.2.7. No association—roving males

In many mammal species, the male and female travel separately much of the time even if they share the same annual range, associating mainly to mate (Wilson 1975, ch. 23). Among primates, some have this typical mammalian system. Several of the nocturnal lorises, the galagines (the bushbabies) are examples (Bearder 1987). So also is the orangutan (ch. 4.2.3.2).

Males of even some group-living species also follow the strategy. We have already mentioned (sec. 2.4.2.6) Packer's study of successfully breeding baboon males that moved to groups with a greater number of estrous females than their present group (Packer 1979). Similarly, in several other species, the social groups of females are invaded during the breeding season by a number of males that leave at the season's end. What is odd about these societies is that one or a very few males continue to associate with the females outside the breeding season. Examples are the ring-tailed lemur, patas monkey, and blue monkey (Carlson and Isbell 2001; Cords 2000; Kappeler 1999).

In primates, males appear least likely to remain with females outside mating in species in which (a) females need no help in caring for offspring, particularly carrying them (species of more than about 1 kg in body mass [sec. 2.5.1]); (b) breeding is highly seasonal and therefore mate-guarding is a benefit for only a limited time (lorises, lemurs [sec. 2.4.2.3]); or (c) females are as large or larger than males and therefore infanticide is less of a danger (lorises, lemurs [sec. 2.4.2.4; 2.4.3]). Additionally, males and females range apart in species in which the costs of increased competition for food that would result from association (sec. 2.2.3) are likely to be greater than the benefits gained by association. An example is the orangutan, which concentrates on ripe fruit, a rare and often small item in the forest (ch. 4.1.8; 4.2.3.2). And finally, the benefits of increased access to females by roving in search of them (Barnes 1982; Dunbar 1988, ch. 13; 2000) could outweigh the cost of the occasional infant killed by an infanticidal male as a result of the absence of a protective father. After all, not all infants are killed by the invading males.

2.4.3. Male-male mating competition and infanticide: Male-female conflict

Females and males potentially have such different survival, mating, and rearing strategies that conflict between the sexes is an almost inevitable aspect of any species' society. The killing of infants by non-father males is one of the most obvious forms of this conflict (Ebensperger 1998; Hausfater and Hrdy 1984; Parmigiani and vom Saal 1994; van Schaik and Janson 2000). However, a continuum of conflict exists, from either sex refusing to mate, to harassment to induce mating (usually by the larger-bodied sex), and finally to infanticide (Clutton-Brock and Parker 1995; Manson 2007; Smuts and Smuts 1993; van Schaik and Dunbar 1990).

In mammals, infanticide's main beneficial payoff when performed by males is to bring the victim's mother into estrus sooner than if the infant had lived (Hausfater and Hrdy 1984; Parmigiani and vom Saal 1994; van Schaik and Janson 2000). In effect, the infanticidal male increases the local number of fertilizable females. Additionally, in species in which the male associates with females and their offspring, and helps protect them, the killing would prevent the male investing extra help in another male's offspring, and would prevent another male's offspring competing with his own (Ebensperger 1998). These latter two arguments are more often made for females killing other females' offspring (Digby 2000; Ebensperger 1998), but the killing could provide a subsidiary benefit for males. Lastly, it has been suggested that infanticide is a demonstration to the female of the male's power, equivalent to a courtship display in the sense that its benefit to the male is to persuade the female to mate with him (Hamai et al. 1992).

Some still argue against any of these benefits of infanticide, against the adaptiveness of infanticide, but the data and hypotheses for its benefits are now well substantiated (box 2.4).

The prevalence of infanticide, and hence of societal responses to it, varies across species. We have already mentioned long lactation in relation to duration of gestation as a risk factor (sec. 2.4.2.4). Infanticide should also be more common in species in which males are much larger than females (such as Old World monkeys and apes). Again, though, reality does not seem to match prediction. Thus, in a phylogenetically uncorrected analysis, sexual dimorphism did not predict incidence of infanticide (Janson and van Schaik 2000a) and, more specifically, the monomorphic lemurs commit infanticide (Jolly et al. 2000).

Box 2.4. Infanticide in primates: Evolved behavior or pathology?

Almost all of the evidence fits the evolutionary explanation that males benefit from using infanticide as a means of competition for mates with other males because the infanticide brings potential mates back into estrus sooner than if their offspring had been allowed to live (Ebensperger 1998; Hausfater and Hrdy 1984; Packer 2001; Parmigiani and vom Saal 1994; van Schaik and Janson 2000). Almost none of the evidence fits the idea that the behavior is merely pathological, that it is an accident from generalized aggression, or that nobody has properly seen the behavior anyway and that the discovered dead infants died of something else (Bartlett, Sussman, and Cheverud 1993; Dagg 1999; Sussman, Cheverud, and Bartlett 1995).

Three large nails seal the coffin of the anti-adaptationist arguments. One, infanticide has been recorded in the same standard situation in scores of mammalian species: a male that is quite a bit more powerful than females is in the presence of infants that were not sired by that male. Two, in a bird, the jacana, in which females are larger and more colorful than males, and in which females compete aggressively with one another over access to males, which sex is it that commits infanticide? It is, of course, the female (Emlen, Demon, and Emlen 1989). It seems highly unlikely that if the behavior were pathological, the appropriate sex would do it in the appropriate situation in a completely different class of animals. Three, captive male mice will remain infanticidal for about two weeks after mating a female; they then turn non-infanticidal for the time it takes to wean pups; and then they turn infanticidal again (Parmigiani et al. 1994; Perrigo and vom Saal 1994). It seems unlikely that infanticidal behavior perfectly timed to miss a male's own offspring is pathological.

Why in the face of such evidence—and a mass more—will some not accept that adaptive infanticide by males occurs? Volker Sommer's (2000) analysis of the infanticide debate hints that the non-acceptance is because the behavior is perceived as offensive. Offensive as the behavior might be, nevertheless, the adaptationist explanation certainly goes a long way to explaining why infanticide in two-parent human families is around 55 times more likely when one parent is a stepparent (Daly and Wilson 1988, ch. 4).

However, in humans the far more common killer than the stepparent is the mother—who cannot cope with the baby and needs to cut her losses to start again another day when she is more capable of raising a healthy child (Daly and Wilson 1988, ch. 3, 4; Hrdy 1999). This adaptationist explanation is surely disliked by some, but again, it explains a lot of infant-killing, and suggests some precise means of alleviating the problem (Daly and Wilson 1988, ch. 3, 4; Hrdy 1999).

Infanticide should also be more common in single-male species or populations (Palombit 2003), and might be (Janson and van Schaik 2000a). Males are often far larger than are females in such species, but in addition and seemingly more importantly, if only one male mates, all other males in the population can benefit from infanticide (Harcourt and Greenberg 2001; van Schaik and Janson 2000).

Finally, the prevalence of infanticide and male-female conflict in any society will be a balance between conditions that might promote infanticide (e.g., one-male breeding groups) and evolved and current adaptations to prevent it (e.g., long estrous periods and concealed ovulation along with promiscuous mating, or (of particular relevance for this section) association with a male (sec. 2.4.4.1).

2.4.4. Female and male responses to mating competition

2.4.4.1. Female responses to harassment and infanticide

As we have said, if a male benefits by preventing infanticide through continued association with females with which he has mated (sec. 2.4.2.4), the females presumably benefit from associating with a protective male for the same benefit (Harcourt and Greenberg 2001; Palombit, Seyfarth, and Cheney 1997; Sterck 1997; Sterck et al. 2005; van Schaik and Janson 2000; van Schaik et al. 1999; Wrangham 1979). Although some disagree (Ebensperger 1998), much evidence now seems to favor this anti-infanticide hypothesis for females associating with a male or males.

The alternative for the females is to mate promiscuously, so converting potentially infanticidal males into acting as if they were potential fathers (Ebensperger 1998; Harcourt and Greenberg 2001; Soltis 2002; van Schaik 2000a; van Schaik, Hodges, and Nunn 2000). The multimale alternative works better, the longer the estrous period (i.e., the more chances to mate with a number of males), the more obvious it is that the female might be in estrus and fertile (i.e., the more males attracted), and the longer the daily distance traveled (i.e., the more chances of meeting a male) (Harcourt and Greenberg 2001; van Schaik 2000a; van Schaik et al. 2000). Thus one finds an association in primates between long estrous periods, obvious sexual swellings, and multimale matings by females (Clutton-Brock and Harvey 1976; Sillén-Tullberg and Møller 1993; van Schaik 2000c; van Schaik et al. 2000; van Schaik, Pradhan, and van Noordwijk 2004).

Promiscuous mating carries costs, however, and it could be cheaper for a female to associate with one male for protection, especially as he might protect her not only against infanticide but also against other forms of harassment, including in some instances disturbance from several males trying to mate (Clutton-Brock and Parker 1995; Harcourt and Greenberg 2001; Matsumoto-Oda and Oda 1998; Smuts and Smuts 1993; Williams et al. 2002; Wrangham 2002).

Nunn, van Schaik, and others argue that females can get the best of both worlds, that is, a powerful protective male and many or most of the local males acting as potential fathers by at least not being harmful to the infants (Nunn 1999a; van Schaik et al. 2000). The females do this by both having a long non-fertile estrous period before ovulation (when they can mate with several subordinate males that are taking the chance that the female has reached maximal swelling and is ovulating), and then a period of obvious maximal swelling when the waiting dominant male(s) takes over and obtains most access to her.

In analyses of sexual swellings and promiscuity as responses to infanticide, the benefits of inducing caring behavior and minimizing harmful behavior are often separated (Paul 2002; Soltis 2002; Stallmann and Froehlich 2000; Wolff and Macdonald 2004). However, if care is simply the absence of harm, or harm the absence of care, the benefits of each become the benefits of the other, and it might not be useful to make separate predictions for care and harm. Rather, we see increased care and decreased harm as simply opposite ends of a gradient of benefit. The argument for swellings as a graded signal (Nunn 1999a; van Schaik et al. 2000) not only treats the hypotheses as complementary but nicely integrates them into a two-step argument, instead of treating them as competing explanations.

Several additional benefits of swellings and promiscuity, and the consequent actual or potential multiple sires, exist (Soltis 2002; Stallmann and Froehlich 2000; Wolff and Macdonald 2004). Some still need to be demonstrated as operating in primates (Stallmann and Froehlich 2000), even if they apparently operate in other taxa, such as improved offspring viability with multiple sires (Madsen et al. 1992; Olsson et al. 1994).

Finally, as said, it could be that some females respond to the increased risk of infanticide that might result from more males being attracted to larger groups of females (sec. 2.4.2.5) by emigrating or forcing other females to emigrate, and so keeping groups small (Crockett and Janson 2000; Steenbeek and van Schaik 2001; Sterck and Korstjens 2000).

2.4.4.2. Male responses to female responses

Some male responses to female responses to male mating competition (the arms race never stops) are perhaps more easily interpreted as physiological and immediately behavioral rather than societal. For instance, in primate societies where females have swellings and mate multiply, males have larger testes for their body mass and inseminate more sperm of higher quality in greater volumes of semen from large secondary glands than in societies in which females usually mate with a single male (Dixson 1998, ch. 8; Gomendio, Harcourt, and Roldan 1998; Harcourt 1997). And males aggressively herd away females that approach other males (Clutton-Brock and Parker 1995; Kummer 1968; Smuts and Smuts 1993), a form of aggression that is one of the more obvious costs to females of promiscuity (Matsubara and Sprague 2004; Williams et al. 2002; Wrangham 2002).

While we present these responses as adaptations to female promiscuity, they are in effect other means of mate-guarding than we have discussed so far (sec. 2.4.2.3). The point is that male and female strategies and counter-strategies are tightly linked, indeed so tightly linked that it can be difficult—maybe even unhelpful—to separate them. How different is herding (Kummer 1968; Wrangham 2002) from longer-term friendships (Palombit et al. 1997; Smuts 1985) from the permanent association that one sees in gorilla society (ch. 4.2)?

2.5. Rearing and Society

2.5.1. Need for help: Male-female association

In some species, the sexes associate because it is difficult, even impossible, for the female to raise the offspring on her own (sec. 2.4.2.1). The many birds whose nestlings depend on two adults feeding them are an obvious example. Among primates, a classic case is the titi monkey, with the adult male and female sitting shoulder to shoulder, tails entwined, with the offspring on the father's back (see Mendoza and Mason 1986; Rowe 1996). The titis are especially classic, because not only does the father do most of the carrying of the infant, but he does so because sometimes the mother appears actively uninterested in the infant (Hoffman et al. 1995; Mendoza and Mason 1986). Part of the reason for the high proportion of primate species with stable male-female association arises from the high proportion (relative to other mammals) with males that actively help by carrying offspring (Kleiman 1977), even if the proportion that is monogamous is less than previously thought (Fuentes 1999).

But why do primates have a high proportion of species with helping fathers? Because male primates can help is one answer: they can carry infants (unlike, say, male sheep). This "because they can" argument explains why in so many more birds than mammals males help, and why among mammals, male canids (which can regurgitate) help, but male felids (which cannot) do not (Emlen 1991).

There is also the issue of need for help. The species in which males actively help by carrying are all small-bodied (≤ c. 1 kg), with offspring that weigh a large proportion of the mother's body mass, especially for the smallest-bodied, which have litters (Goldizen 1987; Leutenegger 1979; Rylands 1993). In these species, a helping male can make a large difference to the female's ability to raise offspring (Koenig 1995), with the result that a male that abandons his mate would often have lower reproductive success than one that stays and helps her raise offspring (Dunbar 1988, ch. 12).

Other animals can help too, and many of the callitrichines (the marmosets and tamarins), have more than one helper, including other males and subadults (Goldizen 1987; Rylands 1993). Although more helpers correlates with more offspring raised (e.g., Koenig 1995), the problem for a male of accepting a male helper is that the helper can also be a mating competitor: even if dominant male tamarins garner most of the matings, the large size of tamarin (but not marmoset) testes relative to their body mass indicates a multi-male mating system (Harcourt et al. 1995).

In groups of the larger-bodied socially monogamous species, such as the gibbons, males do little active care. Instead, it seems that they maintain an association to guard the female and offspring from other males (sec. 2.4.2.4) (Dunbar 1988, ch. 12; van Schaik and Dunbar 1990).

2.5.2. Female competition to rear

Females compete with one another for resources for their offspring. Competition for food between females, perhaps especially lactating females, is rearing competition. The rearing competition continues with females helping their own offspring to win contests against other females' offspring. Such rearing competition can have a profound effect on the nature of society, as when it results in daughters achieving dominance ranks next to those of their mother, so producing a nepotistic class system in the society, with matrilines ranked in respect to one another (sec. 2.2.3.2). Some have even argued that females might time their harassment so as to delay their competitors' estrus or to induce miscarriage (Wasser and Starling 1988).

And the final resort if previous competition has not worked is infanticide (Digby 2000; Digby, Ferrari, and Saltzman 2007). It is not only males that kill competitors' offspring. Thus, among the callitrichines, usually only one female in a group breeds (Goldizen 1987; Rylands 1993). If more than one does so, the dominant female will sometimes kill the rival infant (Digby 1995; Digby et al. 2007). This is rearing/resource competition, and is common in mammals (Digby 2000). The female does not benefit from access to mates but from retaining all the helpers' attention for her own offspring (Digby 2000; Digby 1995; Digby et al. 2007).

Conclusion

As the title of the chapter indicates, this is just a brief introduction to primate socioecology. We present it to illustrate, on the one hand, the nature and breadth of the ideas that we bring to our interpretation of gorilla society, and on the other hand, the theory that we hope our analysis of gorilla society will illuminate. No recent full treatment of the topic is available. Nevertheless, Dunbar's *Primate Social Systems* is still highly relevant (Dunbar 1988), as is the multi-authored 1987 *Primate Societies* edited by Barbara Smuts and colleagues (Smuts et al. 1987), and the new *Primates in Perspective* edited by Christina Campbell and colleagues (2007a). And Karen Strier (2007) has provided an easy introduction to the topic with her *Primate Behavioral Ecology* even if she does, as her title suggests, concentrate on the behavior of individual primates rather than the nature of primate societies.

In all these books, the environment, both physical and social, is a strong influence on the behavior of individuals and hence the nature of societies; in all, male and female strategies are different and can conflict; in all, the females-to-food, males-to-females dichotomy is a foundation of understanding; in all, the operation of the separate survival, mating, and rearing strategies of the sexes is obvious; and in all, one can see emerging the fact that simple rules can produce complexity in society, as they operate to influence the nature of conflict, compromise, and cooperation among the interacting members of the societies.

Statistical Details

Stats. 2.1. Intense competition correlates with residence and cooperation with kin. Ten of 20 species with intense competition stay and cooperate; 0 of 13 species with mild competition do so ($\chi^2 = 9.3, p < 0.01$ uncorrected for taxonomic relatedness; $p = 0.05$ when most rigorously corrected) (Nunn and van Schaik 2000).

Stats. 2.2. Across primate species, larger-bodied species live in larger groups ($r^2 = 0.21, F_{1,89} = 24.4, p < 0.0001$; data from the literature). However, across genera (less double-counting than with species—see box 2.2), no relationship at all exists ($F_{1,41} = 0.4$; $F_{1,39} = 2.0$ [excluding outliers]). And as soon as phylogeny is fully taken into account by the use of comparative analysis by independent contrasts (see box 2.2), or the small-bodied, solitary or pair-living nocturnal taxa are removed, the relationship is completely flat even for species ($r^2 = 0.01$), let alone for genera. The African antelope show a similar relationship: overall, small-bodied species live in smaller groups and hide as an antipredator strategy (and have a higher-quality diet), but when phylogeny is accounted for, no significant relationship exists between group size and body mass (Brashares et al. 2000).

Stats. 2.3. Body size appears not to affect predation rate. With all of the data, and no corrections, body mass correlates with predation rates (Isbell 1994). However, count only diurnal species (to remove the confounding of two antipredator strategies), and remove the two (small-bodied) congeneric tamarin species that have very high predation rates, and no relationship exists in the remaining 16 species ($F_{1,14} = 0.7$, $p > 0.1$). Alternatively, account statistically for phylogenetic relatedness with CAIC (box 2.2), and no significant relation between predation rate and body mass remains in either the total sample or when only diurnal species are counted (All species, $N = 10, 7$; $T = 61$; $z = 0.7$; $p > 0.1$; Diurnal species only, $N = 10, 6$; $T = 59$; $z = 0.5$; $p > 0.1$).

Stats. 2.4. Percent successful defenses by red colobus males by number of males ($N = 8$, $r^2 = 0.7$, $p < 0.05$ [Stanford 2002]).

Stats. 2.5. Larger primate groups occur under conditions of greater predation risk (no phylogenetic correction, $\chi^2 = 9.3$, $p < 0.01$; after correction, $p < 0.05$ [Nunn and van Schaik 2000]). Using the Nunn/van Schaik categories of predation risk with our independently collected data on adult group size, we get much the same answer that they do: species in their higher two categories of predation risk live in groups with about three times as many adults as do those in the lowest risk category (they assumed small-bodied, terrestrial species in open habitat in mainland Africa and South America to be most at risk).

Stats. 2.6. Hill and Lee (1998) did not correct for phylogenetic effects but argued that they were comparing populations, not species, and that because most variation in the data was at the population level, taxonomic effects were minor.

Stats. 2.7. Sexual dimorphism by female body mass in Old World monkeys and apes (Catarrhini, $N = 47$). With no phylogenetic correction, $p > 0.1$; with it, $p < 0.05$ (Smith and Cheverud 2002); but with three influential data points removed (i.e., the minority that strongly influence the relationship), we found that no relationship remains, $p > 0.2$. (Female mass is used in these calculations, not male or average of male and female, because then dimorphism would be in the *x*-axis as well, and in effect, dimorphism would being plotted against dimorphism.)

Stats. 2.8. All terrestrial or partially terrestrial genera are Old World. Terrestrial genera are significantly more sexually dimorphic than are arboreal genera ($N = 7$ terrestrial, 12 arboreal, $df = 2$, $z = 2.37$, $p < 0.02$). Data from Harcourt and Stewart (unpublished). See also Harvey, Kavanagh, and Clutton-Brock (1978).

Stats. 2.9. Unusual proportion primate species with male-female association. Insectivora, 10%; Artiodactyla, 20%; Rodentia, 35%; Carnivora, 35%; Primates, 75%. Data from van Schaik (2000a).

PART 2

GORILLAS, ECOLOGY, AND SOCIETY

Lactating female with 2–3 yr old infant. (Mountain gorilla.) © A. H. Harcourt

CHAPTER 3

Introducing Gorillas: Some Background

Summary

We here present some background information on gorilla biology that is fundamental to understanding of the species' socioecology.

Gorillas live only in Africa. They are in the taxonomic family Hominidae, along with humans and all the other great apes (the orangutan and both species of chimpanzee), and in the subfamily Homininae, which does not include the orangutan. Our knowledge of wild gorillas comes mainly from twelve study sites, eight in west-central Africa and four in eastern Africa. Gorillas are by far the largest of the primates, and one of the most sexually dimorphic, with males at 160 kg—about twice the size of females. Despite their relatively large body size, gorillas have fast life histories, compared to *Pan* and *Pongo.*

3.1. Distribution, Taxonomy, and Study Sites

3.1.1. Distribution: Where gorillas live

Gorillas exist only in equatorial Africa, in two main regions separated by the Congo Basin (fig. 3.1). The western gorilla is the most numerous and widespread form, inhabiting eight countries of west-central Africa, west and north of the Congo River (Butynski 2001) (see also ch. 14.1.2). Its presence in Angola's Cabinda enclave has only recently been confirmed after decades of uncertainty (Ron 2005). Nearly a thousand gorilla-free kilometers to the

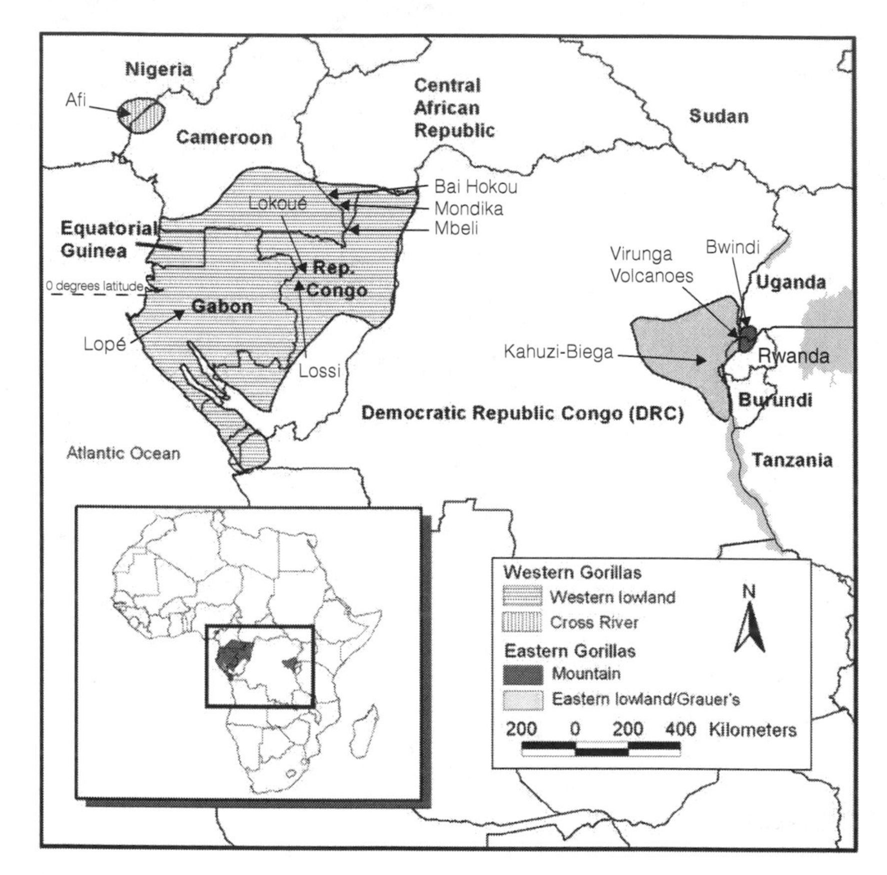

Fig. 3.1. Extent of gorilla distribution in Africa. Arrows indicate sites in table 3.1. Shading encompasses general regions of distribution, but gorillas occur discontinuously within these areas. For more detailed maps of separate regions see Caldecott and Miles (2005, ch. 7, 8). Sources: Data from Butynski (2001), Groves (1970, 1971), Lee, Thornback, and Bennett (1988), Schaller (1963), Wolfheim (1983).

east are the eastern populations, divided into two subspecies, the eastern lowland gorilla, also called Grauer's gorilla, of eastern Democratic Republic of the Congo (DRC), and the rarer mountain gorilla of Rwanda, Uganda, and DRC (fig. 3.1) (Butynski 2001). Grauer's and the mountain gorilla are completely separated from each other, primarily by savanna.

3.1.2. Taxonomy: How many species?

Gorillas belong to the family Hominidae, which includes all the great apes—chimpanzees, gorillas, orangutans—along with humans (Groves 2001a). The orangutans of Asia separated from the common ape-human ancestor earlier than did chimpanzees or gorillas and are classed in their own subfamily, Ponginae. Both genetic and fossil evidence suggest that the line leading to orangutans separated from the line leading to the African apes and humans 12–16 million years ago (Goodman et al. 1998). Humans and the African apes are classed together in the subfamily Homininae, which is divided into the three genera of *Gorilla, Pan,* and *Homo.* Almost all genetic analyses indicate that *Pan* and *Homo* are more closely related to each other than either is to *Gorilla* (Goodman et al. 1998; Ruvolo et al. 1994; Vigilant and Bradley 2004). The lineage leading to gorillas is estimated to have diverged around 6.3–8.5 million years ago, whereas the chimpanzee-human split probably occurred some 5–7 million years ago (Chen and Li 2001; Kumar et al. 2005; Vigilant and Bradley 2004).

Within the genus *Gorilla,* the number of species and subspecies is a subject of debate. The traditional classification, in place for over thirty years, divides one gorilla species into three subspecies, one in western central Africa, the western lowland gorilla (*Gorilla gorilla gorilla*), and two in eastern central Africa, the eastern lowland or Grauer's (*G. gorilla graueri*) and the mountain gorilla (*G. g. beringei*) (Groves 1970; Uchida 1996).

At present, however, gorilla taxonomy, indeed ape taxonomy, is in somewhat of an upheaval. Colin Groves recently reclassified the western and eastern gorilla populations into two species, *Gorilla gorilla* in the west, and *Gorilla beringei* in the east (fig. 3.1 legend) (Groves 2001a; Grubb et al. 2003). Within these divisions, the same subspecies are recognized with one potential addition: the small isolated population on the Nigeria/Cameroon border has been reclassified by some as a separate subspecies, *G. g. diehli* (Groves 2003; Grubb et al. 2003; Oates et al. 2003).

Not everyone agrees with the new classification of the gorilla (see Taylor and Goldsmith 2003). It follows the trend of taxonomic splitting in primates as a whole (Groves 2001a; Grubb et al. 2003). The splitting has been criticized on various grounds (Agapow et al. 2004; Isaac and Purvis 2004) with strong counterargument (Brandon-Jones et al. 2004; Groves 2001b). Indeed, it is not unlikely that current varieties of many understudied tropical species of animal could be true species (Hebert et al. 2004).

While most genetic, morphological, and anatomical studies show a divide between eastern and western gorilla populations, the extent of the difference depends on the exact analysis performed (Uchida 1996). The relatively young field of DNA taxonomy is particularly problematic because interpretation of results can be difficult (Jensen-Seaman, Deinard, and Kidd 2003; Vigilant and Bradley 2004). For example, gorillas' mitochondrial DNA (mtDNA) shows relatively high rates of "mistaken" insertion into nuclear

DNA, resulting in "fake" mtDNA sequences that are difficult to distinguish from the real thing. This means that conclusions based on mtDNA analyses, such as the length of time that western and eastern populations have been separated, are largely suspect (Thalmann et al. 2004).

The wisdom of splitting western gorillas into subspecies is also questionable due to the extensive genetic variation across that group as a whole (Garner and Ryder 1992; Ruvolo 1997). In sum, there is not yet a definitive answer to the question of how many gorilla species or subspecies exist (box 3.1).

We have decided to stick to the old-fashioned taxonomy, and be "lumpers" (one species, three subspecies) and not "splitters." In essence, we are using the ornithologists' concept of the superspecies, or species group (Grubb 2006). However, lumping or splitting makes no difference to what we say about gorillas in the book (ch. 1.3). Most of the data on wild gorillas come from particular populations from each of the still-recognized subspecies, mountain, eastern lowland (or Grauer's), and western gorillas. These names stay the same whatever the taxonomy, and therefore they are the names that we use. Similarly, our usual use of site names for populations, for example, Cross River gorillas or Virunga gorillas, obviates differences of opinion about subspecific status.

3.1.3. Study sites

Full understanding of the socioecology of a long-lived species such as the gorilla entails monitoring individually recognized animals over many years. For this sort of study, the animals must be sufficiently habituated to human observers to tolerate their more or less close presence (fig. 3.2). However, many kinds of information can be gleaned from simply following gorillas' trail in the forest or examining nightly nesting sites. Trail evidence provides crucial data on diet, ranging, group size and composition, and population size (e.g., Oates et al. 2003; Remis 1997a, b; Tutin and Fernandez 1984; Yamagiwa and Mwanza 1994).

The main sites for gorilla studies differ in levels of habituation of the animals, visibility of animals, and the ease with which trails can be followed. Thus, sites vary in the types of information they yield. We provide brief descriptions of the major long-term (at least two years) study sites, separated by subspecies (table 3.1). While these are the primary sites for long-term research, numerous other shorter studies, including surveys and censuses, have provided valuable demographic and ecological information on gorillas and are referred to often throughout this book.

Box 3.1. Gorilla taxonomy: One species or two?

> *"Non-arbitrary taxonomy is flatly impossible"* (p. 173). *"To insist on an absolute objective criterion* [of what constitutes a species] *would be to deny the facts of life, especially the inescapable fact of evolution."* (p. 152). **George Gaylord Simpson (1961, ch. 5)**
>
> *"[A] taxonomic scheme is a hypothesis . . . and so is absolutely bound to change. . . ."* **Colin Groves (2001a, p. vii)**
>
> *"Our taxonomy is not definitive; taxonomy never is."* **Peter Grubb et al. (2003, p. 1302)**

We have written this book as if only one species of gorilla exists, although we recognize that some classify the gorilla as divided into two species (and four subspecies) (Groves 2001a; Grubb et al. 2003). Taxonomy is to some extent in the eye of the beholder, a sentiment reflected in the quotes at the top of this box. What biologists term a *species* can sometimes be a more or less arbitrary break in a continuum of variation, except with, for example, good evidence that two forms exist in the same site and rarely interbreed. Indeed, it might even be sensible to change definitions of what constitutes a species depending on the question being asked and the sort of organism being studied (Hey et al. 2003).

Groves is the world's premier primate taxonomist. The recent massive revision of primate taxonomy, its accompanying 40% increase in number of primate species, and its doubling of the number of gorilla species is largely his work (Groves 2001a), and largely a result of raising subspecies to the level of species (Isaac and Purvis 2004). These changes are based less on new findings from, for example, genetic or anatomical or biogeographic research, and more on new decisions about where to break the continuum of variation. Groves himself was explicit about the reasoning behind his reclassifications. In order not to miss a difference, he assumed one existed until similarity was demostrated (Groves 1993, p. 38; 2001b). This approach to defining species is part of what has been termed the *phylogenetic species concept,* in which the guesstimation of potential for interbreeding of the classic biological species is dropped in favor of identification in any way of a distinguishable population of individuals (Groves 2001a, ch. 3).

Taxonomy, like all of science, is producing and substantiating interesting ideas about the form of the world and how it works (Hey et al. 2003). Testable hypotheses and interpretation of data are as much a part of taxonomy as they are of any branch of science. Beyond a certain point, a choice between two taxonomic hypotheses depends on judgment, not on demonstrable rightness or wrongness of one or the other. As Charles Darwin wrote in a letter to Asa Grey in 1857, "When I was at systematic work, I know I longed to have no other difficulty . . . than deciding whether the form was distinct enough to deserve a name; & not to be haunted with undefined and unanswerable question whether it was a true species" (Burkhardt 1996, p. 183).

Fig. 3.2. Observer (Kelly Stewart) studying well-habituated gorillas (Group 5 of the Karisoke Research Center; see table 3.1). (Mountain gorilla.) © A. H. Harcourt

3.1.3.1. Mountain gorillas

Until recently, generalizations about most aspects of socioecology of wild gorillas were based on the small population of approximately 360 mountain gorillas in the Virunga Volcanoes of Rwanda, Uganda, and DRC (table 3.1). It was here that George Schaller's pioneering study in 1959 laid the groundwork for the most famous gorilla researcher, American zoologist Dian Fossey, and all others that followed her (Fossey 1983; Schaller 1963). In 1967 she pitched a tent at around 3,000 meters in the Virunga Volcanoes in Rwanda, thereby establishing the Karisoke Research Center. It was the start of one of the longest studies of any wild primate anywhere in the world. The descendents, down to three generations, great grand-offspring, of the gorillas that Fossey first studied are still being observed today (Robbins et al. 2005). The Virunga gorillas are still the population from which comes most information on social relationships, mating and reproduction, group dynamics and demographic processes, and hence the structure of the society (Stewart, Sicotte, and Robbins 2001).

The other population of mountain gorillas, only 30 km from the Virungas, in the Bwindi Impenetrable National Park of southwest Uganda, has been observed since the late 1990s. Living at altitudes about 1,000 m lower than the Virunga Volcanoes, the Bwindi gorillas have enabled valuable comparisons with the Virunga population in diet, ranging, and group sizes and compositions (Goldsmith 2003; Nkurunungi et al. 2004; Robbins and McNeilage 2003; Sarmiento, Butynski, and Kalina 1996).

Table 3.1. Long-term study sites for mountain, eastern lowland, and western gorillas in order of establishment of research program. If research is continuing, dates indicate year that studies began. Type of study: T = data from gorillas' trails; P = study groups partially habituated; H = study groups habituated; Clearing = all data from observations at swampy clearings. NP = National Park.

	Country	Protected area	Study site	Type of study	Start date
Mountain *G. g. beringei* *(G. b. beringei)*	Rwanda, Uganda, Dem. Rep. Congo (DRC)	Virunga Volcanoes NPs	**Karisoke**	H	1967
	Uganda	Bwindi Impenetrable NP	**Bwindi**	T, H	Late 1990s
Eastern lowland	DRC	Kahuzi-Biega NP	**Kahuzi**	H	1970s, 1980s
(Grauer's) *G. g. graueri* *(G. b. graueri)*			**Itebero**	T	1987–1991
Western lowland *G. g. gorilla*	Nigeria	Afi River Forest Reserve	**Afi Mountain**	T	1996–1999
	Gabon	Lopé National Park	**Lopé**	T, P	Early 1980s–early 2000s
	Central African Republic (CAR)	Dzanga-Sangha Reserve	**Bai Hokou**	T, P	Early 1980s
		Dzanga-Ndoki NP	**Mondika**	T, P	Mid-1990s
	Republic of Congo	Odzala NP	**Lossi**	H	1994
		Nouabalé-Ndoki NP	**Mbeli**	Clearing	1995
		Odzala NP	**Maya Nord**	Clearing	1996–early 2000s
		Odzala NP	**Lokoué**	Clearing	2001

3.1.3.2. Eastern lowland, or Grauer's gorillas

Researchers have been studying Grauer's gorillas in Kahuzi-Biega National Park in eastern DRC intermittently since the 1970s (table 3.1) (Casimir 1975). Of particular value are the observations of four habituated groups, conducted since the early 1980s, and tragically curtailed by recent war and deliberate killing of the animals (Yamagiwa 1999b; Yamagiwa et al. 2003a, b; Yamagiwa and Kahekwa 2001). The Kahuzi-Biega National Park is separated into two distinct sectors. One consists of higher altitude montane forest, the Kahuzi sector. The other is a far larger lowland forest region, the Itebero sector. The habituated animals inhabit the highland area and are the source of all long-term behavioral and demographic data on this subspecies. The gorillas of Itebero are not habituated, and data from here consist primarily of information gleaned from trail evidence during regular censuses (Hall et al. 1998; Yamagiwa et al. 2003b).

3.1.3.3. Western gorillas

Western gorillas have proved much more difficult to habituate and observe than eastern populations. Part of the problem lies with their habitat and diet. In tropical lowland forest with its abundant fruit but relatively sparse undergrowth, foraging gorillas (often searching for fruit) travel further than do those in montane habitats, but leave a less obvious trail as they go (ch. 4.1.4). They are therefore challenging subjects for researchers to find and observe consistently. In addition, western gorillas have always been more widely hunted for meat than have mountain gorillas (ch. 14.1.1), and are consequently far more fearful of humans.

Nevertheless, over the last twenty-five years, several long-term study sites have provided an ever clearer picture of western gorillas' socioecology (table 3.1). The longest of these studies was based at Lopé National Park in Gabon. It was initiated by Caroline Tutin and Michel Fernandez in the early 1980s following their amazing countrywide census on foot (782.8 km of transects) of gorillas and chimpanzees (Tutin 1996; Tutin and Fernandez 1984). While observations of habituated western gorillas are still rare, these data are finally starting to emerge. In ongoing studies at Bai Hokou and Mondika in Central African Republic (CAR), researchers can now find and follow gorillas through the forest at distances close enough to observe their behavior. At both sites, this achievement took one to two years of dedicated daily follows of a single group (Cipolletta 2003; Doran-Sheehy et al. 2004).

Only ten years ago, estimates of western gorilla numbers more than doubled when surveys uncovered high densities of animals in the swamp forests of west-central Africa (ch. 4.1.2). There are three main sites situated at swampy forest clearings, all in the Republic of the Congo (Rep. Congo) (table 3.1) (Gatti et al. 2004; Levréro et al. 2006; Magliocca, Querouil, and Gautier-Hion 1999; Parnell 2002). At these sites, large numbers of gorillas feed completely out in the open while observers watch them from raised platforms set up on the edge of the clearings (box 4.1). These clearing studies (table 3.1) enable researchers to collect demographic and behavioral data on large numbers of individually recognized and repeatedly observed gorillas.

3.1.4. Comparison with *Pan* and *Pongo*

Chimpanzees, the closest relatives of gorillas, are confined to Africa and live in all forests inhabited by gorillas except the Virunga Volcanoes of DRC, Rwanda, and Uganda. Chimpanzees, however, inhabit more diverse habitats than do gorillas, and their distribution covers more of Africa, stretching across the Congo Basin to Sierra Leone in the west, and down to western Tanzania in the east (Boesch, Hohmann, and Marchant 2002; Groves 2001a; Stumpf 2007).

Primate taxonomists currently recognize two species of chimpanzee: *Pan troglodytes,* the common chimpanzee, and *Pan paniscus,* the bonobo, previously known as the pygmy chimpanzee, which is restricted to the Congo Basin in DRC. There are three traditionally recognized subspecies of common chimpanzee that are geographically distinct, *P. t. verus* in western Africa, *P. t. troglodytes* in central Africa, and *P. t. schweinfurthii* in eastern Africa, but as for the gorilla, these long-standing taxonomic divisions are now being reconsidered (Groves 2001a; Uchida 1996).

In our general inter-ape comparisons, we usually distinguish between the chimpanzee species but not subspecies. In many instances, however, we refer to separate chimpanzee study sites, because different populations show considerable diversity in ecology, behavior, and the nature of their societies (Doran et al. 2002a; Uchida 1996). Long-term studies at six sites, four in eastern Africa and two in western Africa, have provided years, and in some cases, decades, of data on habituated chimpanzees. There are two such sites for bonobos (Stumpf 2007).

Orangutans, confined to the Asian islands of Borneo and Sumatra, are the least well known of the great apes (Knott and Kahlenberg 2007). Until recently, most researchers recognized two subspecies, *Pongo pygmaeus abelii* on Sumatra, and *P. p. pygmaeus* on Borneo, although many now argue that these should be considered separate species (Groves 2001a; Wich et al.

2004). In this book we will refer to them as the Sumatran and Bornean orangutan. Most of the information we have on wild orangutans come from two long-term study sites on Sumatra and three on Borneo (Delgado and van Schaik 2000).

3.2. Life History and Reproduction

3.2.1. Introduction

The socioecology of a species is closely linked to its life history. *Life history* refers to the speed at which animals live their lives, for example, the rates at which they mature and reproduce. In mammals, the most important life history traits are gestation length, size and number of newborns, age at weaning and first reproduction, interbirth interval, and life span (Kappeler, Pereira, and van Schaik 2003). While individuals and populations vary in many of these parameters (e.g., Bentley 1999; Harcourt 1987a; Tutin 1994), life history traits generally typify a taxon. Species can thus be characterized as having fast or slow life histories (Charnov 1991; Harvey, Promislow, and Read 1989; Judge and Carey 2000; Promislow and Harvey 1990; Read and Harvey 1989).

Life history traits set the boundaries within which males' and females' survival and reproductive strategies evolve (ch. 2) and are therefore fundamental influences on the nature of societies. For example, slow reproductive rates lower the number of sexually active females at any one time relative to the number of sexually active males. This lowered ratio will tend to increase mating competition between males, and so have a major impact on the mating system (ch. 2.4.1.2) (Mitani et al. 1996b).

Probably the most important single feature for understanding interspecific differences in life histories is body size (ch. 2). Body mass correlates closely with virtually all life history traits, as well as with the size of all major organs, including the brain (see review in Leigh and Blomquist 2006). As a general rule, larger animals live slower lives than do smaller animals (Harvey, Martin, and Clutton-Brock 1987; Kappeler et al. 2003; Pagel and Harvey 1993; Read and Harvey 1989).

Of course, body size is not the whole story, and mammals differ significantly in the speed of life after body size is controlled for. As a taxon, primates have among the slowest life histories for their size of any mammalian order (Charnov and Berrigan 1993; Harvey et al. 1989; Kappeler et al. 2003; Lee 1996), a fact that is ultimately tied to their relatively low mortality rates.

The risk of untimely death appears to be closely linked to life history schedules across mammals (Charnov 1991; Charnov and Berrigan 1993; Harvey and Pagel 1989; Promislow and Harvey 1990; Purvis et al. 2003). When the probability of reaching and surviving adulthood depends heavily on factors such as disease or predation, over which individuals have little control, then it is beneficial for females to wean their offspring early, and for those offspring to mature quickly before they die. In short, and with exceptions to the rule (Abrams 2004), the higher the risk of dying, the faster animals live their lives. When mortality rates are low and life expectancy high, animals can take their time to mature and reproduce. Thus bats, almost immune to predation, can live about thirty years, sparrows three years, and mice often less than a year. The long life enjoyed by primates might correlate with their arboreal lifestyle and the safety from predators this may, in some cases, afford (van Schaik and Deaner 2002).

While body size and mortality risk are key determinants of life history, many other factors exert an influence. Among the primates, numerous ecological and social variables, such as diet or non-maternal care, correlate in sometimes complex ways with life history variation (Bentley 1999; Kappeler et al. 2003; Lee and Kappeler 2003; Ross and Jones 1999).

The following section summarizes information on gorilla body size and life history, presents related data on reproduction and causes of mortality, and contrasts the findings with information from the other great apes.

3.2.2. Body size and sexual dimorphism

The great apes are the largest of all primates, and gorillas are the largest of the apes (ch. 1.3). Along with orangutans, they are also among the most sexually dimorphic primates (fig. 3.3A, B) (Mitani et al. 1996b; Plavcan 2004; Smith and Cheverud 2002): adult male gorillas weigh around 165 kg (365 lb), while adult females weigh about half this (fig. 3.3A, B).

Male gorillas are not only far larger than females, but develop distinctive secondary sexual characteristics, that is, traits more obviously associated with mating success than with survival or rearing of offspring (fig. 3.3A). Across mammals, marked sexual dimorphism results primarily from sexual selection, that is, male mating competition and female choice that select for traits useful in defeating or intimidating rivals and impressing mates (ch. 2.4.1). In male gorillas, these features include huge sagittal (down the middle) and nuchal (across the top of the back) crests on their skulls for attachment of jaw muscles, and relatively large canines (Harvey et al. 1978).

Fig. 3.3A. Male gorillas are twice the size of females (the second-largest animal in the picture), and have distinctive morphological traits. Note the male's silver back and the large crest (for jaw muscles) on his head. The smallest two animals are a juvenile and infant. (Mountain gorilla.) © A. H. Harcourt

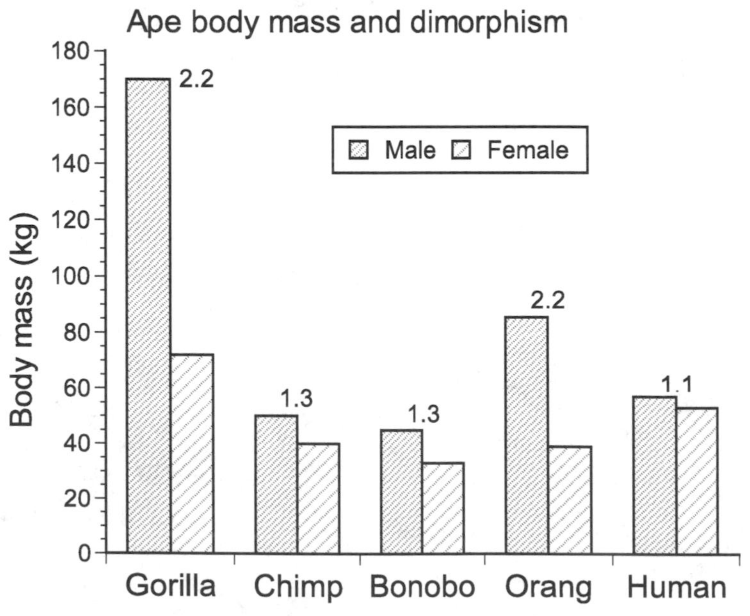

Fig. 3.3B. Mean body mass, in kilograms, of male and female great apes and, for the sake of comparison, humans. Numbers next to male histograms indicate ratio of M/F weights, that is, degree of sexual dimorphism. Data are for wild animals when available but also include some data from captivity (Smith and Jungers 1997). The values for humans come from four hunter-gatherer societies (Kaplan et al. 2000).

Not all the gorilla's sexually selected characters involve weaponry. As individuals approach full maturity, males develop a saddle of grey-white hair on their backs, hence the title silverback (Schaller 1963). This saddle may be a feature designed to make males look bigger and more imposing: a famous optical illusion shows that breaking up a black rectangle with a central white area makes the rectangle appear longer. A silver back is also a visual signal of full maturity. Another trait that differs between males and females is the quality of their chest-beats. While all gorillas beat their chests in times of excitement, the behavior is most developed in adult males and is an important component of their aggressive displays. A silverback's chest-beat has a unique resonant sound like that of a gigantic champagne cork popping, and is amplified by inflatable air pouches that are extensions of the larynx (Schaller 1963, p. 225).

While sexual dimorphism reflects a species' mating system (ch. 2.4.1), it can in turn influence society by affecting males' abilities to control the behavior of females (ch. 2.4.1) (Smuts and Smuts 1993).

3.2.3. Maturation and reproduction

3.2.3.1. Infants to subadults

The striking differences between the sexes in body form and behavior take years to emerge. At birth, both males and females weigh a little over 2 kg and grow at similar rates until about five to seven years of age. Then males' growth begins accelerating relative to females' (Leigh and Shea 1995; Smith and Jungers 1997).

Most of our data on maturation in wild gorillas come from the Virunga population of mountain gorillas. However, the available information on eastern lowland and western gorillas indicates comparable growth rates. The following age class definitions are based on mountain gorillas (Fossey 1979; Harcourt, Fossey, and Sabater Pi 1981a; Schaller 1963; updated in Watts and Pusey 1993) but are applicable to other populations and subspecies (Robbins et al. 2004; Yamagiwa and Kahekwa 2001).

Infants are animals from birth through thirty-six months, by which time they are usually weaned and their mother has returned to sexual cycling (sec. 3.2.3.3). Young gorillas, however, continue to share their mother's night nest until her next infant is born (Fletcher 2001; Fossey 1979; Stewart 1988). **Juveniles,** aged three to six, have been weaned but have not yet reached puberty. **Subadults** are between six and eight years of age, that is, between puberty and fertility. At this point, male and female developmental paths diverge (Watts and Pusey 1993).

3.2.3.2. Maturity and reproduction in males

It is difficult for observers to assess the sexual development of subadult males. Unlike chimpanzees, they lack conspicuous testes that show measurable accelerated growth at the onset of puberty. The first completed copulation is the clearest indication of puberty, which, in wild mountain gorillas, occurs between eight and nine years, when males enter adolescence (Watts 1990b, 1991b).

Blackbacks are young adolescent males between eight and about twelve years. During this period, the hair on their backs becomes shorter and gradually turns a silvery white. In addition, their other secondary sexual traits develop (discussed earlier). **Young silverbacks** are males between about twelve to fourteen years, by which time they are considered mature **silverbacks.** Fourteen is the youngest known age at which a male (mountain gorilla) became a dominant breeder (Watts and Pusey 1993). Observers of wild mountain and Grauer's gorillas estimate that the males reach full physical maturity at around fifteen years (Watts and Pusey 1993; Yamagiwa, Kahekwa, and Basabose 2003a), although analyses of weights from captive western gorillas indicate that they stop growing at around thirteen years (Leigh and Shea 1995).

The age at which males in the wild are physiologically fertile is not known. In captivity, gorillas as young as eight to nine years have produced viable sperm (Groves and Meder 2001). In the wild, although the youngest known father (revealed by DNA analysis) was just under twelve (Bradley et al. 2005), males are unlikely to be successful breeders before becoming silverbacks due to mating competition from older dominant males (ch. 4.2.1.2; ch. 11.1.4).

3.2.3.3. Maturity and reproduction in females

Female mountain gorillas reach puberty between the ages of six to eight years. During this interval, they show their first labial swellings (relatively inconspicuous and visible only in adolescent females), first estrous behavior, and first complete copulation. Although females are considered sexually mature (**adults**) from eight years on, they do not normally give birth until about ten years old, which is also the age at which they reach full adult size (Gerald 1995; Harcourt, Stewart, and Fossey 1981b; Stewart, Harcourt, and Watts 1988; Watts 1990b, 1991b).

The data for eastern lowland gorillas are very similar (Yamagiwa and Kahekwa 2001). Our only comparable data on sexual maturation in western gorillas comes from captivity, where the mean age of first birth is 9.4 years, and menarche occurs at around 7 years (Groves and Meder 2001; Tutin

1994). We discuss the surprising similarity of these findings to those from the wild in section 3.2.7.

The time lag between female sexual maturity and first reproduction is known as a period of "adolescent sterility" and is common to all the great apes and humans (Harcourt et al. 1980; Harcourt et al. 1981b; Short 1980). It may be important as a time during which females can learn about sexual interactions, search for the right mate or breeding environment, or complete physical development before producing an infant (Short 1980).

In gorillas, estrus coincides more or less closely with ovulation. Observers of mountain gorillas judge estrus by the females' behavior, mainly their seeking and accepting copulations with males. Fertile gorilla females are in estrus during just one to two days every twenty-eight or so days, and usually take two to four cycles to conceive. Pregnant females sometimes copulate, but their cycle lengths are erratic (Harcourt et al. 1980; Sicotte 2001; Watts 1991b).

In wild mountain gorillas, the interval between births when the first infant survives is about 4 years, similar to the 4.6 years for eastern lowland gorillas (fig. 3.4) (Harcourt et al. 1980; Harcourt et al. 1981b; Stewart et al. 1988; Stewart 1988; Watts 1991b; Yamagiwa and Kahekwa 2001). There are few published data from western gorilla sites. Six intervals between surviving births from Mbeli and Lossi in Rep. Congo ranged from four to six years (Robbins et al. 2004).

The birth interval is made up of gestation (pregnancy), lactation, and the time it takes a female to conceive after resuming sexual cycling, usually two to four months for gorillas (Harcourt et al. 1981b; Watts 1991b). Gestation length is relatively fixed at 255 days, or about 8.5 months (Groves and Meder 2001; Harcourt et al. 1981b). Lactation, during which females nurse their infants, makes up the greatest part of the interbirth interval. Breastfeeding suspends females' ovarian function and sexual receptivity through hormonal changes caused by an infant's sucking and consequent milk production (Short 1984; Stewart 1988). This condition is known as lactational anovulation or anestrus. Mothers do not resume sexual cycling until they have weaned their offspring, usually when the infant is between two and a half to three years of age (Fletcher 2001; Stewart et al. 1988).

The contraceptive effect of lactation is highlighted when a young infant dies. With the cessation of sucking and thus lactation, mothers quickly return to sexual cycling. Hence, the intervals between births following the death of a young infant are far shorter than those between surviving infants. When a mountain or eastern lowland female loses an infant, she gives birth after one to two years rather than four (fig. 3.4). Western gorillas do the

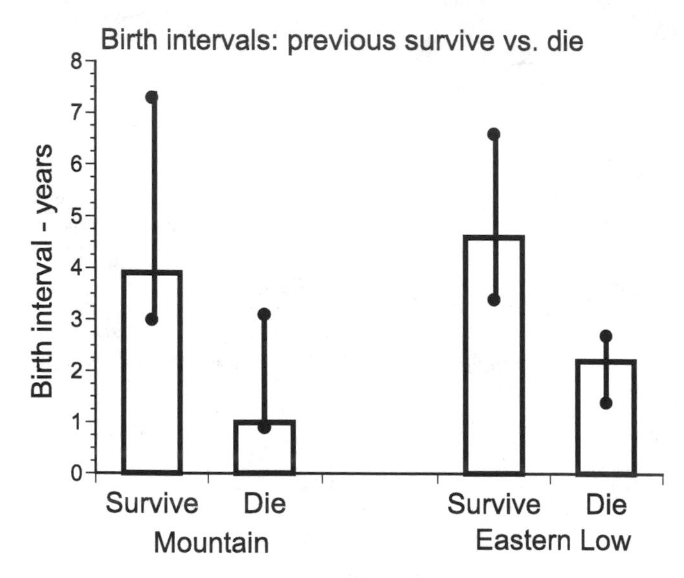

Fig. 3.4. The interval (in years) between births when the previous infant survived and when it died, for wild mountain and eastern lowland gorillas. Bars indicate medians, lines show range of values. *Details, including data sources, at end of chapter.*

same (Robbins et al. 2004). This fact of life has a major impact on male and female reproductive strategies with consequently large effects on the nature of society (ch. 7.2; ch. 11.1.2 and 11.2.4).

3.2.4. Mortality

3.2.4.1. Mortality rates

As in most animals, infancy is the most dangerous time of life for gorillas, the time when an animal is most likely to die. In the Virunga gorillas, 26% of infants die during their first year and 34% before they reach three years (65 births, Watts 1991b). Estimates from other populations of mortality rates in the first years vary from 26% in Grauer's gorillas of Kahuzi-Biega (46 births, Yamagiwa and Kahekwa 2001) to 22% and 65% for western gorillas at Lossi and Mbeli, Rep. Congo (12 and 32, Robbins et al. 2004) (see table 3.1 for further site details).

Once a gorilla makes it past weaning, its chances of living to adulthood are good, at least in the Virunga population. Approximately 60% of the Virunga gorillas survive to adulthood (Robbins 2006). In fact, after the age of four, mortality rates are low, varying between zero and 3% per year until animals reach their twenties (sec. 3.2.5) (Harcourt et al. 1981a; Robbins and Robbins 2004, and see next section).

3.2.4.2. Causes of mortality

Data on causes of mortality are hard to come by, especially for infants. One day a baby is in its mother's arms and the next day it is gone and its tiny body is never found. In the Virunga gorillas, however, infanticide by unrelated males is a well-documented cause of infant death (ch. 4.2.1.3).

Other recorded causes of death for all age groups are various diseases, predation by leopards and humans, and fighting between adult males. The relative impact of each factor varies with region. For example, in the high, cold Virungas, respiratory diseases are often fatal to mountain gorillas (Mudakikwa et al. 2001; Watts 1998b), whereas in western gorillas, the viral disease Ebola has devastated some populations (ch. 14.1.1) (Huijbregts et al. 2003; Walsh et al. 2003). And while leopards seem no longer to be a problem in the Virungas, they probably have more of an impact in the west (ch. 7.1).

Human predators, usually hunting for meat, are an important cause of death in west-central Africa, and have recently become so in eastern Africa, with the anarchy, human starvation, and proliferation of guns associated with the wars there (ch. 14.1.1). For example, in the Virungas, between 1992 and 2000, at least 12 mountain gorillas of 255 monitored animals were shot. While some of these animals were killed by poachers trying to capture infants, most were victims of militia crossfire (ch. 14) (Kalpers et al. 2003). In Kahuzi-Biega in DRC, conservation workers and researchers estimate that half the gorillas in the national park were killed by poachers in 1999 after the outbreak of war. This slaughter included 60 gorillas in the three study groups that have provided most of the information on eastern lowland gorillas (Yamagiwa and Kahekwa 2001).

Finally, adult males sometimes wound one another so severely during fights that they die. Lethal aggression between silverbacks has been documented in western and mountain gorillas (Robbins 1995; Tutin 1996).

3.2.5. Life span

The life span of wild primates is difficult to document. Death itself is usually not seen, so the record shows simply a disappearance, which could also have been an emigration. Furthermore, primates, especially the great apes, live a long time and so it takes many years of study to record the life span of even one animal, let alone a decent sample. Most wild female mountain gorillas probably die by their mid-twenties to early thirties (Robbins and Robbins 2004). Six females observed as adults for at least ten years that were not killed by poachers lived to a median age of at least thirty-four years (Fossey 1983; Sicotte 2001). We say "at least," because some were breeding females when first seen and were given a minimum age of ten years; they could well have been older. The oldest known wild gorilla female so far, Effie, from one of the Virunga study groups, was probably in her early to mid-forties at death (Robbins et al. 2005).

It is usual for the males of sexually dimorphic mammals to have shorter lives than the females due to the wear and tear of mating competition, which includes, for mammals, emigration from the natal group (Clutton-Brock 1988; Dunbar 1988). In mountain gorillas, males are likely to die by their late twenties (Robbins and Robbins 2004). Of seven well-studied males that died in adulthood of natural causes (i.e., were not killed by humans), the median minimum age of death was twenty-six years (Robbins 1995), but some males in the Virunga study groups are still alive in their thirties (Bradley et al. 2005). The oldest estimate from the wild was for an eastern lowland male from Kahuzi-Biega, guessed to be about forty-three years when he died (Yamagiwa 1997). The oldest known gorilla was a captive western male that died at age 54 years (Groves and Meder 2001).

3.2.6. Lifetime reproduction

Many differences between the sexes are tied to the fact that successful males in polygynous species (males mate several females) can produce more offspring than can successful females, in other words, have a higher reproductive rate than do the females, and hence expend more energy on competing for mates than on caring for offspring (ch. 2.1.2; ch. 2.4).

Lifetime reproductive success for male gorillas varies widely depending on their access to fertile females. We discuss the variation in chapter 11.1. For a female, lifetime reproduction depends on the age at which she first reproduces (10 years), length of intervals between surviving births (3.9 years, fig. 3.4), her reproductive life span (on average, 14 years) (Robbins 2006), and the number of her offspring that die during infancy (34%, sec. 3.2.4). In the Virunga study population, the average female produces in her lifetime two to three surviving offspring (Robbins 2006; Robbins and Robbins 2004). A female that rears four or five is doing very well, indeed. The highest number of known surviving offspring is six, born to Effie, the gorilla that has lived the longest (Watts 1991b). While there is much less

Box 3.2. Bimaturism in male orangutans

Sexually mature orangutan males come in two forms, developed and undeveloped; a state of affairs known as "bimaturism." Fully developed adults are heavy and possess secondary sexual characteristics such as long hair, throat pouches that enable the production of loud long calls, and cheek flanges. This last characteristic has led to use of the terms *flanged* and *unflanged* to refer to developed and undeveloped males, respectively.

Unflanged males (in the past referred to as subadults) weigh less than flanged males and are missing the secondary sexual characteristics, although they are sexually active and may even sire as many offspring in the wild as do the larger males (Knott and Kahlenberg 2007; Utami and van Hooff 2004; Utami et al. 2002). A male can be undeveloped for most of its life before changing into a developed male (te Boekhorst, Schürmann, and Sugardjito 1990). Being unflanged, therefore, lies somewhere between an alternative strategy and arrested development. There is evidence that the timing of the change from undeveloped to developed is linked to social factors. Specifically, the presence of a fully flanged male can retard the development of unflanged individuals (Maggioncalda, Czekala, and Sapolsky 2000; Utami and van Hooff 2004). On the other hand, if social conditions permit, undeveloped males can apparently change to full adulthood in a matter of months (Galdikas 1985a, b).

information for eastern lowland and western females, the data so far suggest comparable reproductive prospects (Robbins et al. 2004; Yamagiwa and Kahekwa 2001).

3.2.7. Comparison with *Pan* and *Pongo*

Despite considerable variation among chimpanzee and orangutan populations in life history and reproductive parameters, one fact stands out in a comparison with gorillas: wild gorillas live faster lives than do the other great apes (see box 3.2 for extreme variation among male orangutans in age of full maturation, a condition known as bimaturism). Female gorillas start breeding earlier and have shorter interbirth intervals than do *Pan* or *Pongo* females (table 3.2A, B). And while data on mortality rates and longevity are few, estimates suggest that gorillas also have generally shorter life spans than do chimpanzees or orangutans (Hill et al. 2001; Wich et al. 2004).

Table 3.2A. Developmental and reproductive milestones in wild apes. Western gorillas omitted due to lack of data, but see text for comparable interbirth intervals. *Details, including data sources (superscripts), at end of chapter.*

	Gorilla		**Chimpanzee**		
Age or interval in years	Mountain	Eastern lowland	Common	Bonobo	**Orangutan**
Weaning	3.0–4.0[1,2]	—	5[3]	—	~6[4]
Menarche	7.0–8.0[5,6]	—	10.6–11.1[7]	~11[8]	10–11[9,10]
First birth	10.2[11]	10.6[12]	14.5[7]	~13–15[8,13]	15–16[7,9]
Interbirth interval (survive)	3.9[6,11]	4.6[12]	5.8[7]	4.8[14]	7–9[7,9]
First ejaculation	~9[6]	—	9[15]	—	~8–10[4]
Full male size	~15[16]	~15[12]	~14[17]	—	21–35[9]

Table 3.2B. Infant mortality and life expectancy for females (F) and males (M) in wild apes. Mtn = mountain gorilla; East = eastern lowland (Grauer's) gorilla; West = western gorilla. *Data sources (superscripts) at end of chapter.*

	Gorilla		Chimpanzee[3]		Orangutan[4]	
Infant mortality in first year: % live births	26% (Mtn)[1], 20% (East)[2]		20%		8%	
	F	M	F	M	F	M
Life expectancy (years) at 15 years	32 (Mtn)[5] 27 (West)[6]	25[5] 20[6]	30.4	29.2	39.6	46.4
Estimated oldest age (years)	~40 (Mtn)[7]	35+ (Mtn)[8] ~40 (East)[9]	55	46	42	58

These findings do not match simple predictions based on the body size/ life history relations discussed previously. Being the largest of the great apes, gorillas should, on the face of it, be the slowest maturing and reproducing. What accounts for the apparent contradiction?

One explanation might lie in relative brain size. Species with large brains compared to their body size generally have slower life histories than do those with smaller brains; and of all the great apes, gorillas have the smallest relative brain size (Harvey et al. 1987; Judge and Carey 2000).

Relative brain size, in turn, is linked to diet: within taxonomic families, folivores have smaller brains than do frugivores (Clutton-Brock and Harvey 1980; Deaner, Barton, and van Schaik 2003; Milton 1988), and gorillas are far more folivorous and less frugivorous than either of the two chimpanzee species or orangutans (ch. 4.1). But brain size might not be the only variable linking ecology to life history.

Differences among the great apes in life history traits may reflect the different environmental and nutritional conditions associated with their diets. As any livestock raiser knows, better fed animals breed sooner and faster than do hungry ones (Sadleir 1969). Primates are no exception (Bentley 1999; Harcourt 1987a). The gorilla, being relatively folivorous, has a more consistently available food supply than do the more frugivorous apes, whose food can vary considerably with season (ch. 4.1.3). Thus, gorillas live a generally easier life, with less likelihood of suffering from food shortages than do chimpanzees or, especially, orangutans, and may be able to afford faster growth and reproduction (ch. 4.1) (Knott 2001; Knott and Kahlenberg 2007).

A comparison of rates of maturation and reproduction between wild and captive gorillas and chimpanzees (the species for which we have the most data) illustrates this point. Animals in captivity generally have easier access to high-quality food than do those in the wild. In addition, they spend no energy looking for mates or escaping predators. As a consequence, captive animals usually breed earlier and reproduce faster than do their wild counterparts (De Lathouwers and Van Elsacker 2005; Knott 2001; Lee and Kappeler 2003; Tutin 1994).

Nevertheless, wild gorillas mature and reproduce at rates remarkably similar to those for captive animals (fig. 3.5A, B). For instance, the interbirth interval for wild mountain gorillas is 3.9 years, for wild Grauer's gorillas 4.6 years, and for captive western gorillas (almost all captive gorillas are western) 4.2 years. By contrast, wild chimpanzees do the expected, maturing and reproducing at slower rates than do captive chimpanzees (fig. 3.5A, B). While the data for orangutans are few, they also show that captive animals reach sexual maturity sooner and have shorter interbirth intervals than do those in the wild (Knott and Kahlenberg 2007). In short, wild gorillas, at least those of eastern populations, appear to be growing and reproducing closer to their evolutionary potential than do wild chimpanzees or orangutans.

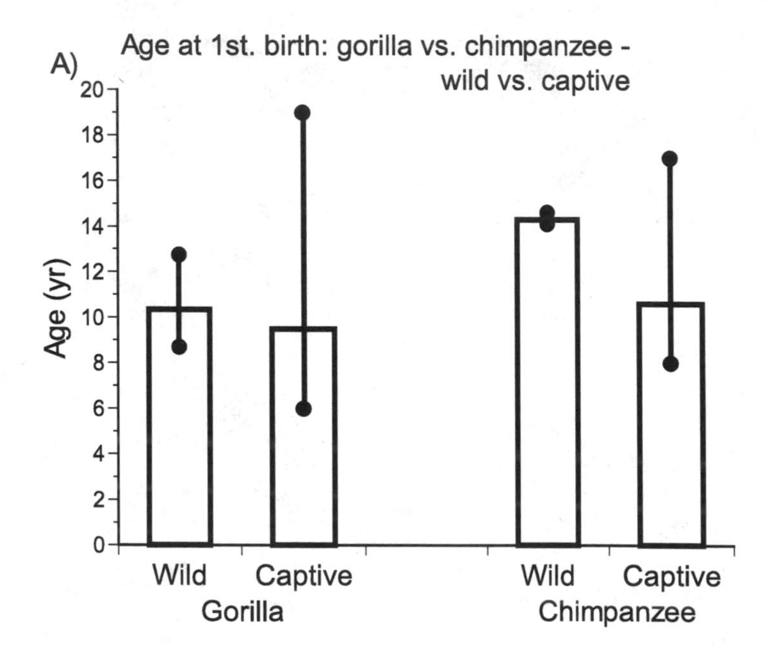

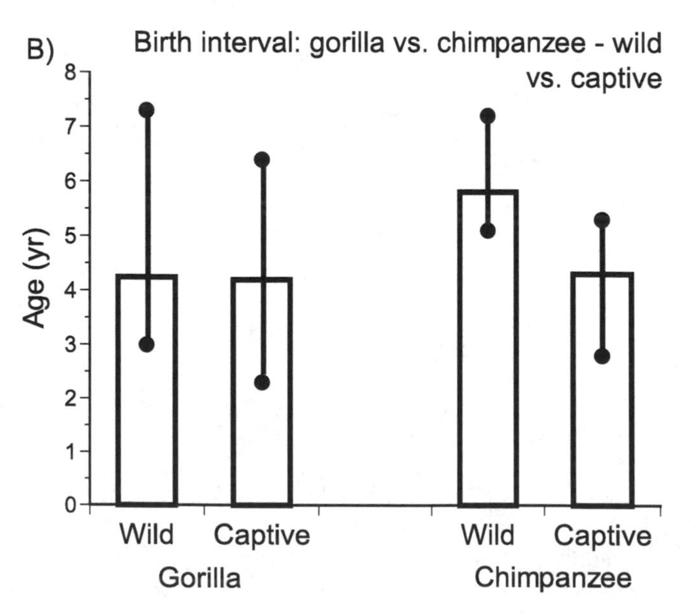

Fig. 3.5. (A) Age at first birth and (B) intervals between surviving births for wild and captive gorillas and chimpanzees. Bars indicate medians; lines show range of values. *Data sources: Wild, see details for table 3.2A, B at end of chapter; captive, see Figure Details at end of chapter.*

Another factor contributing to variation in reproductive rates among the apes could be differences in mortality rates and life expectancy at sexual maturity (sec. 3.2.4; 3.2.5). Orangutans, with their extraordinarily long interbirth intervals, appear to have lower infant mortality rates than do wild chimpanzees or gorillas (table 3.2B). In fact, orangutans have significantly lower mortality rates than chimpanzees at all ages (Wich et al. 2004). While similar differences in mortality rates between the two African apes are not apparent (table 3.2B), estimates of life span and maximum ages in the wild are generally lower for gorillas than for either orangutans or chimpanzees (table 3.2B) (Gerald 1995; Hill et al. 2001; Judge and Carey 2000; Wich et al. 2004). Taken together, these data support the general relation between living long and living slowly (sec. 3.2.1).

Finally, the nature of the society itself may influence differences among the apes in growth and reproductive rates. For example, Tutin (1994) suggested that the more cohesive social groups of gorillas compared to chimps and orangutans (ch. 4.2) might favor earlier weaning and shorter birth intervals because the gorilla mother is not the sole protector and socializer of her offspring. Along similar lines, the help in rearing infants that human mothers receive may explain the unusually short birth interval that humans can attain (Hawkes et al. 1998; Hrdy 1999; Kaplan et al. 2000).

Conclusion

Our discussion concerning reasons for differences in life history traits among the apes is largely speculative. Clearly these differences are not well understood. What is certain is that life history variation must be considered in any comparisons of different animal societies.

Among primates, body weight still explains a large proportion of life history variation. This factor underlies animals' vulnerability to ecological influences, primarily food availability, and predation risk, and partially explains the impact of these forces on species' social systems (ch. 2.2; 2.3). Chapter 4 describes the ecology of the gorilla, the basis of which is its folivorous diet, and begins with a consideration of the influences of body size.

Table Details

Table 3.2A. Life history parameters. Menarche for gorillas and orangutans is defined as regular sexual cycling and full copulation; for chimpanzees, as first adult-sized sexual swelling.

For full male size: Leigh and Shea (1995) show that captive western gorillas attain full growth by ~ 13 yr, and chimpanzees the same. Captive bonobos are full grown at 11–12 yr, but there are no estimates from the wild.

Sources: Data from [1]Stewart (1988); [2]Fletcher (2001); [3]Pusey (1983); [4]Delgado and van Schaik (2000); [5]Harcourt et al. (1980); [6]Watts (1991b); [7]Knott (2001, table 18.1 data from several long-term sites); [8]Furuichi (1987); [9]Wich et al. (2004); [10]Galdikas (1995); [11] Gerald (1995); [12]Yamagiwa and Kahekwa (2001); [13]Kuroda (1989); [14]Furuichi et al. (1998); [15]Nishida, Takasaki, and Takahata (1990); [16]Watts and Pusey (1993); [17]Goodall (1986).

Table 3.2B. Infant mortality and life expectancy in wild apes. Sources: Data from [1]Watts (1991b); [2]Yamagiwa and Kahekwa (2001); [3]Hill et al. (2001); [4]Wich et al. (2004); [5]Gerald (1995); [6]Groves and Meder (2001); [7]Robbins et al. (2005); [8]Bradley et al. (2005); [9]Yamagiwa (1997).

Figure Details

Fig. 3.4. Birth intervals are short when the previous infant dies. Data for mountain gorillas are 26 intervals when previous infant survived, 15 when it died, for 12 females (Watts 1991b). More recent data for three times as many intervals and females show almost exactly the same result: median intervals when survived = 3.88 yr compared to 1.41 yr if died, $N = 62, 39$ intervals for 42 females (Gerald 1995). Data for eastern lowland gorillas, $N = 9, 3$ intervals for 9 females (Yamagiwa and Kahekwa 2001). With no or almost no overlap between survived and died intervals, the contrasts are highly significant.

Fig. 3.5A, B. Gorillas and chimpanzees, wild versus captive. For wild gorilla and chimpanzee data, see table 3.2A and references. For data from captivity: (A) Gorillas, International Studbook, records for 44 females, cited in Tutin (1994); Chimpanzees, median from seven captive facilities, data on 51 females (De Lathouwers and Van Elsacker 2005; Tutin 1994). In the only statistical comparison available, between wild chimpanzees at Taï (with relatively young first parturition ages), and first-time mothers in captivity, the captive mothers gave birth at significantly earlier ages ($U = 0$, $N_1 = 13$, $N_2 = 7$, $p < 0.001$) (De Lathouwers and Van Elsacker 2005).

(B) Gorillas, $N = 16$ intervals, 13 females in American and European zoos (Sievert, Karesh, and Sunde 1991); Chimpanzees, median from four zoos and one captive facility; data from 56 intervals and 32 females (Courtenay 1987; De Lathouwers and Van Elsacker 2005; Tutin 1994). A comparison between wild data from Taï and those from four European zoos showed that captive females had significantly shorter intervals ($U = 242$, $N_1 = 33$, $N_2 = 31$, $p < 0.001$) (De Lathouwers and Van Elsacker 2005; Tutin 1994).

A group of western gorillas feeding in a swampy clearing. © A. H. Harcourt
A group of mountain gorillas resting. © A. H. Harcourt

CHAPTER 4

Gorilla Ecology and Society: A Brief Description

4.1. Gorilla Ecology

Summary: Gorilla Ecology

Gorillas prefer fruit to foliage. Their large body size, however, enables them to survive on more abundant, but lower-quality food such as pith, leaves, and woody stems, when preferred fruit is unavailable. This dietary flexibility helps to explain variation across gorilla populations in diet, ranging behavior, and some aspects of social cohesion. Thus, they are most frugivorous in west-central Africa, where fruit is more abundant than in eastern gorilla habitat; they travel longer distances in the fruiting season than the non-fruiting season, as they search for scattered fruiting trees; and groups in western populations spread out more when feeding than do those in eastern populations. Nevertheless, gorilla groups across Africa are relatively cohesive and stable compared to groupings of chimpanzees that live in the same forests and eat much of the same food. The gorilla's ability to switch diets and foraging strategies when fruit is scarce contributes to this contrast. Similarly, the striking differences between all the great apes in the nature of their societies are closely tied to their contrasting diets, and to related variation in how they find food, that is, their foraging strategies.

Introduction: Gorilla Ecology

The food that animals live on, and the energy expended in its acquisition (foraging effort) have major impacts on species' social systems (ch. 2.2).

Foraging effort depends on three interrelated factors: the costs of locomotion, which are largely the outcome of body size, morphology, and locomotory mode; the abundance and distribution of food in time and space; and the size of the group that is foraging (Chapman, Wrangham, and Chapman 1995; Chapman and Chapman 2000; Clutton-Brock and Harvey 1977b; Janson and Goldsmith 1995). As group size increases, so too does feeding competition, which means that group members must increase their foraging effort to maintain their food intake. Beyond a certain level of effort, animals must either switch to a less expensive resource or decrease the size or cohesiveness of groups. Thus, differences between species and populations in the nature of resources and foraging effort are intimately linked to differences in the nature of their society (ch. 2.2).

Among primates, fruit-eating requires more foraging effort than does foliage-eating because fruit is less abundant than is foliage, and more widely and patchily distributed. Hence, frugivores, but not folivores, tend to increase their daily travel distance (commonly used as an indication of foraging effort) as groups get larger (ch. 2.2.3.1) (Dunbar 1988; Isbell 1991; Janson and Goldsmith 1995). Nevertheless, primates favor fleshy fruit because it is packed with energy and easily digested. For example, at Lopé National Park in Gabon, seven of eight diurnal primate species, including gorillas and chimpanzees, are mainly frugivorous in the fruiting season (Tutin et al. 1997).

The problem with a dependence on fruit is that it varies widely in availability, fluctuating between abundance and scarcity within and between years. Primates must have more abundant, if less preferred, foods to fall back on during fruit shortages.

To understand the influence of diet and feeding competition on society, it is crucial to consider fallback foods—the foods that animals survive on when times are lean. Studies of several different primate communities show that species' diets are most divergent and niche separation most obvious during periods of fruit shortage (Gautier-Hion 1979; Terborgh 1983; Tutin et al. 1997; Wrangham, Conklin-Brittain, and Hunt 1998). Going back to Lopé, all the eight diurnal primate species changed their diets in fruit-poor months to incorporate less preferred, but more abundant foods such as seeds, leaves, and herbaceous pith. Species differed, however, in the extent to which they gave up fruit, which seemed to affect their ability to cope with seasonal changes. At Lopé, those species with a more diverse folivorous diet lived at higher densities than did more dedicated frugivores (Tutin et al. 1997).

In general, primates that rely heavily on fruit may have a hard time of it during ecological crunches (Oates et al. 1990), a constraint that is especially pertinent to apes. While most frugivorous primates eat both unripe and ripe fruit, apes appear to have a strong preference for the latter (Doran-Sheehy, Shah, and Heimbauer, in press; Remis 2003; Wrangham 1979). This may be due to a relatively low tolerance for chemical antifeedants, such as tannins, that are present in unripe fruit (Wrangham et al. 1998). Such specialization limits the amount of preferred foods available, thus making frugivory relatively expensive for apes, in terms of foraging costs, and acting as a constraint on gregariousness (Wrangham et al. 1996).

While initial studies of mountain gorillas earned the genus the label of consummate folivore (Fossey and Harcourt 1977; Schaller 1963, ch. 5), subsequent research on other populations has revealed more diversity in what gorillas eat (Doran et al. 2002b; Rogers et al. 2004). Diet varies not only across populations, but between seasons and years within the same forest (Doran and McNeilage 2001; Tutin et al. 1991b). This dietary flexibility is a key to understanding gorilla socioecology and the variation between populations. From a broader perspective, differences in dietary flexibility between the great apes helps explain some of the striking contrasts in their social systems. Because seasonality in food supply affects virtually all aspects of foraging effort, it is a key factor linking environment to society, and therefore constitutes an important theme throughout this chapter.

Before going any further in this consideration of gorilla ecology, we need a word here about vegetation. A major component of gorillas' diet is the vegetative parts of plants that are neither fruit, seed, nor flower. This component includes stems, bark, pith, roots, shoots, and leaves, all of which vary in their nutritional quality and digestibility but contain more fiber and less sugar than does fruit (Remis 2003; Waterman 1984; Watts 2000b). Many terms are used to encompass this salad mix, including *herbaceous terrestrial vegetation, fibrous foods,* or *foliage.* These terms all refer to the non-reproductive parts of plants, whether they belong to vines, tree leaves, shrubs, or ground herbs.

4.1.1. Body size and diet

Gorillas are built to be ground-living folivores. They are one end of a general trend across primates in which increasing body size correlates with greater terrestriality and folivory (Clutton-Brock and Harvey 1977a, b).

Being large enables gorillas to exploit relatively low-quality resources. In general, larger-bodied mammals can cope with a slower rate of energy intake than can smaller-bodied ones, even though they need absolutely more food (Harvey et al. 1987; McMahon and Bonner 1983, ch. 2; McNab 1999). This slower intake means that large animals can survive on relatively low-quality foods (e.g., woody plant stems) that take a long time to digest, that is, a long time to extract energy from (Jarman 1974; Lambert 1998; Ross 1992) (ch. 2.2).

Larger animals also tend to have a large colon and cecum surface area, and long food retention times compared to smaller animals, which helps maximize absorption (Chivers and Hladik 1980; Martin et al. 1985; Milton 1984; Remis 2003). Gorillas have no other specific adaptations for digesting fibrous, low-energy foods. In contrast, the smaller-bodied folivorous colobines, which do not have a large hind-gut surface area, have evolved specialized gut anatomy and physiology for digesting foliage (Lambert 1998; Messier and Stewart 1997).

4.1.2. General habitat and food preferences

Gorillas live in a variety of habitats, from the montane forest of the Virungas to the lowland tropical forests of west-central Africa (ch. 3.1). In all regions, they generally favor open canopy forest with its dense cover of ground vegetation, over closed canopy forest with its sparser undergrowth (Bermejo 1999; Schaller 1963, ch. 2; Tutin and Fernandez 1984; Yamagiwa et al. 2003a). This preference means that they thrive in secondary forest, often associated with human activities like slash-and-burn agriculture. For example, Lossi Reserve in Rep. Congo (see table 3.1), with its unusually dense gorilla population, is the site of an abandoned village and is associated with especially dense ground cover (Bermejo 1999). In some regions, gorillas frequent swamp forests or swampy clearings where they feed intensively on abundant aquatic vegetation (see box 4.1).

Despite their essentially terrestrial nature, gorillas in some areas spend considerable amounts of time in trees, frequently as high up as 35 m (Doran and McNeilage 1998; Remis 1995; Remis 1999). The use of trees varies with the nature of the habitat. For example, in the tree-sparse Virungas, mountain gorillas spend over 90% of their foraging time on the ground, where they also build the vast majority of their night nests (Fossey and Harcourt 1977; Schaller 1963, ch. 5). In contrast, in Gabon's Lopé National Park, gorillas not only harvest much of their food arboreally but also build 35% of their night nests in trees (Tutin and Fernandez 1993; Tutin et al. 1995).

While gorillas can subsist on herbaceous vegetation, and prefer habitats where this resource is abundant, they eat fruit when they can get it. In fact, taste experiments in captivity showed that gorillas consistently chose fruit over vegetables (Remis 2003). Whether feeding on fruit or foliage, wild gorillas tend to be highly selective in what they eat, as nutritional analyses from both eastern and western sites have shown. They select herbaceous

Box 4.1. West-central Africa's swampy clearings, bais

Some forests in west-central Africa contain large swampy clearings. They are especially common in the 170,000 km^2 of northern Rep. Congo (Fay and Agnagna 1992), where satellite imagery has documented at least 100 such clearings (known locally as *bais*), ranging in size from 4–40 hectares (Magliocca, Querouil, and Gautier-Hion 1999). These sites attract many species of large mammals, including gorillas, buffalos, elephants, and giant forest hogs, all apparently drawn to the sites by the high densities of aquatic herbs that grow there (Vanleeuwe, Cajani, and Gautier-Hion 1998).

Human observers are attracted as well, because of the extraordinary concentrations of animals and excellent observation conditions. While the vegetation is dense, it is very low-growing, giving a typical clearing the aspect of a large meadow in the middle of the forest. During one 17-month study of a clearing just 4 hectares in size, gorillas visited the area on 95% of 380 observations days, with an average of three different groups showing up each day. Researchers identified 377 individual gorillas at this small clearing (Gatti et al. 2004). Three studies at forest swamp clearings have run for at least three years (see table 3.1) (Gatti et al. 2004; Magliocca et al. 1999; Parnell 2002; Robbins et al. 2004).

Gorillas in the swamp clearings feed for three-quarters of their time, spending hours walking or sitting in water that is at least wrist-deep, sometimes waist-deep. They pull up a handful of aquatic plant, swish it in the water with an expert flick of the wrist to wash off the mud, and eat the lot, with little other preparation. At two sites, mineral content appears to influence the choice of food, since the species eaten are particularly high in sodium, potassium, and calcium (Magliocca and Gautier-Hion 2002; Nishihara 1995).

Little is known about the general role of the clearings in gorilla foraging because the animals have rarely been followed through the forest away from the swamps. At one site, researchers calculated that each gorilla group spent, on average, about 1½ hours per month at the swamp clearing (Magliocca et al. 1999). Whether or not these groups spent the rest of their time visiting other clearings or in more common forest habitat is still a mystery.

plants that are low in fiber and high in protein (such as bamboo shoots), and prefer fruits with a relatively high sugar and low tannin content, such as succulent ripe fruit rather than unripe fruit (Calvert 1985; Nishihara 1995; Remis 2003; Rogers et al. 1990; Waterman et al. 1983).

Unlike chimpanzees, wild gorillas have not yet been seen to eat vertebrates, even though we have seen curious young animals handle mice and chameleons. However, they do occasionally eat insects, primarily ants or termites (Harcourt and Harcourt 1984; Tutin 1983; Watts 1989b). Studies of Grauer's and western gorillas have found insect remains (of numerous species) in about a third of fecal samples from six sites (Deblauwe et al. 2003; Yamagiwa et al. 1991). In the Bwindi forest of Uganda, one group of mountain gorillas consumed ants on an average of 18% of days (judged from fecal samples) and in some months, on as many as 43% of days (Ganas and Robbins 2004). We know little to nothing about the nutritional significance of insects to gorillas, but given the small proportion of the diet that they constitute, they probably have little effect on foraging and therefore on the nature of the society.

4.1.3. Diet, altitude, and season

Researchers use three methods to determine what gorillas eat. The majority of dietary data come from fecal analyses in which gorillas' dung is collected and examined for plant and insect remains. This method provides good information on the general frequency and diversity of fruit consumption, since gorillas swallow most fruit seeds whole, and excrete them intact, enabling researchers to find and identify the species being eaten. It is a less accurate method for identifying herbaceous species, which do not leave behind such convenient signatures as seeds (Doran et al. 2002b; Rogers et al. 2004; Tutin and Fernandez 1985). Nevertheless, fecal sampling allows researchers to estimate roughly the relative consumption of fruit versus non-fruit foods and to monitor changes over time (e.g., Remis 1997b).

A second type of data comes from the characteristic feeding remains, such as discarded peelings from stems that gorillas leave behind as they forage. Trail evidence is useful for identifying species of herbaceous foods and the general frequency of their consumption (e.g., Doran et al. 2002b; Goldsmith 1999; Oates et al. 2003; Yamagiwa et al. 2005). While these indirect methods provide a good picture of dietary diversity and changes over time, they tell us little about the actual amounts of different foods that gorillas eat (Doran et al. 2002b; Robbins et al. 2006). For this, direct observations are needed, which require at least semi-habituated animals. Systematic, quantitative studies of gorilla feeding behavior come only from mountain gorillas in the Virungas and Bwindi (Fossey and Harcourt 1977; Robbins et al. 2006; Watts 1984), eastern lowland gorillas in Kahuzi (Yamagiwa et al. 2005), and from one western site, Mondika in CAR (Doran-Sheehy et al., in press).

Ecological research over the past twenty-five years, using varying combinations of these methods, has shown that gorillas' diets differ markedly from region to region, with one of the most striking contrasts being in degrees of frugivory. In terms of number of species consumed, western gorillas eat the most fruit, mountain gorillas the least, and eastern lowland gorillas are intermediate between the two (Doran et al. 2002b; Robbins et al. 2006; Rogers et al. 2004; Yamagiwa et al. 2005).

Both altitude and seasonality affect the contrasts between the sites, with altitudinal effects being the most obvious. With rising altitude, mean annual temperature drops, and with it the number of plant species, and hence the amount and diversity of fruit in the forest and the diet (fig. 4.1) (Ganas et al. 2004; Nkurunungi et al. 2004; Yamagiwa and Basabose 2006). Even within gorilla populations, altitude affects diet, as in the Bwindi population of mountain gorillas in Uganda (B in fig. 4.1). The fact that the variation within Bwindi falls pretty much on the line across populations emphasizes the influence of altitude, as opposed to merely site, on the gorilla's diet (Robbins et al. 2006).

The decline with altitude in diversity of fruit does not mean that gorillas living at high altitudes have less food than those in lower fruit-rich forests. Indeed, abundance of gorillas' staple foods, terrestrial herbs, tends to increase with altitude (Goldsmith 2003; Nkurunungi et al. 2004; Watts 1984).

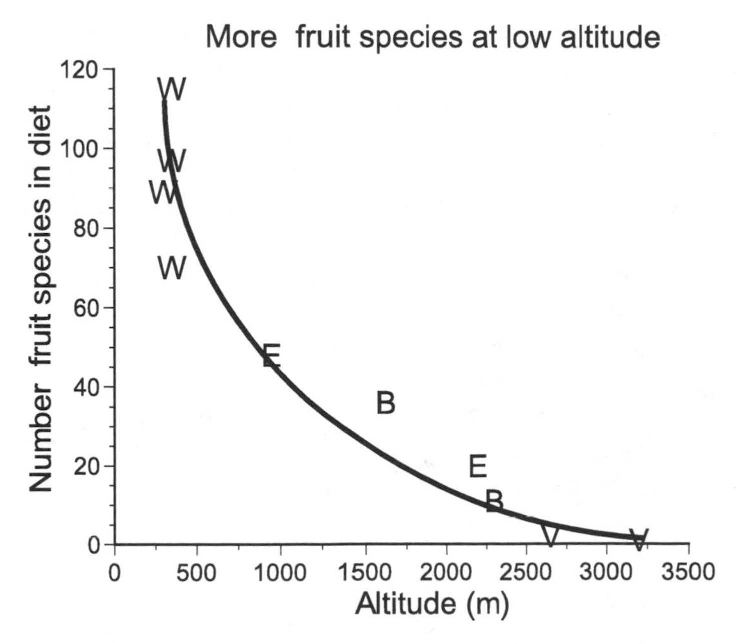

Fig. 4.1. Number of fruit species in diet by altitude. Lower altitude populations eat more fruit species. Each letter represents a study site (see table 3.1) for western gorillas (W), eastern lowland gorillas in Kahuzi-Biega National Park (E), and mountain gorillas in Bwindi (B) and the Virungas (V). *Details, including data sources, at end of chapter.*

Fruit-eating populations show the most marked seasonality in diet, because the availability of fruit varies more over time than does the availability of herbaceous foods. When preferred fruit is available, gorillas eat it, but when it is scarce, they switch to fibrous foods.

4.1.3.1. Western gorillas

At all study sites, western gorillas regularly eat succulent fruit throughout the year. The diversity of dietary fruit ranges from 70 to at least 115 species (fig. 4.1). During the fruiting season, fruit remains are found in 90%–100% of gorilla fecal samples (Doran et al. 2002b; Goldsmith 1999; Kuroda 1992; Remis 1997b; Rogers et al. 2004; Williamson et al. 1990; Yamagiwa 1999a). In fact, at some sites during certain times of the year, gorillas eat about as many species of fruit as do sympatric chimpanzees (sec. 4.1.8).

While these data show that fruit is clearly important to western gorillas, the degree of frugivory in their diet has been overemphasized in the past, due to the biases inherent in indirect sampling (see sec. 4.1.3). In a recent study at Mondika in CAR, direct behavioral observations revealed nearly three times as many leaf species in the diet as did trail evidence (Doran-Sheehy et al., 2006). The study (admittedly based on just one silverback in a group) concluded that fruit accounted for about 22% of food species identified and that gorillas spent only about 14% of their feeding time consuming fruit versus non-fruit foods. These data indicate a lower degree of frugivory in western gorillas than all previous feeding studies based on indirect sampling or limited behavioral observations (Remis 1997b; Rogers et al. 2004).

It is now clear that throughout their range, western gorillas eat large quantities of herbaceous vegetation as well as fruit. Most importantly, their staple foods, that is, the foods that are eaten year-round, are all non-fruit species, specifically, high-quality herbs (monocotyledonous species). Across six well-studied western sites, the gorillas have very similar dietary staples, the most common being species in the families Zingiberaceae (ginger) and Marantaceae. We know of no common name for the African genera of Marantaceae, but those from some other regions are termed prayer plant or arrowroot (Rogers et al. 2004).

Western gorillas change their diet according to seasonal availability of fruit. When fruit is plentiful they add it to their herbaceous staples, dropping out the less nutritious and less favored fibrous foods. In the non-fruiting season, however, not only do the gorillas continue consuming herbaceous staples, but they fall back on lower-quality herbs and woody vegetation, as well as non-preferred fruits with relatively high tannin and fiber content (Doran and McNeilage 2001; Oates et al. 2003; Remis 1997b, 2003; Rogers et al. 1990; Tutin et al. 1997).

This seasonal variation in diet is exemplified by corresponding changes in fruit-to-fiber ratios (Doran et al. 2002b; Kuroda et al. 1996; Nishihara 1995; Remis 1997b; Tutin et al. 1997). For example, at Bai Hokou in CAR, as fruit production in the forest increased, so too did the average number of fruit species per month found in gorilla dung samples, whereas measures of fiber intake did the opposite (fig. 4.2).

4.1.3.2. Eastern lowland, or Grauer's gorillas

The eastern lowland gorillas of Kahuzi-Biega in DRC eat fewer species of fruit than do western gorillas, and consume it less often. For example, at western sites, 90%–100% of fecal samples contain fruit remains compared to 53% of fecal samples from Kahuzi (Yamagiwa et al. 2005). The diet also varies considerably with altitude (fig. 4.1). The animals of the Kahuzi region, which live at relatively high altitude, eat about half the fruit species that the population lower down in Itebero eats. Gorillas in both areas, high and low, vary their diet with season (Yamagiwa 1999a; Yamagiwa et al. 2005). Like western gorillas, when fruit is scarce, they increase their consumption of foliage, bark, and woody vines. The Kahuzi population also experiences seasonal fluctuation in bamboo shoots, high in protein, and a favored food

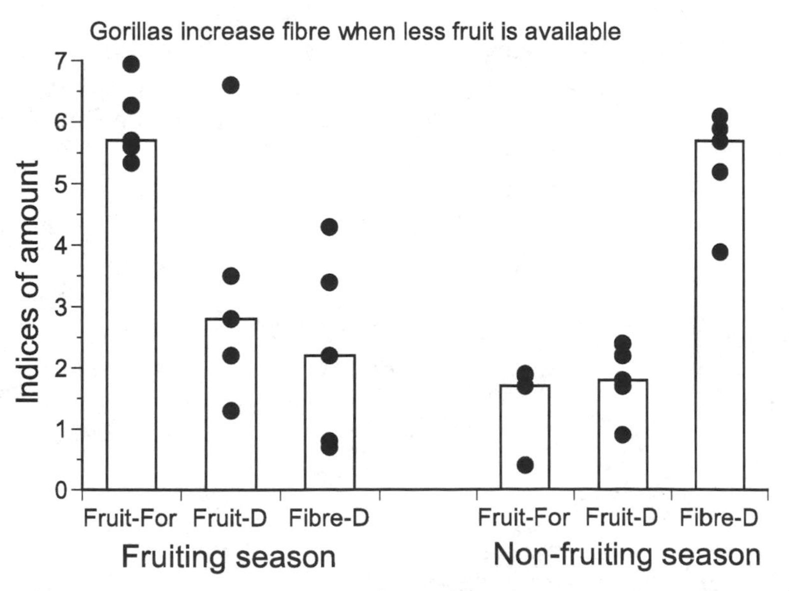

Fig. 4.2. Measures of fruit availability in the forest (Fruit-For) compared to amount of fruit and fiber remains in gorilla dung samples (Fruit-D, and Fiber-D). Fruiting and non-fruiting seasons are defined as the five most extreme months in the study. • = monthly values; bar = median. *Details at end of chapter.* Source: Data from Bai Hokou in CAR, collected over two years (Remis 1997b).

when it is available (Casimir 1975; Goodall 1977; Yamagiwa et al. 2005; Yamagiwa et al. 1996).

4.1.3.3. Mountain gorillas

Until well into the 1980s, our knowledge about the diet of wild gorillas, as for most aspects of gorilla socioecology, came primarily from the population of mountain gorillas in the Virunga Volcanoes (ch. 3.1.3). The region is at the upper limit of gorillas' altitudinal range (fig. 4.1), with most of the forest lying between about 2,400 to 3,800 m (Fossey and Harcourt 1977; Schaller 1963, ch. 2; Watts 2000b). The habitat is primarily moist montane forest, characterized by a dense ground cover of herbaceous vegetation, open canopy, and very little fruit. The gorillas here are among the most folivorous of any primate (fig. 4.1) (Clutton-Brock and Harvey 1977b), and the most terrestrial of any gorilla population.

The Virunga gorillas live on leaves, pith, and stems of herbs and vines, and in some areas, seasonal bamboo shoots (Fossey and Harcourt 1977; McNeilage 2001; Vedder 1984). Compared to most other primates, the Virunga population inhabits a relatively uniform environment in which food is evenly and abundantly dispersed. Throughout most of this range, almost anywhere a gorilla stops, it will find something to eat (Watts 2000b). Nevertheless, habitat quality does vary, and gorillas clearly favor the richest areas (Waterman et al. 1983; Watts 1990a, 2000b). Because all the major foods in the region are available year-round, the only Virunga gorillas that experience obvious seasonal variation in diet are those with access to bamboo forest, which they use heavily when the bamboo is shooting (Fossey and Harcourt 1977; Vedder 1984; Watts 2000b).

The other population of mountain gorillas in the Bwindi Forest of Uganda lives at lower altitude, from 1,160–2,607 m, where there is greater plant diversity and more species of fruit than in the Virungas (Ganas et al. 2004; Goldsmith 2003; Robbins and McNeilage 2003). The Bwindi gorillas' diet reflects this difference in fruit availability (fig. 4.1). The fruit component of their diet changes seasonally, although to a much lesser extent than in western gorillas, and fruit consumption varies inversely with consumption of several important herb species (Ganas et al. 2004; Nkurunungi et al. 2004; Robbins et al. 2006).

4.1.4. Diet and daily ranging: Variation in foraging effort

Because fruit is more sparsely distributed than herbaceous vegetation and is therefore more difficult to find, frugivorous primate species tend to invest more foraging effort, that is, travel farther per day, than do folivorous spe-

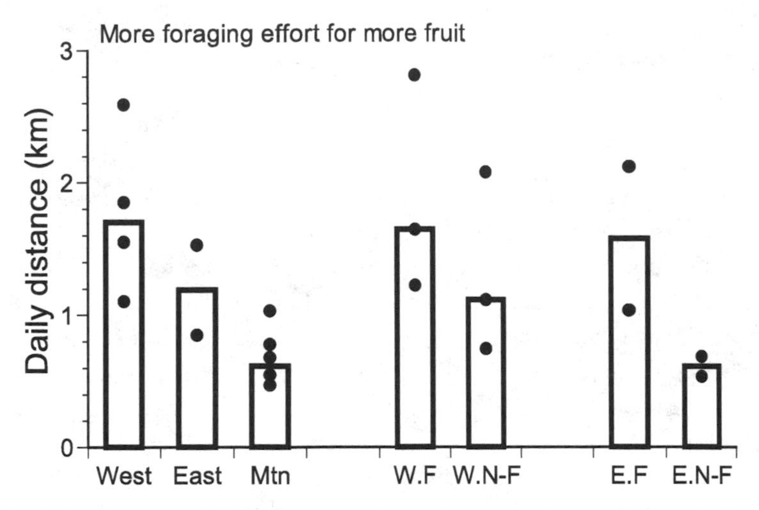

Fig. 4.3. Daily distances traveled by site and amount fruit in diet. The three bars on left show daily distances for western (West), eastern lowland (East), and mountain (Mtn) gorillas. Pairs of bars on right show daily distances in western and eastern lowland gorillas when fruit in diet was high (W.F and E.F) and low (W.N-F and E.N-F). • = mean values for different populations (different groups for Mtn); bar = median. *Details, including data sources, at end of chapter.*

cies (ch. 2.2.2 and sec. 4.1, Introduction). This contrast is evident between the more frugivorous western gorillas and the less frugivorous eastern lowland and mountain populations (fig. 4.3). In fact, all gorillas inhabiting fruit-rich lowland forest, including the Grauer's gorillas in the Itebero region of Kahuzi-Biega (represented by the higher value for East in fig. 4.3) travel farther each day than do those in highland forest (Ganas and Robbins 2005; Goldsmith 2003; Yamagiwa 1999a; Yamagiwa et al. 2003b).

A further indication that fruit-eating increases foraging effort is the seasonal variation within populations. At all sites except the Virungas, studies consistently show that gorillas travel farther each day during fruiting periods than non-fruiting periods (fig. 4.3). This difference was significant in seven populations (Doran-Sheehy et al. 2004; Goldsmith 1999; Kuroda et al. 1996; Tutin 1996; Yamagiwa et al. 1996), including the mountain gorillas of Bwindi, although the difference here was very small, on the order of a few hundred meters (Ganas and Robbins 2005). These data indicate that gorillas are more than just casual fruit-eaters, that is, consuming fruit only when they happen upon it. When fruit is available, gorillas actively seek it out (Rogers et al. 2004).

Contrasts between regions in foraging effort will depend not just on fruit abundance in the forest but also on the density of herbaceous vegetation, particularly the gorillas' staples, high-quality terrestrial herbs. We have already pointed out that a drop in number of fruiting species with altitude does not mean a drop in amount of food, because the density of terrestrial herbs tends to increase with altitude (sec. 4.1.3). A greater density of a favored food will mean that animals need go less far each day to find it (Ganas and Robbins 2005).

In addition to amount and distribution of food, group size can also influence the distance that a group travels each day. The more animals foraging at any one spot, the less food there is for the average animal, and therefore the more the group needs to move for all to find enough food (ch. 2.2.3.1). The relationship between foraging effort and group size has been examined closely in only mountain gorillas. In Bwindi, gorilla group size had a positive effect on the daily distance traveled (Ganas and Robbins 2005), while in the Virungas, results were more equivocal. Watts's study showed a positive correlation only at the smallest or largest groups sizes (Watts 1991a).

4.1.5. Diet and home range size

In much the same way that diet and habitat correlate with daily travel distance, so they might be expected to correlate with home range size. Across gorilla populations, however, this relationship is not clear. While the mountain gorillas of the Virungas, with their abundant, evenly distributed, non-seasonal food supply, have the smallest annual home ranges of any population, between 4 and 12 km^2 (Fossey and Harcourt 1977; McNeilage 2001; Vedder 1984; Watts 2000b), all other populations, eastern and western, have comparable annual home ranges, varying between 16–31 km^2 (summarized in Ganas and Robbins 2005). This lack of correlation between home range and habitat is probably because a group's annual home range encompasses all areas where a group has traveled in a year and reflects the general distribution of different foods in the habitat rather than daily foraging effort. Thus, a gorilla group in Lopé National Park, Gabon, spent most of its time within a 10 km^2 core area but made periodic safaris outside its normal stomping ground to feast on fruit bonanzas in far off trees (Tutin 1996). Similarly, Grauer's gorillas in the montane forests of Kahuzi-Biega, DRC, dramatically shift their range during September to December to eat seasonal bamboo shoots (Casimir 1975; Goodall 1977; Yamagiwa 1983). While this shift to the bamboo zone increases the total size of their annual home range, they do not travel longer distances each day when feeding on bamboo (Yamagiwa et al. 2003b).

Monthly home range is probably more closely linked to foraging effort than is yearly range size. For example, one group of Bwindi gorillas increased its monthly home range when feeding on fruit. Furthermore, as with daily travel distance (see earlier discussion), its monthly home range size was positively correlated with group size (Ganas and Robbins 2005).

Social as well as ecological factors influence gorillas' ranging behavior. Because gorillas' daily distances are too short for them to defend a territory (Mitani and Rodman 1979), their home ranges overlap extensively, and groups and lone males encounter one another more or less frequently (Fossey and Harcourt 1977; Tutin 1996; Vedder 1984; Watts 1990a, 1991a; Yamagiwa 1999a). Males' attempts to acquire females sometimes result in pursuit and flight between social units that can alter ranging abruptly and dramatically (Cipolletta 2004; Watts 1994c). We discuss the nature of these encounters in section 4.2.1.4 and their influence on ranging in chapter 11.1.3.

4.1.6. Ecology and group cohesion

While traveling farther is one way to cope with sparsely distributed food, the other way is for group members to spread out more, or to break up into smaller foraging parties so that individuals interfere less with each other's feeding (Watts 1998a; Wrangham et al. 1996; Wrangham, Gittleman, and Chapman 1993). Given that western gorillas are more frugivorous than are Grauer's or mountain gorillas, are western groups correspondingly more spread out, that is, less cohesive?

At western gorilla sites, including Lopé in Gabon and Bai Hokou and Mondika in CAR, group members not uncommonly spread over several hundreds of meters (Bermejo 2004; Doran and McNeilage 2001; Goldsmith 2003). This spread is greater than for eastern lowland or mountain gorillas, where inter-individual distances during feeding are normally within 15 m, and maximum group spread is usually between 100–200 m (Fossey and Harcourt 1977; Goldsmith 2003; Yamagiwa 1983).

In the past, researchers at Bai Hokou and Mondika have reported that gorillas occasionally split into temporary subgroups (e.g., sleeping 800 to 1,000 m apart) (Remis 1997a) for one or two days at a time, and tend to do so during the fruiting season (Doran and McNeilage 2001; Goldsmith 2003; Remis 1997a). It is unclear, however, how common this behavior is. Much of the evidence for it is based on following trail and counting nests of unhabituated animals, a method that sometimes leads to errors about the identity of the groups being sampled. In other western gorillas, subgrouping occurs either rarely or not at all (Bermejo 2004; Doran and McNeilage 2001), and is equally infrequent at eastern sites (Goldsmith 2003; Yamagiwa et al. 2003b). It remains to be seen whether or not subgrouping is a habitual behavior in at least some western populations, and therefore whether it represents a fundamental difference in social flexibility. Only direct observations of known gorilla groups can answer this question.

4.1.7. Ecology and population density

If food affects ranging and grouping, it also affects how many gorillas a region can support, in other words, the density of the population. For gorillas the most crucial factor affecting population density is probably the density of high-quality terrestrial vegetation (Doran et al. 2002b; Goldsmith 2003; Rogers et al. 2004; Rogers and Williamson 1987). In general, gorillas are most abundant in secondary forest, which is open with a dense ground cover (Bermejo 1999; Poulsen and Clark 2004; Schaller 1963, ch. 4; Tutin and Fernandez 1984; Yamagiwa 1999a; Yamagiwa et al. 1993).

Thus Tutin and Fernandez (1984) recorded variation in density across Gabon from less than 0.5/km^2 to over 7/km^2 in what they called thicket. A recent overview of six well-studied western sites showed that gorillas live at the highest densities in Marantaceae forest and swamp forest where gorillas' staple foods are most abundant (Rogers et al. 2004). For example, while the median density in three regions of mature forest, with its sparse undergrowth, was 0.5 gorillas/km^2, the median density from four areas dominated by either Marantaceae or swamp forest was 4.3 gorillas/km^2, with a remarkable 11.3 animals/km^2 in the Marantaceae forest of Odzala National Park in Rep. Congo. (Box 4.2 gives details of how fieldworkers count gorillas and arrive at estimates of densities.)

Density varies not only across regions but also within sites, a fact that poses challenges to anyone trying to estimate gorilla densities over a large area (Walsh et al. 2003). Thus the average density in Odzala in the 1990s was 5.4 gorillas/km^2 (Bermejo 1999), meaning that some areas will have had far less than the maximum of 11.3/km^2. Similarly, in the Virunga Volcanoes, densities in the Karisoke Research Center study area are upward of 3 gorillas/km^2, but the average for the Virunga region as a whole is 1 animal/km^2 (Kalpers et al. 2003).

To put these density figures into a broader perspective, across primates, population density correlates quite well with body mass: large primates live at lower densities than do smaller ones, most likely because they require more food and therefore need more land to supply it (Clutton-Brock and Harvey 1977a). The correlation means that for a species of a particular body mass, its expected density, were it an average primate, can be calculated. It turns out that, based on data for diurnal African primates, the gorillas' predicted density is 1/km^2 in good habitat (stats. 4.1). We say "good" habitat

Box 4.2. Counting gorillas

An extraordinary fact about gorillas is that not only can they be counted without a single animal ever being seen, but the census worker can even get a fairly good idea of the ages and even the sex of animals in the group or population. We did a census of gorillas in Nigeria and obtained an estimate of not just numbers but also composition of the population (Harcourt, Stewart, and Inahoro 1989a, b). And yet in the whole of our three months in Nigeria, one of us saw only one gorilla for perhaps three seconds.

How did we do it? Gorillas leave two obvious signs of their presence: their nests (platforms of bent-in branches that they sleep in at night) and their dung (Schaller 1963). The nests give an idea of how many animals there are in a group. The dung in and around a nest closely corresponds to the size of the animal that made it (box fig. 1) and thus indicates the age of the animal.

The ideal place to count gorillas is the Virunga Volcanoes of Rwanda, Uganda, and DRC. There, it is so cold at night that once the gorillas are bedded down, the animals do not move. One nest therefore equals one nest-making animal, in other words, one weaned individual. And if tiny dung is in or near the nest, then census workers know they have found the nest of a mother and infant. The largest nests and dung of all are produced by silverbacks. So, we can both count the animals and estimate the sex of several (Harcourt and Groom 1972).

The other reason that the Virungas are ideal is that the population is small, just 300 animals or so, and has been intensively studied. Consequently, with good organization, censuses of almost the entire population are possible, and several have been done (Harcourt and Groom 1972; Harcourt et al. 1983; Kalpers et al. 2003; Weber and Vedder 1983). These censuses are why we know that the Virunga mountain gorilla population is increasing.

In lower-altitude areas, the gorillas can confuse the census worker by making more than one nest at night. Also, with larger populations less intensively studied, censuses have to be samples. Not all of the population can be counted, nor all of the area covered. This is the normal situation for censuses.

Perhaps the most systematic yet comprehensive gorilla census to date was that conducted by Caroline Tutin and Michel Fernandez of gorilla and chimpanzee populations across the whole of Gabon (Tutin and Fernandez 1984). They used the long-established transect method (see, e.g., Buckland et al. 1993). Transects of known length are walked (the transects preferably placed randomly in respect to the area sampled), all sign of the species of interest noticed, and various calculations performed to convert the sightings to density of animals. Sightings are

(*continued*)

Box 4.2. continued

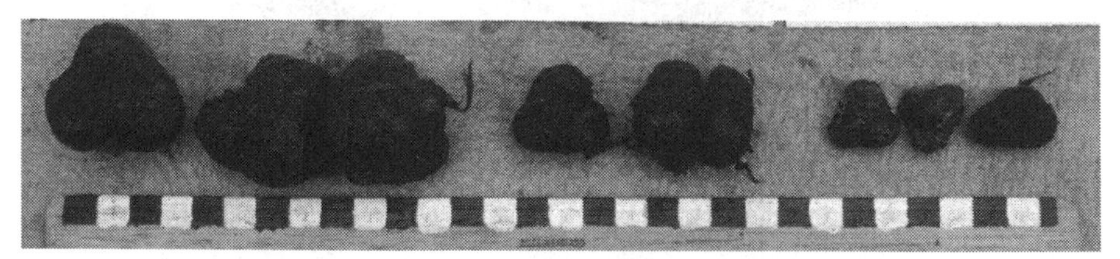

Box 4.2, Fig. 1. Gorilla dung: size of bolus corresponds to size of individual producer. (A) Adult male and female; (B) Infant of 1 yr, adult male; (C) Immatures of 3.5, 2.5, and 1.5 yr. © A. H. Harcourt

sometimes recorded as groups of animals, and number of groups multiplied by average size of group to produce number of animals. In the case of the apes, who all build night-nests, nests are usually counted rather than animals, with corrections made for how long it takes nests to decay (Tutin and Fernandez 1984). With numbers obtained, length of transect is multiplied by the perpendicular distance from the trail within which observers are likely to see any nests to produce the sample area covered. And so the density is calculated.

Immense error in measurement occurs at all stages. Usually transects cover not even 1 percent of the total area for which the biologist calculates density of the species. Visibility varies depending on the nature of the vegetation. Observability of animals varies hugely depending on the species, the age and sex of the animals, and the behavior of the animals at the time. While sophisticated computer programs and statistics are available now to calculate densities from transects (e.g., Buckland et al. 1993), they are only as good as the data entered, and there is little escape from the error of the data. Densities from sample transects are almost always extremely imprecise. If an estimate is given, and error of less than 30% is suggested, the writer is probably being optimistic.

All conservation management that depends on knowledge of numbers obtained from transects must incorporate knowledge of the imprecision. There is little that can be done about it: precise estimates are extremely difficult, perhaps especially in forest (Plumptre 2000).

because, of course, most values for density are obtained by fieldworkers who watch their study species in areas where they occur at high density—there is no point in studying in an area where there are hardly any study animals.

Despite this general relation between gorilla abundance and the nature of their habitat, there are numerous non-ecological factors that influence gorilla numbers, including human disturbance, such as hunting, or epidemic disease such as Ebola (ch. 14.1.1) (Kalpers et al. 2003; Leroy et al. 2004; Walsh et al. 2003). All of these must be taken into account when considering the relation between ecology and population density.

4.1.8. Comparison with *Pan* and *Pongo*

4.1.8.1. Pan

The ability of gorillas to free themselves from ripe fruit distinguishes them from both species of chimpanzee. Chimpanzees and bonobos live in a wide variety of habitats, from primary forest to wooded savanna. At all sites, herbs and other fibrous vegetation are important parts of the diet, just as they are for gorillas. However, all chimpanzees and bonobos depend on fruit far more than do gorillas (Boesch et al. 2002; Tutin et al. 1997; and see refs. in following discussion).

As we have said (ch. 3.1.1), chimpanzees occur everywhere that gorillas do except the Virunga Volcanoes. A comparison between the two apes inhabiting the same forest highlights the key ecological differences between them. The home ranges of sympatric chimpanzees and gorillas overlap extensively (Kuroda et al. 1996; Stanford 2006; Tutin and Fernandez 1993; Yamagiwa and Basabose 2006). Nevertheless, their habitat preferences differ. While gorillas are most abundant in open secondary forest, chimpanzee densities are highest in closed canopy, primary forest (Bermejo 1999; Tutin and Fernandez 1984; Yamagiwa 1999a).

Both eastern lowland and western gorillas are more folivorous than are chimpanzees inhabiting the same forest, eating proportionally more fiber and less fruit, as well as a greater diversity of herbaceous foods, at all times of year (fig. 4.4) (Stanford 2006). Yet, in terms of fruit species eaten, the diets of the two apes are more similar than different. At Lopé, Gabon, and Kahuzi-Biega, DRC, for example, chimpanzee and gorilla diets have 60%–80% of fruit species in common (fig. 4.5). Despite this overlap, there appears to be little overt feeding competition between the two apes, possibly due to mutual avoidance. Even when they encounter each other in the same fruiting trees, antagonism, such as threats and chasing, is rare. In fact, the two species sometimes seem unconcerned by each other's presence (Kuroda et al. 1996; Stanford 2006; Tutin and Fernandez 1993; Yamagiwa et al. 1996).

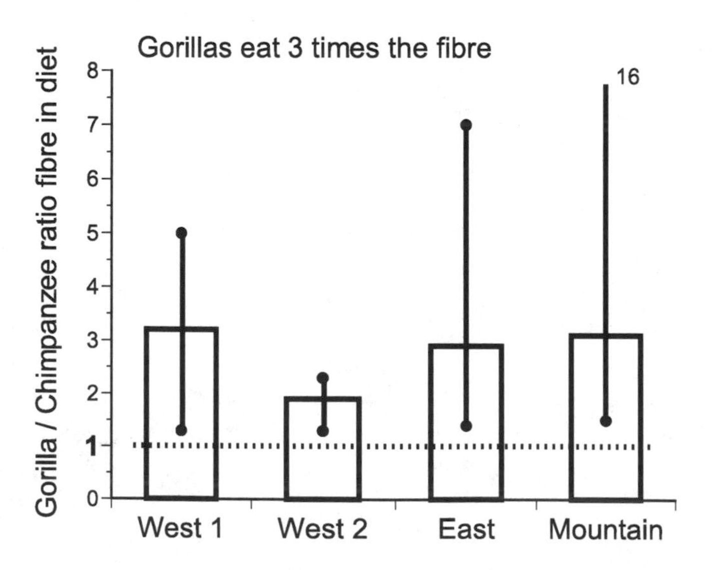

Fig. 4.4. Gorilla/chimpanzee ratio of fiber content in diet, analyzed from dung samples. Range and median (bar) of monthly fiber scores. Sources: Data from Ndoki, Rep. Congo (West 1) (Kuroda et al. 1996), and Lopé, Gabon (West 2) (Tutin and Fernandez 1993); Kahuzi-Biega, DRC (East) (Yamagiwa et al. 1996); and Bwindi, Uganda (Mountain) (Stanford and Nkurunungi 2003).

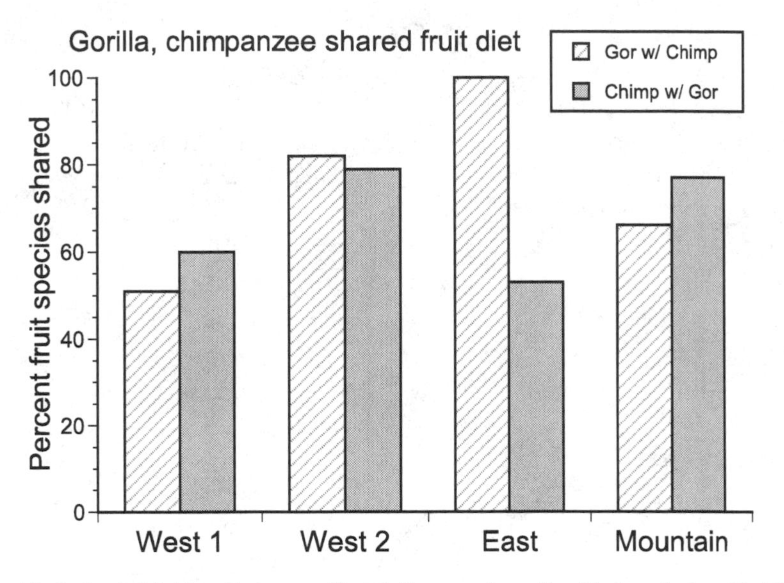

Fig. 4.5. Fruit part of diet shared between gorilla and chimpanzee. Proportion of fruit species in gorillas' diet that chimpanzees eat (Gor w/ Chimp) and proportion of chimpanzees' fruit diet shared by gorillas (Chimp w/ Gor). *Details at end of chapter.* Sources: Data from same sites and sources as in figure 4.4.

One explanation for this apparent lack of competition is that their diets are most similar in the fruiting season, when there is plenty of food. Also, their markedly different foraging styles might prevent contest competition. Chimpanzees have been characterized as "focused" frugivores. They move rapidly between food sources, often spending hours in the same fruiting tree and returning to it for consecutive days until the crop is depleted. In contrast, gorillas employ an eat-as-you-go style of fruit foraging. They spend less time in any one tree than do chimpanzees, for example, minutes rather than hours, and move slowly between fruit sources, feeding along the way on ground vegetation (Doran 1996; Kuroda et al. 1996; Rodman 2002; Tutin and Fernandez 1993; Yamagiwa and Basabose 2006).

The foraging styles and diets of the two apes are most divergent at times of fruit shortage. Then, gorillas contract their daily travel and switch to a predominantly herbaceous diet, including lower-quality foods such as woody stems and bark (sec. 4.1.3). In contrast, sympatric chimpanzees expand their range and persist in searching for fruit (Basabose 2005; Kuroda et al. 1996; Stanford and Nkurunungi 2003; Tutin and Fernandez 1993). While they increase the amount of foliage that they eat when fruit is scarce, they never stop being primarily frugivorous. Indeed, they will eat otherwise less preferred fruits rather than switching to foliage (Basabose 2002; Wrangham et al. 1996; Yamagiwa and Basabose 2006).

Chimpanzees in general travel farther each day than do gorillas. For example, the average daily distance for gorillas varies from 550 m in the Virungas (Fossey and Harcourt 1977; McNeilage 2001; Watts 1998a) to just over 2 km at Mondika, CAR (Doran-Sheehy et al. 2004). By contrast, chimpanzees (including those at sites where gorillas do not occur) routinely travel 2–5 km per day (Boesch and Boesch-Achermann 2000; Goodall 1986; Wrangham 1977). In Taï forest, Côte d'Ivoire, the chimpanzees can average as much as 11 km per day in some months (Boesch and Boesch-Achermann 2000). This is an almost inconceivable distance to anyone who has strolled slowly along after a habituated, feeding gorilla group.

Their longer daily journeys mean that chimpanzees can cover a larger percentage of their home range in a shorter period of time than do gorillas (Basabose 2005; Stanford 2006; Watts 1991a; Yamagiwa et al. 1996), and therefore have the option of being territorial, which they are at most sites (ch. 2.2.3; sec. 4.2) (Doran et al. 2002a; Mitani and Rodman 1979; Williams et al. 2004).

The ecological differences between *Gorilla* and *Pan* translate into differing costs of being gregarious and help explain the striking contrasts in the nature of their societies (Wrangham 1979, 1986; Wrangham et al. 1996; Yamagiwa 1999a). Bonobos and chimpanzees have socially flexible, fission-fusion societies (sec. 4.2.3.1). This flexibility is best explained as an adaptation to fluctuations in the supply of ripe fruit. It allows the animals to respond to fruit scarcity and rising foraging costs by reducing the size of foraging groups, thereby reducing feeding competition (Chapman, White, and Wrangham 1994; Wrangham 2000; Wrangham et al. 1996). At all well-studied chimpanzee sites and in one community of bonobos, fruit availability, measured in the general habitat or in individual patches, is one of the most important factors determining the size of subgroups, with smaller parties forming when food is less abundant (Boesch 1996; Chapman et al. 1995; Doran 1997; Matsumoto-Oda et al. 1998; Mitani, Watts, and Lwanga 2002a; Sakura 1994; White 1996; Wrangham 1986).

Variation in food supply, assessed by seasonality of rainfall, also helps explain some of the differences between *Pan* populations in levels of gregariousness (Chapman et al. 1994). Eastern chimpanzees, the least sociable, face the greatest number of dry months per year and, presumably, the greatest fluctuations in fruit availability (Doran et al. 2002a). The more gregarious chimpanzees in western Africa experience less variation in rainfall, while the two bonobo populations, the most gregarious, face the fewest dry months (Doran et al. 2002a; White 1996). By comparison, gorillas enjoy the equivalent of the fewest dry months because their primary food, herbaceous vegetation, is less susceptible than is fruit to seasonal fluctuations.

4.1.8.2. Pongo

Of all the great apes, orangutans are the most constrained by the problem of finding enough to eat. Like chimpanzees, they are primarily frugivorous, depending heavily on ripe, succulent fruit (Galdikas 1988; Knott 1998a; Rodman 1988b). Unlike the African apes, however, orangutans are predominantly arboreal. This arboreality, combined with their large body size (ch. 1.3; ch. 3.2.2; they are the heaviest arboreal mammal on earth) affords them near immunity from predation but presents serious foraging challenges (Horr 1975; Mitani et al. 1991). Orangutans require a lot of food to maintain their large bulk, but their size prevents fast movement through the trees. They are slow and cumbersome, especially the heavy adult males. They do not travel as rapidly as chimpanzees, and so cannot cover as much distance in their search for fruit (Rodman 1984).

On top of this, the forests of Borneo and Sumatra are less productive than are African forests, and subject to the phenomenon, unique to Southeast Asia, of widespread fruit "masts." These are brief intervals of fruit bonanzas when many species over a large area produce huge fruit crops (Corlett and Primack 2006; Terborgh and van Schaik 1987). Unfortunately, the periods of superabundance occur only once every two to ten years, punctuating much longer intervals of low fruit production. Orangutans are therefore subject to extreme fluctuations in food availability, on a much larger scale than the African apes (Delgado and van Schaik 2000).

During fruit masts, orangutans feed well and put on weight (Knott 1998a; Leighton et al. 1995). As fruit declines, they continue to search for it, increasing their daily range to do so. But they cannot keep up this strategy during severe shortages. Then, they reduce their ranging and eat lower-quality, fibrous vegetation. Bark is one of their most important fallback foods (Delgado and van Schaik 2000; Rodman 1988b). While their dietary flexibility enables them to survive, they do not thrive on fibrous foods the way gorillas do. Indeed orangutans in Borneo approach starvation during fruit-poor periods as they metabolize their fat reserves, a process detected by high levels of breakdown products, ketones, in the urine (Knott 1998a). These periodic times of very low food availability might be one reason for the species' slow reproductive rates (ch. 3.2.7) (Knott and Kahlenberg 2007).

Their dependence on fruit and its distribution in the forest is a key to understanding the solitary nature of orangutans, the least gregarious of the great apes (sec. 4.2) (Delgado and van Schaik 2000; Galdikas 1979; Horr 1975; Rijsken 1978; Rodman 1988b; van Schaik 1999). At all sites, it is clear that ripe fruit is not produced regularly enough (in time or space) or in large enough quantities to allow more than infrequent, intermittent association with others. Being gregarious, even in small parties, means an increase in foraging effort that orangutans cannot afford (Knott 1999).

Conclusion: Gorilla Ecology

Gorillas are herbivores that eat ripe fruit when it is available. The dietary staples, even of "frugivorous" western gorillas, are high-quality terrestrial herbs, the abundance of which correlates with gorilla densities. These herbaceous foods differ from fruit in being more evenly distributed and available throughout the year. Gorillas do not, therefore, face food shortages to the same extent as do the other great apes, which depend far more heavily on ripe fruit. As a consequence, gorillas are not as constrained by foraging costs as are *Pan* or *Pongo*. They can afford to live in permanent, cohesive groups, whereas the other apes must be able to respond to variation in food availability with flexible sociality.

The fundamental differences between the great apes in ecological constraints on grouping have led to pervasive differences in the nature of their societies.

4.2 Gorilla Society

Summary: Gorilla Society

Gorilla society is based on cohesive groups usually containing one adult male, several breeding females (median of 3) and their offspring. The society is characterized by long-term association between males and females, sometimes lasting for years. Both sexes normally leave the group of their birth. Females immediately join a silverback, either group-living or solitary, while dispersing males usually wander alone, or sometimes associate with other bachelors, until they attract females and form their own breeding group. Competition between males for females is intense and sometimes involves infanticide.

The different subspecies and populations of gorillas have a similar social structure, despite large differences in habitat and diet (sec. 4.1). Societal details such as the nature and timing of dispersal by both sexes, processes of group formation and change, and the nature of intergroup encounters are comparable at most sites.

The most notable difference across populations is that multi-male groups are far more common in mountain gorillas than in eastern lowland or western gorillas. Other reported contrasts are less well documented, but possibly important. Specifically, infanticide appears to be less predictable in other populations than it is in mountain gorillas; and at some western sites, peaceful encounters between groups (including lone males) are relatively frequent.

Ecological differences probably underlie some of these contrasts in the nature of societies. For example, the mountain gorillas' rich, year-round food supply could allow larger maximum group sizes and hence multi-male groups. In western gorillas, high densities in some regions (such as swamp clearings) and correspondingly high encounter rates could make aggression simply too costly to be the most common response to another social unit.

The social relationships between males and females emphasize the importance of the male, his dominance over females, and their attraction to him. For instance, females (except when in estrus), tend to approach or follow the silverback more than vice versa, and even compete with each other to be near him. Among females, most of which are unrelated, relationships are at best merely tolerant. Close kin are characteristically friendly, however. For immature animals, the silverback is usually the most important adult in the group, after their mother, as indicated by the amount of time they spend near him.

The stable group membership of the gorilla contrasts with the parties of continually changing composition that characterize the society of the two chimpanzees, and the solitary society of orangutans. At a fundamental level, this contrast reflects varying ecological constraints on the costs of grouping.

Introduction: Gorilla Society

In this part of the chapter we describe the basic structure of gorilla society: the size and composition of groups; the social processes, such as dispersal, that lead to changes in the structure of groups, including their formation and demise; and the general nature of interactions between groups and between individuals within groups. This description is the raw material for later chapters that interpret gorilla society within the socioecological framework discussed in chapters 1 and 2.

We summarize data only from studies in the wild. This is not to say that captive research has not contributed to our knowledge of gorilla behavior, providing data, for example, on reproductive physiology and mating (Nadler 1975a, 1981), maternal behavior and infant development (e.g., Beck and Power 1988; Hoff, Nadler, and Maple 1981, 1983), intragroup social interactions (Enisco, Calcagno, and Gold 1999; Meder 1990, 2005; Scott and Lockard 2006), and some aspects of life history such as interbirth intervals (see refs. in ch. 3.2.7). Findings from captivity generally support the picture from the wild, although behavior of captive animals may, of course, be influenced by factors specific to captivity, such as boredom or inability to evade group members (e.g., Mitchell 1989). It is sometimes difficult, therefore, to make generalizations from captivity about the selection pressures that have shaped gorilla society, which is what socioecology is all about. Studies on captive animals are better suited to address questions about proximate rather than functional aspects of behavior, for example, the experiments on taste preferences referred to in section 4.1.2.

4.2.1 Social structure and social processes

4.2.1.1. Group size and composition; population structure

Group size. Almost all gorillas in all populations live in cohesive groups consisting of adults of both sexes and their offspring. Despite variation in degrees of frugivory and associated foraging effort (sec. 4.1.3), median group size does not differ significantly between gorilla subspecies (fig. 4.6). From the Atlantic coast of Gabon to the montane forest at 3,000 m in the Virunga Volcanoes, gorillas live in groups of about eight animals (Robbins et al. 2004).

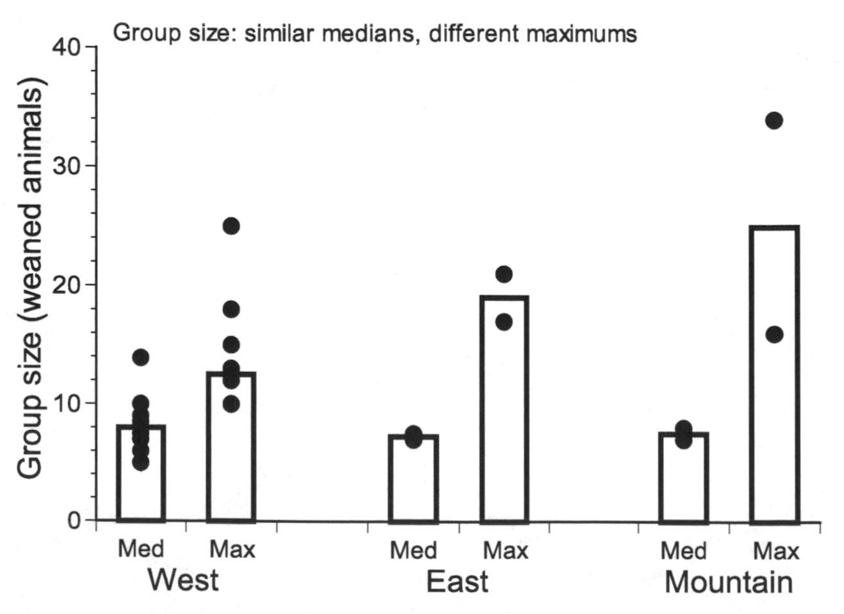

Fig. 4.6. Gorilla group sizes. Median (Med) and maximum (Max) group size for different gorilla populations (•), and medians for three gorilla subspecies (bar), western (West), eastern lowland (East), and mountain. Only weaned individuals are considered because infants can be missed in censuses. For each population only data from one census are shown. *Details, including data sources, at end of chapter.*

The largest groups recorded have consistently been found in the montane forests of Kahuzi-Biega National Park, DRC, and in the Virunga Volcanoes of DRC and Rwanda. There, groups can reach 20, even 30 weaned animals (fig. 4.6) (Doran and McNeilage 2001; Kalpers et al. 2003; Schaller 1963; Yamagiwa 1983; Yamagiwa et al. 1993). Mountain gorillas in the Virungas hold the record, with one group of 34 weaned (47 in total) animals, and still growing (Kalpers et al. 2003). In western gorillas, only at Lossi in Rep. Congo has weaned group size exceeded 20 members (Bermejo 1999).

The contrast between the regions in maximum group size could relate to environmental differences, in particular, the abundance of high-quality herbaceous ground vegetation (4.1.3). Regions in which terrestrial herbs are dense, such as the Virunga Volcanoes, the highland forest of Kahuzi-Biega, or Lossi's Marantaceae forest, provide abundant, permanently available food (Bermejo 1999; Watts 2000b; Yamagiwa et al. 2003a). These conditions increase the likelihood that extra-large groups can arise and persist (Goldsmith 2003; Yamagiwa et al. 2003a).

Group composition. Just as the size of groups is similar across all populations, so also is their composition (Robbins et al. 2004). Gorilla groups typically have one fully adult male, three to five adult females, and immature animals of various ages (table 4.1). In all well-studied populations, groups are remarkably stable, with adult males and females living together for years (Sicotte 2001; Stokes, Parnell, and Olejniczak 2003; Tutin 1996). For example, in mountain gorillas, the typical adult female will reside in the same group for about 10 years, while the average silverback's breeding tenure is 9–10 years (Watts 1996, 2000a). One female, Effie, stayed with the same male, Beethoven of Group 5, for at least 18 years, from the time Dian Fossey first saw them in 1967 until Beethoven's death in 1985.

The only striking contrast in composition between populations is the proportion of breeding groups with more than one silverback. Multi-male groups are rare among eastern lowland and western gorillas (table 4.2). By contrast, over 40% of mountain gorilla groups are multi-male. The contrast has been evident ever since the first studies in each region, several decades ago (Harcourt et al. 1981a; Kalpers et al. 2003; Schaller 1963). Mountain gorillas, however, are the least numerous subspecies. We calculate that, across all gorilla populations, about 97% of breeding groups of gorillas in Africa have only one fully adult male (fig. 4.7). It is fair to conclude, then, that in general, gorilla society is based on a single-male system.

This difference between mountain gorillas and others in the proportion of multi-male groups may relate in part to the differences in maximum group size. Mountain gorillas have the largest groups on record, as we have described previously, and within mountain gorillas, those groups with

Table 4.1. Composition of breeding groups in different populations. Values are mean number of individual per age-sex class per group. Numbers in parentheses next to site name are numbers of breeding groups from which mean was obtained. SB = silverback; BB = blackback; Ad F = adult female; Inf = infant; Inf/Ad F = ratio of infants to adult females in each group; Juvs + Subads = juveniles and subadults. See ch. 3.2.3 for age class definitions.

	Age-sex class					
Site	**SB**	**BB**	**Ad F**	**Inf**	**Inf / Ad F**	**Juvs + Subads**
Western						
Lokoué (37)	1	0.2	3.2	2	0.6	1.8
Maya (29)	1	0.1	4	2.4	0.6	3.9
Mbeli (12)	1	0.8	3.5	1.8	0.5	2.3
Lossi (4)	1	2.1	7.1	4.5	0.6	7
Lopé (4)	1.3	0.9	2.9	1	0.2	3.25
Western median	**1**	**0.8**	**3.5**	**2**	**0.6**	**3.25**
East Lowland						
Kahuzi (11)	1.1	0.3	4.2	1.3	0.2	2.9
Mountain						
Virunga '00 (15)	2.1	1	5.9	4.6	0.7	2.9
Bwindi (28)	1.8	1.7	4.5	2.7	0.6	2.2
Mountain median	**1.95**	**1.3**	**5.2**	**3.65**	**0.65**	**2.5**

Sources: Lokoué (Gatti et al. 2004); Maya (Magliocca et al. 1999); Mbeli (Parnell 2002); Lossi (Bermejo 1999); Lopé (Tutin 1996); Kahuzi (Yamagiwa et al. 1993); Virunga '00 (i.e., in 2000) (Kalpers et al. 2003); and Bwindi (McNeilage et al. 2001).

multiple males are significantly larger (weaned group size) than groups with just one silverback. The difference is not significant when only adult female group size is considered (stats. 4.2, 4.3) (Kalpers et al. 2003; McNeilage et al. 2001; Robbins 1995).

The relation between weaned group size and number of males, however, does not extend to other populations. For example, the largest recorded group of western gorillas, at Lossi (25 weaned individuals), and the largest eastern lowland groups with over 20 weaned animals, each had only one adult male (Bermejo 1997; Yamagiwa and Kahekwa 2001).

The development and persistence of multi-male groups must depend on a host of ecological, demographic, and social factors all interacting to affect

Table 4.2. Proportion of breeding groups that have more than one fully mature male (silverback).

	Total groups (*N*)	Multi-male (*N*)	Multi-male (%)
Western			
Lokoué	37	0	0
Maya	29	0	0
Mbeli	12	0	0
Mikongo	4	0	0
Lopé	8	2	25
Western median			**0**
East Lowland			
Itebero	10	0	0
Kahuzi	24	2	8.3
East Lowland median			**4.2**
Mountain			
Virunga '60	10	3	30
Virunga '81	27	11	41
Virunga '00	16	9	56
Virunga median			**41**
Bwindi	28	13	46
Mountain median			**43.5**

Sources same as for fig. 4.6 for western, eastern lowland, and Bwindi groups (see Figure Details at end of chapter). For Virunga Volcanoes: Virunga '60 (Schaller 1963); Virunga '81 (Harcourt et al. 1983); and Virunga '00 (Kalpers et al. 2003).

male and female dispersal and reproductive strategies. We come back to these in some detail (ch. 8; ch. 11).

Lone males and non-breeding groups. In addition to breeding groups, solitary males, usually silverbacks, are always detected in some numbers in any long-term study (table 4.3) and are an influential part of gorilla society (e.g., ch. 11.1.3). Since lone animals are easy to overlook in censuses, it can be difficult to be sure of exactly how many there are in any population. Nevertheless, their lower representation in mountain gorillas compared to western or eastern lowland populations fits the greater proportion of multi-male groups in mountain gorillas.

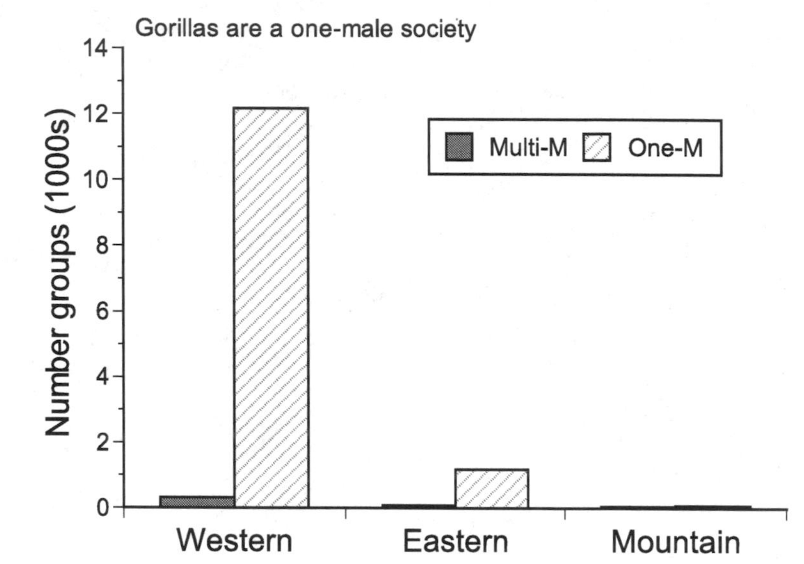

Fig. 4.7. Total numbers of one-male and multi-male groups in the western, eastern lowland, and mountain gorillas. *Details at end of chapter.*

Table 4.3. Proportion of social "units" that are solitary males in western, eastern lowland, and mountain populations. Solitary % = percentage of all social units that are solitary silverbacks (SBs) or blackbacks (BBs).

	Groups + solitaries(*N*)	Solitary SBs (*N*)	Solitary BBs (*N*)	Solitary (%)
Western				
Lokoué	76	20	11	41
Maya	47	16	0	34
Mbeli	21	7	0	33
East Lowland				
Itebero	16	6	0	38
Kahuzi	34	8	1	27
Mountain				
Virunga '00	42	10	0	22
Bwindi	35	7	0	20

Sources same as for fig. 4.6 (see Figure Details at end of chapter).

And finally, most populations have a small proportion of non-breeding groups, also known as "bachelor" groups. While these occasionally include immature females, they consist primarily of young males, from juveniles to blackbacks, and almost always at least one silverback (Gatti et al. 2004; Levréro et al. 2006; Robbins 1995; Yamagiwa and Kahekwa 2004). The best-studied non-breeding group, in the Virunga Volcanoes, was observed for fourteen years, and for eleven of these, it consisted of males only (Robbins 1996). In a study of non-breeding groups in the very dense population at Lokoué clearing in Rep. Congo, 81% of the animals living in non-breeding groups ($N = 54$ animals of known sex, 14 groups) were males (Levréro et al. 2006). Both these studies of mountain and western gorillas revealed similar group dynamics. Membership changed frequently, as males left at maturity to become solitary, while immatures and blackbacks immigrated from other groups. Nevertheless, some male associations were quite long-lasting. For example, in the non-breeding mountain gorilla group studied by Robbins, five of the individuals resided together for six years (Robbins 1996).

4.2.1.2. Dispersal; group formation, transition, and demise

The size and makeup of a group are the result of births, deaths, immigration, and emigration. In chapter 3, we talked about birth and death. Immigration and emigration are included in the term *dispersal,* our current topic. Across primates, young animals on the brink of adulthood are the individuals that most commonly disperse, doing so voluntarily, in search of a social group or territory in which to breed. Less regularly, breeding animals also disperse (ch. 2.4.2.6). Sex differences in the timing, nature, and contexts of dispersal are a key to understanding the nature of a society (ch. 2.4.2.6), and in gorillas, the two sexes differ markedly in these aspects (Harcourt 1978a; Harcourt, Stewart, and Fossey 1976; Sicotte 1993, 2001; Watts 1991b, 1996; Yamagiwa 1999a).

Female dispersal and transfer. In all well-studied gorilla populations across Africa, most females leave the group of their birth, their "natal group" (Harcourt 1978a; Harcourt et al. 1976; Sicotte 1993; Stokes et al. 2003; Watts 1991b, 1996; Yamagiwa and Kahekwa 2001). Natal dispersal by the majority of females means that the breeding females in gorilla groups are mostly unrelated (Stewart and Harcourt 1987; Watts 1994a), with, of course, many potential ramifications for the nature of gorilla society (ch. 2.2.3.2).

Most females leave their natal group between the ages of sexual maturity (seven to eight years old) and first breeding at ten years (details on maturation in ch. 3.2.3.3). In the Virunga Volcanoes and Kahuzi-Biega,

50%–70% of females observed since birth emigrated before breeding, depending on exact dates and sample analyzed (Sicotte 2001; Watts 1991b, 1996; Yamagiwa and Kahekwa 2001). While there are fewer long-term data for western gorillas, natal dispersal is clearly common there too. At Mbeli, Rep. Congo, for example, over half of 35 females that dispersed had not reached breeding age (Stokes et al. 2003).

In the Virunga Volcano population, the largest sample indicates that while 69% (20 of 29) of females emigrated before giving birth, 31% (9 of 29) had their first infant in their natal group (Watts 1996). Of these 9, 4 (or 14% of the total) subsequently emigrated before giving birth again. Of the remainder, 3 never dispersed, all in the famous Group 5 of the Karisoke Research Center. (It is not known what happened to the other 2). One difference between those that emigrated before giving birth and the others lies in the availability of a second adult male in the group. All females that reached maturity in a group with only one silverback dispersed before breeding. We consider the implications of this contrast for avoidance of inbreeding in chapter 8.2.

Once females have dispersed, they often transfer again. Thus, in the Virungas, 46% of the 24 females observed for at least ten years as adults transferred more than once, while among western gorillas at Mbeli, of 17 dispersing adult females, 47% were moving for at least the second time (Sicotte 2001; Stokes et al. 2003). The typical female probably transfers once or twice in her life after leaving her natal group, but some individuals have moved up to five times during their adulthood (Watts 1996). Even from the relatively short study of six years at Mbeli clearing, females are known to have transferred as many as three times and bred in four different groups (Stokes et al. 2003).

If young nulliparous females (meaning they have not yet had an infant) are the most likely to disperse; mothers carrying suckling infants are the least likely. The primary exceptions to this difference involve females in one-male groups whose silverback dies. In such cases, the widowed females, no matter what their reproductive state, do not wander alone, or stay as a male-less group, but instead almost always join another male (Harcourt 1978a; Stewart and Harcourt 1987; Stokes et al. 2003; Watts 1989a). We describe and discuss the few exceptions to this rule of death-induced, involuntary dispersal a little later in the section "Group transition and demise."

Females usually transfer on their own, but occasionally move together. In mountain and western gorillas, the most common context of multi-female transfer is the death of a silverback in a one-male group. At these times, preadolescent females (juveniles and older infants) follow dispersing adults—

often their mothers—into another group. These immature immigrants sometimes mature in the new group to become eventual mates of the resident male (Sicotte 2001; Stewart and Harcourt 1987; Stokes et al. 2003).

Among Grauer's gorillas of Kahuzi-Biega, however, females transfer together more often than in other populations, regardless of whether or not their silverback has died (Yamagiwa and Kahekwa 2001). The co-dispersal of relatives is another way, in addition to females remaining in their natal groups, through which closely related adult females may come to reside in the same group (Bradley et al. 2005). Additionally, it is not impossible that females might be more likely to immigrate into a group that already contains kin, if kin can recognize one another after an absence, and are less aggressive to one another than to non-kin (sec. 4.2.2.2; ch. 5.2) (Cheney and Seyfarth 1983).

While the background and the explanation for most emigrations and transfers are obvious enough, when a female transferred voluntarily, it always took us by surprise. A female waits until another male is close by before dispersing. She slips from one male to another in a matter of minutes during interunit encounters, while the silverbacks are busy exchanging their flamboyant displays (Harcourt 1978a; Sicotte 1993). With no change of behavior over the previous weeks or even days to warn us, she suddenly takes off. Why this male, why this interunit encounter? Perhaps most surprising are the females that leave behind young juvenile offspring when they transfer (Harcourt 1978a; Stewart and Harcourt 1987).

Male dispersal. In most gorilla populations, the majority of males, like females, disperse before reaching full adulthood. They leave as blackbacks or young silverbacks, usually between the ages of 10 and 15 years (Robbins 1995; Watts and Pusey 1993) with an average age of 13.5 years (Robbins 1995). Dispersal before full maturity explains in part why most gorilla groups have only one fully adult male (sec. 4.2.1.1).

The process of emigration appears to be voluntary, and unlike for the females, is often preceded by many months during which the young male spends increasingly more time farther and farther away from his group, punctuating ever longer absences with ever more temporary visits, until finally he has to be counted as having emigrated (Fossey 1983; Harcourt 1978a; Schaller 1963; Tutin 1996).

Males that leave a breeding group usually wander alone, sometimes for years, before acquiring mates through female transfer. Attracting females is the only way for a lone adult male to gain access to mates, since resident silverbacks intensely resist outside males. No fully adult male has ever been recorded to immigrate into a breeding group, and none has ever been recorded to take over a group by ousting the resident silverback (mountain gorillas: Harcourt 1978a; Watts 1991b, 1996, 2000a; eastern lowland gorillas: Yamagiwa and Kahekwa 2001; western gorillas: Gatti et al. 2003; Stokes et al. 2003; Tutin 1996). Blackbacks rarely join breeding groups: they have been recorded to do so in western gorillas, at Lossi, Rep. Congo, and once in the Virunga mountain gorillas (Robbins 1995; Watts 2000a).

Sometimes males disperse while still immature, that is, less than eight years old, particularly after the death of their group's silverback. While immature males occasionally manage to join breeding groups (Fossey 1983; Robbins et al. 2004; Stokes et al. 2003), even young juveniles can meet aggressive resistance from resident silverbacks (Robbins 1999; Sicotte 2000; Watts 1990b; Yamagiwa and Kahekwa 2001). An alternative to a solitary life for "homeless" immature males, as well as dispersing blackbacks, is to join a non-breeding group or a solitary silverback. Young males usually reside in bachelor groups until they mature, when they emigrate to become solitary (Levréro et al. 2006; Robbins 1995; Yamagiwa 1987a) (and see previous discussion).

Males can find themselves maturing in a non-breeding group by default, if all females of a breeding group leave and the remaining males simply stay together. As a result, the males of bachelor groups are sometimes close relatives. Genetic analyses of four non-breeding western groups at Lokoué, for example, found that three of the four silverbacks were closely related to at least one of the immatures in their group (Levréro et al. 2006).

Male philopatry. Philopatry is a convenient jargon in socioecology to mean staying in a place, usually the place of one's birth. Just as some female mountain gorillas are philopatric (i.e., they remain to breed in their natal group), so also are some males. In the Virungas, of 11 individuals for which there are adequate records, 7 (64%) stayed in their natal group to become dominant breeding silverbacks, a process known as succession. This occurs either because the dominant male dies or the younger silverback usurps the alpha position from the older silverback (Robbins 1995; Watts 2000a). The combined observations of no immigration, no takeover, and some male philopatry means that the mature males in multi-male groups are likely to have known each other since at least one of them was immature, and be more or less close relatives, for example, uncle-nephew, cousins, half siblings, or even father-son or full siblings. While all these combinations are known or suspected from the Virunga study groups, recent genetic analyses indicate that father-son or full sibling pairs are less common than initially expected (Bradley et al. 2005). This is partly due to the slow reproductive rate of females (producing, on average, one surviving son every 8 years), the time it takes males to mature fully (about 12 years), and the likelihood that they will die in their late twenties (ch. 3.2.3–3.2.5).

At another level of philopatry, some dispersing males stay close to their natal group in an adjacent range. This is indicated by genetic evidence from western and mountain gorillas showing that the silverbacks in neighboring groups are often close relatives (Bradley et al. 2004; Nsubuga 2005, cited in Robbins 2006).

Group fission. A maturing male has a third route, somewhere between emigration and succession, to become a breeder. This route is group fission. Instead of emigrating and waiting to attract females, he emigrates accompanied by one or more females (and sometimes their offspring) from his group. In two well-documented cases, one in eastern lowland and one in mountain gorillas in the Virungas, fission took several months, during which temporary subgrouping gradually became permanent separation (Robbins 2001; Yamagiwa and Kahekwa 2001).

Group fission is uncommon in gorilla populations because it requires a multi-male group and one that is relatively large. Of seven records of group fission, five in the Virungas, and two in Kahuzi-Biega, the median group size at the time of split was at least 21 (including infants), over twice the median group size for each population (Kalpers et al. 2003; Robbins 2001; Yamagiwa and Kahekwa 2001). An eighth case of fission was recently documented in Bwindi (Ganas and Robbins 2005). The phenomenon will necessarily be more common in mountain gorillas than other populations, because mountain gorillas have more multi-male groups.

Group transition and demise. A gorilla group can be described as having a life span: it is newly born when a female joins a male, maturing as more females join and offspring are born, and dying as animals either leave or die (Parnell 2002; Robbins 1995; Yamagiwa 1987b, 1999a). The same description has been applied to Thomas langurs, a primate with a very similar society to that of gorillas (ch. 1.3) (Steenbeek et al. 2000).

The death of a group can occur relatively suddenly when a silverback dies. In western and mountain gorillas, single-male groups usually disintegrate catastrophically after a male's death, either disappearing completely or dwindling to a non-breeding group, following dispersal of all females (see earlier section, "Female dispersal and transfer").

However, some eastern lowland groups in Kahuzi-Biega have survived a leading male's death. In one case the group of females and offspring remained intact for 29 months with no mature male. In the end, a male joined them to become the resident silverback. This degree of cohesiveness in the absence of silverbacks has never been observed anywhere else (Yamagiwa and Kahekwa 2001). Widow groups must be uncommon in Kahuzi-Biega, since censuses of the population reveal no groups without mature males, and indeed, only three such groups have been recorded in 26 years of observation (Yamagiwa et al. 1993). Nevertheless, these records are notable because they contrast so markedly with the data from western populations and mountain gorillas. We come back to them later when we discuss both female and male strategies (ch. 7.1.2; ch. 7, Conclusion; ch. 11.3.2).

Multi-male groups are buffered against disintegration in the event of a silverback's death, because a surviving resident male can take over the group in a seemingly smooth transition. In five well-documented cases in the Virungas, when the older or dominant male of a multi-male group died, all females and immature group members remained with the younger silverbacks. In one case, the group fissioned after the leading male's death, with the females and immatures partitioning themselves between the two remaining males (Robbins 2001). The inheritance of a group by young resident silverbacks after a dominant male's death has also been observed in western gorillas at Lopé, and Grauer's gorillas in Kahuzi-Biega (Tutin 1996; Yamagiwa 1983).

It appears that fully mature males are more able than younger males to retain females after a dominant silverback's death. For example, in two groups of mountain gorillas that lost their leading silverbacks, all females deserted the surviving males that were sexually mature, but not yet fully silvered (Fossey 1983).

To summarize, the processes of dispersal, group formation, transitions, and decline are similar for gorillas across Africa (Levréro et al. 2006; Parnell 2002; Robbins 2001; Robbins and Robbins 2004; Yamagiwa 1987b; Yamagiwa and Kahekwa 2001). New groups form most commonly through acquisition of females by solitary males, or occasionally through group fission (documented at Kahuzi-Biega, the Virungas, and Bwindi). As groups mature, they will contain varying proportions of infants, juveniles, subadults, and adults. Breeding groups can alternate in structure between single male and multi-male through maturation of natal males and their dispersal, or through group fission. Groups can dwindle to a non-breeding structure when all adult females disperse, or groups can disappear completely if males die or disperse as well (Mbeli, Virungas). Bachelor groups can change in size through immigration or dispersal of males (Lokoué and Virungas) and can switch back to breeding units through the immigration of females (Virungas, Kahuzi-Biega).

4.2.1.3. Infanticide

Gorilla males use the tactic of infanticide to increase their breeding opportunities, and possibly to avoid expending energy or risking injury protecting unrelated infants (ch 2.4.2.4; ch. 2.4.3; box 2.4; ch. 3.2.3.3; fig. 3.4; ch. 11.1.2).

Most infanticides occur after the silverback of a one-male group dies. In the course of ensuing group disintegration, all females, including those with unweaned infants, disperse to other groups or solitary males (Watts 1989a). Around the time of transfer, their infants are killed by the unfamiliar males. Less often, females may lose their infants during interunit encounters when their silverback is still alive (Watts 1989a; Yamagiwa and Kahekwa 2004).

Infanticide in gorillas is usually associated with female transfer, a situation that contrasts with that in many other primate species in which invading males kill infants after ousting the resident males, as in the classic case of Hanuman langurs (Hrdy 1977). As previously mentioned, takeovers by outside males have not been observed in gorillas. These different scenarios, however, are merely variations on the same theme: males kill infants they have not sired to increase their access to fertile females.

How do males assess paternity? Silverbacks, of course, do not run genetic tests on infants, nor are they likely to recognize any familial resemblance. Instead, degree of familiarity with the mother, including sexual contact with her, are probably the cues used (Watts 1989a; Yamagiwa and Kahekwa 2004). Two cases illustrate these points. In a fascinating recent account from Kahuzi-Biega National Park, DRC, two pregnant females transferred to a male and gave birth within three months of their move. The new male killed their infants a few days after their birth. Importantly, in the same month, the same male spared the newborn of a female with which he had resided for at least a year (Yamagiwa and Kahekwa 2004). Unfortunately, there are no data on the interactions between the pregnant immigrants and the new silverback, so we don't know whether it was their short residence with him or a lack of sexual interaction in particular that prompted his infanticidal response to their infants (Yamagiwa and Kahekwa 2004).

The crucial importance of mating per se was demonstrated in the Virunga Volcanoes by a young silverback (Beetsme) who had immigrated into a breeding group when he was a blackback—the only known case of a sexually mature male joining a breeding group. Following the death of the dominant silverback, Beetsme killed the young infant of an adult female with whom he had never been seen to mate during his residency in the group, but spared (indeed was quite friendly toward) the infant of a female with whom he had mated (Fossey 1984; Watts 1989a). The influence of male-female familiarity (including mating) on paternity assessment by males and their subsequent behavior toward infants has shaped male and female reproductive strategies and, consequently, gorilla society (ch. 2.4.4; ch. 7.2).

In the Virunga population, infanticide has been a major cause of death of infants, accounting for over a third of infant mortality during certain periods ($N = 50$ infants, 19 deaths; Watts 1989a). Infanticide is a predictable, if relatively rare, event, because of the narrow range of circumstances in which it occurs: females with unweaned infants that are widowed or, less often, separated from their group silverback, are virtually doomed to have their infants killed by another male. The reason it does not happen more often is that the context is relatively unusual.

There have been no documented cases of infanticide in the Virungas since the mid-1980s, the reason being that all study groups since that time have been multi-male (Kalpers et al. 2003). Consequently, the females do not disperse to other males following the death of the mature silverback, and hence the infants of these groups are effectively buffered against infanticide (sec. 4.2.1.2; ch. 7.2.2.1; ch. 11.2) (Robbins and Robbins 2005; Watts 2000a).

Until recently, infanticide had not been observed in eastern lowland gorillas, despite numerous occasions in which it was predicted (Yamagiwa and Kahekwa 2001). Three instances have now been described from the Kahuzi population (see earlier discussion), each occurring in precisely the same situation as in the Virungas, namely, the transfer of females with infants to a non-father male (Yamagiwa and Kahekwa 2004).

Nobody has yet seen infanticide in any western gorilla population. However, researchers at Mbeli strongly suspect it in at least two instances when the disappearance of an infant coincided with the death of a leading silverback and the subsequent transfer of the mother to a new group (Stokes et al. 2003).

While infanticide might occur in all populations, it nevertheless seems to be less predictable in western and eastern lowland gorillas than in mountain gorillas. At Mbeli and Maya Nord in Rep. Congo, and in Kahuzi-Biega in DRC, females with unweaned infants have transferred to apparently new males with no harm to their babies (Stokes et al. 2003; Yamagiwa and Kahekwa 2001), a situation that has not been observed in mountain gorillas. These differences may relate to variation in levels of male-male competition, or to varying degrees of familiarity between males of different groups, and are discussed later in the context of male mating strategies (ch. 11.3).

4.2.1.4. Interunit encounters

The nature of encounters. Encounters between groups or between groups and solitary males are potentially important affairs in gorilla society as they are the times when females transfer and when males compete most intensely. We refer to both types of interactions, those between groups or between a group and lone silverback, as "interunit" encounters, where "unit" means a group or a solitary male.

Fig. 4.8. Silverback hooting and chest-beating in response to displays by an adult male from another group. (Mountain gorilla.) © A. H. Harcourt

Gorilla researchers have usually defined encounters as occurring when units come within 500 m of one another, that is, within more or less easy hearing range of the loud hoots and chest-beats that competing silverbacks perform (fig. 4.8). Most recorded encounters, however, take place at closer quarters, with males and other group members from different units within 50–300 m of each other (Harcourt 1978a; Sicotte 2001; Tutin 1996).

Interunit encounters last from around thirty minutes to one or two days (Sicotte 2001; Tutin 1996), but can go on for much longer, up to a week or more (Watts 1994c). They normally end with the participants going their separate ways, but sometimes one party supplants or even chases the other (Bermejo 2004; Levréro 2001; Tutin 1996; Watts 1994c). Intensely agonistic encounters can influence daily and monthly ranging of groups or lone males as they flee from or pursue each other (ch. 11.1.3) (Watts 1998a).

Across sites, encounters vary in their nature from peaceful to lethally aggressive. However, at four out of five sites for which there are data (western gorilla: Lopé, Lossi, Mondika; mountain: Virunga [see table 3.1]), most encounters involved some form of agonistic response, either avoidance/flight, or aggression, with or without physical contact (Bermejo 2004; Doran-Sheehy et al. 2004; Harcourt 1978a; Sicotte 1993; Tutin 1996). It should be said that the data from western gorillas, where one or more units in an encounter are likely to be unhabituated, may overestimate encounters characterized by flight or avoidance, because the gorillas may be fleeing the observers rather than each other.

During interunit conflicts, opposing silverbacks typically exchange aggressive displays. These flamboyant shows of force include hooting, chest-beating, stiff-legged strutting during which lips are pursed, and spectacular runs, ground thumps, and foliage whacks (Harcourt 1978a; Schaller 1963, ch. 5; Sicotte 1993). In the swampy clearings, adult males make impressive use of water in their displays, performing them in streams, causing great splashes and sprays (Parnell and Buchanan-Smith 2001). While silverbacks are the primary protagonists in these contests, other group members, especially subadult males and blackbacks, sometimes join in with struts, runs, or chest-beats.

Although most interunit contests involve only threat displays, some escalate into physical aggression between males, occasionally leading to serious injury or death (Fossey 1983; Harcourt 1978a; Tutin 1996; Watts 1994c; Yamagiwa and Kahekwa 2001). In the Virungas, for example, 24% of encounters within 50 m involved males hitting or biting one another (Sicotte 1993).

At the other end of the spectrum, the members of different social units, including the mature males, sometimes totally ignore one another and occasionally even mingle as if one group (Bermejo 2004; Doran-Sheehy et al. 2004; Harcourt 1978a; Levréro 2001; Sicotte 2001). Such peaceful encounters seem to be particularly common among western gorillas, at least at some sites (Doran and McNeilage 2001) (ch. 11.3.2). At Lokoué in Odzala National Park, for example, intergroup encounters without any displays by males are more frequent than those involving displays (Levréro 2001).

Frequency of encounters. The frequency with which social units encounter each other varies across sites. While estimates are available from only a few studies, they suggest that a high density of gorillas or the use of rare but highly attractive food sources (such as swampy clearings) may lead to relatively high rates of interunit encounters (Doran and McNeilage 2001).

At Mbeli clearing, for example, 45% of the times that a gorilla group visits the swamp, at least one other group is present (Stokes et al. 2003).

At the Lokoué clearing in gorilla-dense Odzala National Park, researchers recorded 125 encounters (simultaneous occurrence of units in the swamp) in 67 days, involving 47 groups or lone males (Levréro 2001). These data translate roughly into 1.2 encounters per group per month. This estimate, of course, applies to only one location in a forest that contains numerous clearings and many gorillas (box 4.1). At Mondika in CAR where gorillas are abundant (13 groups and two lone males in the 15 km^2 study area) and have been observed throughout their range in both swamp and "terra firma" forest, one group encountered other groups or lone males an average of four times per month (Doran-Sheehy et al. 2004).

In contrast, mountain gorillas in the Virungas have no equivalent of swamps—a highly localized food patch with a magnetic attraction for gorillas—and live at an average density of one animal per square kilometer (sec. 4.1.7). The Virunga groups appear to encounter each other less frequently than do western gorillas at Mondika or Odzala. Watts (1989a) estimated a mean rate of 8.2 encounters per group per year—less than one a month—using data from the entire yearly ranges of eight social units.

The sheer frequency of interunit encounters at clearings like Lokoué may in part account for their relative peacefulness (mentioned earlier). Gorillas visit these sites to exploit the abundant high-quality food that grows there. If the animals engaged in conflicts every time they met other groups or lone males, they would have little time left to feed.

Sources of conflict. Researchers have speculated that interunit conflicts over access to food may be more common in western gorillas than in less frugivorous populations because fruit, a relatively clumped resource, is more likely to be contested than is abundant, evenly spread foliage (ch. 2.2.2) (Doran and McNeilage 2001; Tutin 1996). Tutin has documented aggressive encounters at fruiting trees in Gabon, while at Lossi, Bermejo (2004) noted that many intergroup encounters occurred at or near fruit trees, although most of these meetings were not aggressive. Aside from these observations, there are few data with which to consider the issue.

In fact, the bulk of theory and evidence suggests that conflicts between gorilla units represent reproductive competition between males, rather than contest competition over food (ch. 2.2.2) (Doran and McNeilage 2001; Harcourt 1978a; Sicotte 1993; Watts 1996; Yamagiwa and Kahekwa 2001). Fully adult males are almost always the main contestants, with females taking little part. Furthermore, the intensity of encounters does not depend on location in the home range or quality of food at the encounter site. Instead, the nature of encounters (e.g., aggressiveness, duration) tends to correlate with composition of the social units involved (e.g., number of and reproductive state of females). We address this topic further in the chapter on male-male competition, where we treat in detail the mating system of gorillas, and how males' reproductive strategies influence the structure of gorilla society (ch. 11.1.3; 11.3.2).

4.2.2. Social relationships

The pattern of social interactions between group members reflects the competitive or cooperative nature of their relationships, the patterning of which, in turn, contributes to the nature of the society (Hinde 1976). Friendly behavior such as close proximity or grooming, and supportive intervention in fights, indicate a cooperative relationship. Aggressive behavior indicates the opposite.

Almost all our information on social interactions and social relationships comes from the mountain gorilla research groups of the Karisoke Research Center in the Virunga Volcanoes (Robbins et al. 2001). What data we have from eastern lowland and western gorillas support the general picture of intragroup social behavior that has emerged from the Virungas (Stokes 2004; Yamagiwa 1983).

The description that we provide here of social relationships comes from long-term observations, some of them spanning thirty years, primarily from four breeding groups: Groups 4, 5, Nunki (Nk), and Beetsme (Bm), and three bachelor groups, Pn, Bm, and Bilbo.

4.2.2.1. Relationships between adult males and females

From early nineteenth century descriptions of gorillas by explorers and hunters, to present-day detailed studies of social behavior, everyone has noticed how central silverbacks are to the lives of adult females.

Most adult females spend more time near mature males than they do near any other adult except close female kin. The attractiveness of silverbacks is most obvious during rest periods, when group members congregate near the adult male (see ch. frontispiece 2) (Harcourt 1978b; Yamagiwa 1983). Some females spend as much as 50% of their resting time within 5 m of silverbacks, almost touching distance. In small groups, mothers with unweaned infants spend more time near the silverback than do females with older or no infants (Harcourt 1979a), but this effect of young offspring does not occur in large groups (Watts 1992).

Females commonly take the initiative in maintaining close proximity with males in that they approach them more, follow them more, and leave them less than vice versa (Harcourt 1979a; Watts 1992). This situation reverses when females are in estrus, at which time some males are far more active in keeping close to the female (Sicotte 2001).

Females without adult relatives in the group not only favor the silverback for close proximity, but have more friendly contact with him than with any other adult. For example, he is often the females' primary adult grooming partner and tends to receive more of the grooming than he gives (Harcourt 1979a; Stewart and Harcourt 1987; Watts 1992).

All silverbacks, even young subordinate ones, are dominant to females. Females avoid even the non-aggressive approaches of adult males, sometimes almost fearfully, while the silverback never avoids a female, is never submissive to a female, and very rarely is threatened by any female. This complete dominance comes about only when males reach twelve years or so, by which time they are much larger than females, and beginning to develop their silver back (see fig. 3.3A).

By contrast, blackback males (aged 8 to about 12 years) are usually outranked by adult females. As the males mature, they begin to challenge females, displaying near them increasingly frequently. Initially, females either ignore the show-offs, or retaliate. But as the difference in body size increases, females respond with progressively less aggression and more submission (Harcourt 1979a; Stewart and Harcourt 1987; Watts 1992; Watts and Pusey 1993).

Although mature males frequently threaten females, they rarely wound them, even mildly (Watts 1992). Most male threats are in the form of displays. Typically, a silverback struts toward a female or chest-beats close by, sometimes thumping vegetation or the ground, sometimes thumping the female. To human observers, these acts often seem unprovoked. For example, a silverback will display at a female for no apparent reason as he approaches her. Males commonly display near females as they initiate group travel at the end of a rest period (Harcourt 1979a; Stewart and Harcourt 1994). In these contexts, the males appear to be using displays as a general show of strength to control group members, and indeed, the usual response to such acts is to follow or at least remain near the silverback (Harcourt 1979a; Sicotte 1994; Watts 1992).

Non-display aggression by males toward females includes cough grunt vocalizations (the most common threat), lunges or runs, and rarely, grabbing, hitting, and biting. Silverbacks direct non-display aggression to females most frequently during feeding or when they intervene in conflicts between females (ch. 6.2, 6.3) (Harcourt and Stewart 1987; Watts 1992, 1997). Females normally respond by cowering or moving away, or by making appeasement gestures such as touching or embracing the male, and giving intense grumbling or humming vocalizations (Harcourt et al. 1993; Watts 1995).

The appeasement gestures can be more than just momentary submission. Over the half hour after females have been threatened by a male, they spend more time near him, grooming him more than normal (Watts 1995). Such increased friendliness after aggression has received a lot of attention since Frans de Waal (1989a) highlighted it, termed it *reconciliation,* and suggested that it functioned to repair relationships in societies in which stability of the group was important to members' survival. Whether in fact the concept of repairing relationships is necessary is debatable. Perhaps the behavior is more simply interpreted as pacification for immediate benefit (Silk 2002b). In any case, the partner that benefits more often takes the initiative, as is the case for gorillas (Watts 1995).

When more than one male is in a group, females may have distinctly different relationships with each of the males (Robbins 2003; Watts 2003). Typically, females interact more closely with the top-ranking silverback than with subordinate males, which parallels the females' mating preferences (Harcourt 1979a; Robbins 1999; Watts 1990b, 1992). Some females, however, frequently associate and groom with younger, lower-ranking males, in which cases, the males often take an uncharacteristically active role in the interactions (ch. 11.1.4–6).

Females' relatedness to a silverback can also influence their relationship, regardless of the male's rank or mating status. For example, six females that remained in their natal group as adults were far friendlier with the older, deposed male, which was possibly either their father or grandfather, than with the younger dominant silverback, even though they solicited mating only from the younger male. Furthermore, the older silverback consistently supported his presumed daughters and granddaughters in conflicts against unrelated females (ch. 6.2, 6.3) (Watts 1992).

4.2.2.2. Relationships between adult females

Unrelated female gorillas, which include most adult gorillas in a social group (sec. 4.2.1.2), can be described as mostly tolerating one another. They rarely interact in any way, and especially not in a friendly manner. In fact, their most common overt social interaction is probably mild threat (Harcourt 1979b; Watts 1994a). Close relatives, on the other hand, such as mothers and daughters, or full sisters, can spend a lot of time near each other, groom one another frequently, and support each other in fights (Stewart and Harcourt 1987; Watts 2001). In general, paternal relatives, for example, half sisters sired by the same male, show levels of friendliness and antagonism intermediate between those of unrelated females and those of maternal relatives (Watts 1994b).

Researchers have often had difficulty discerning clear dominance relationships between female gorillas, either because there were so few agonistic interactions within a dyad, or because the direction of their interactions was not consistent (Stewart and Harcourt 1987). In every Karisoke study group, however, there have always been some pairs in which non-aggressive supplants (approach-retreat interactions) were consistently in the same direction, indicating a stable dominance relationship between those females (Harcourt and Stewart 1989; Stewart and Harcourt 1987; Watts 1994b). In such cases, higher-ranked animals were usually older but also had lived longer in the group (Harcourt and Stewart 1987; Watts 1985, 1994b).

Recently, Robbins and co-workers analyzed long-term data spanning 30 years on non-aggressive supplants from 51 females in six groups (Robbins et al. 2005). This larger sample shows stable, linear rank differences among most females in most groups. For example, 80% of dyads had the same dominant female across successive sample periods, and four females held top rank in their groups for 15–25 years. Only 4% of all supplants were of a dominant by a subordinate. Rank was positively correlated with age, and after that, with duration of group residence (tenure), except when enough difference in tenure existed (> 7 years), in which case, time in the group became more important than age.

Nevertheless, female gorillas are clearly not ruled by social status in the same way as are, say, baboons or macaques (Cheney 1977; Seyfarth and Cheney 1984; Sterck et al. 1997; Thierry 2007). When aggressive encounters are considered, the outcomes of contests are sometimes unpredictable and non-linear. In fact, the most common response to aggression from another female is either to ignore it or retaliate (Harcourt 1979b; Watts 1994b). Furthermore, female gorillas do not show appeasement behavior to each other like they do toward males (sec. 4.2.2.1), and Watts (1995) found no evidence for reconciliation between female opponents following aggression.

While fighting between females is uncommon in small groups (Harcourt 1979b), in larger groups it can be quite frequent, often escalating beyond the original disputing pair as other females join the skirmish (Watts 1994b, 1997). Female conflicts sometimes escalate beyond just vocal threats. For example, in one study, out of 46 documented fights, one of the participants suffered bite wounds in 22, or 48% (Watts 1997). This is a far higher injury rate than that received from adult males (Watts 1994b).

The most common context for female disputes is feeding, a fact that we will consider in greater detail in chapter 5. In large groups, females also apparently compete to be near the mature male, judged by the fact that a not uncommon context of aggression between females is proximity to a silverback (Watts 1994b).

In chapter 5 we discuss in detail the significance for understanding gorilla society of these contrasts between kin and non-kin, and of the intensity and nature of competition between females.

4.2.2.3. Relationships between adult males

In groups with more than one silverback, the males are likely to have known one another for many years (sec. 4.2.1.2). As with all animals, social interactions between male gorillas can differ depending on the partners' personalities and histories (Harcourt 1979c; Watts and Pusey 1993). Underlying all their relationships, however, is a tension that increases as younger males move into adulthood.

Immature males, before reaching adolescence, associate closely with the dominant silverback, sometimes grooming him frequently, for example (sec. 4.2.2.4). As the young males become blackbacks, they spend less and less time near the older male, and receive increasingly more aggression from him (Harcourt 1979c; Watts and Pusey 1993). All males older than about 11–12 years have generally tense relationships, and manage to live in the same group by avoiding or, at best, tolerating each other. Except at times when co-resident males jointly repel outside silverbacks or defend the group against predators (ch.7.1; ch. 11.2.2), cooperative behavior, such as grooming or coalitions, between adult males in the same group is very infrequent. Nevertheless, close relatives such as fathers and sons are more tolerant of each other than are less closely related pairs, such as half brothers or uncles and nephews, and occasionally support one another in aggressive encounters within the group (Harcourt and Stewart 1981; Robbins 1996; Watts 1997).

Most conflicts between males are mild. The males display at one another, or just cough grunt or lunge without touching. For example, in two multimale groups, one with three silverbacks and one with two, Robbins (1996) recorded 250 episodes of displaying or mild aggression between silverbacks (about 3 per hour in close proximity), but only 21 cases of aggression with physical contact, or less than 1 every four hours in proximity.

Co-resident adult males establish clear dominance relationships, often based on age. Young silverbacks are outranked by fully mature males, among which the older are commonly dominant to younger. As males age, they may lose rank to the next silverback in the hierarchy. In one group, such a rank reversal between an aging father, Beethoven, and his presumed son, Icarus, took several months during which time it was not clear which male was dominant. Then, after one particularly serious fight in October

1980 that Icarus won, he consistently dominated his father (Veit 1983). Rank reversal of younger males over older has been documented in two other pairs in the Karisoke Research Center study groups, and recently also in mountain gorillas in Bwindi (Robbins 2006).

Bachelor groups. All our information on social behavior in non-breeding groups comes from two bachelor groups of mountain gorillas (Robbins 1996; Yamagiwa 1987a), although all-male groups also form in western gorilla populations (Levréro et al. 2006). Males in bachelor groups, including the silverbacks, are friendlier with each other than are those in breeding groups. They spend more time near each other during both resting and feeding periods, and play together relatively frequently. Indeed in one bachelor group, the extra friendliness extended to frequent homosexual behavior, in which the younger male completely adopted female courtship and copulatory behaviors (Yamagiwa 1987a).

Robbins (1996) found that kinship between bachelor males had no influence on their social interactions, although the closest relatives in her sample were half siblings. As in the breeding groups, older males dominated younger. However, between males of similar age, it was more difficult than in breeding groups to discern rank differences because the bachelor males were generally less competitive.

The reason for these relatively relaxed relationships between bachelors becomes immediately apparent with the arrival of adult females, and hence, something to compete over. When several females joined one of the Karisoke all-male groups, the two silverbacks, which had lived peacefully together for six years, began fighting seriously. After about a week of escalating aggression, two new groups formed. The dominant silverback, a blackback, and all the females went one way; the subordinate silverback with all the other younger males went another way (Watts 2001).

4.2.2.4. The social relationships of immature animals

Immature gorillas interact primarily with their mothers during their first year (see ch. 3 frontispiece), after which their social network expands rapidly. They spend much of their time playing with other immature animals, especially similarly aged peers (Fletcher 2001; Fossey 1979). Their associations with adults are similar to those of their mothers. Thus, infants, juveniles, and subadults spend more time near and have more friendly interactions with close maternal relatives, such as older sisters or aunts, than with unrelated females (Robbins 2006; Watts and Pusey 1993).

However, by far the most important adult to them aside from their mother is the dominant silverback (fig. 4.9), which is likely to be their father. Infants in their third year develop an attachment to the mature male,

Fig. 4.9. Silverbacks are remarkably tolerant and protective of infants in their group. (Mountain gorilla.) © K. J. Stewart

spending more time near him than any other adult except their mother or sometimes a closely related female (Fossey 1979; Stewart 2001). Males are tolerant and protective of immature animals, especially orphans. Infants and young juveniles whose mothers die or emigrate increase their association with the male, which allows them to share his night nest (Stewart 2001; Watts and Pusey 1993).

Many immature males and females develop strong grooming relationships with the dominant silverback. The advantages to immatures of associating with the male include his intervention in their aggressive encounters with older animals (Stewart 2001; Watts and Pusey 1993). That benefit may continue into adulthood: one silverback supported his adult offspring in fights more than he did unrelated adults (Watts 1992).

All adults dominate subadults, which outrank juveniles, which dominate infants (Harcourt and Stewart 1989). This ranking is based on non-aggressive supplants or mild threats, since immatures rarely receive serious aggression from older animals (Stewart 2001; Watts and Pusey 1993). If they do, especially if they are infants, other individuals will intervene. Mothers, other adult female kin, and the silverback are the most ardent protectors of infants (Harcourt and Stewart 1989; Watts 1997).

Male and female immature animals do not obviously differ in the nature of their social relationships. While males might play more than do females, statistical tests do not show a difference, possibly due to small sample sizes within age classes (Fletcher 2001; Watts and Pusey 1993). Males and females begin to differentiate around eight years of age when, as we have described (sec. 4.2.2.3), males become progressively more peripheral, as they receive increasing aggression from adults, especially males, and initiate more aggression with adult females (Watts and Pusey 1993).

The nature of animals' early social relationships may have some impact on their later decisions about dispersal (e.g., Harcourt and Stewart 1981), but with all the other factors that influence dispersal of both sexes (ch. 8; ch. 11.2, 11.3), it is going to be a long time before anyone has enough data to test the idea.

4.2.3. Comparison with *Pan* and *Pongo*

The great apes differ strikingly in the nature of their societies, which is why comparison among these closely related taxa is so illuminating. If anyone had to name one factor that distinguished the societies, it would surely be the cohesiveness of their social groups. The contrast in cohesiveness is tied to differences between the species in their diets and habitats (sec. 4.1), and the constraints these put on sociality (Delgado and van Schaik 2000; Doran et al. 2002a; Rodman 1984; Wrangham 1986; Yamagiwa 1999a).

4.2.3.1. Pan

Common chimpanzees and bonobos live in socially distinct communities within which individuals associate in temporary subgroups, or parties, whose membership changes frequently (Boesch 1996; Ghiglieri 1984; Goodall 1986; Hohmann and Fruth 2002; Mitani, Watts, and Muller 2002b; Nishida 1979; Reynolds 1965; Sugiyama and Koman 1979; White 1996; Wrangham et al. 1996). "Frequently" means on the order of minutes or hours, not days, and certainly not weeks. The mean time that parties lasted in four chimpanzee and two bonobo communities ranged from 14 minutes in Budongo, Uganda, to 126 minutes in Bossou, Equatorial Guinea (Boesch and Boesch-Achermann 2000). Any two members of a community can go for hours, days, or even weeks without seeing each other.

Pan populations differ in how gregarious they are. Western chimpanzees associate with others more frequently than do eastern chimpanzees. Thus, in Taï Forest, Côte d'Ivoire, only 5% of common chimpanzee units (subgroups and lone animals) were lone animals, whereas at three eastern sites (Gombe and Mahale in Tanzania, and Kibale in Uganda), 18%–21% of units were lone animals (Boesch and Boesch-Achermann 2000). Bonobos, in turn, are even more gregarious than are western chimpanzees and even less often on their own (Chapman et al. 1994; Doran et al. 2002a; Hohmann and Fruth 2002; White and Wrangham 1988; Wrangham 1986).

Numerous factors influence the size of *Pan* subgroups (Anderson et al. 2002). The most important are food and estrous females. More food in larger patches attracts larger groups, and estrous females attract males (Boesch 1996; Matsumoto-Oda et al. 1998; Mitani et al. 2002a; Sakura 1994; Wrangham 1977).

The society of common chimpanzees can be described as male-bonded and male-dominated (Mitani et al. 2002b; Stumpf 2007). Males typically spend their lives in the community in which they were born (Boesch and Boesch-Achermann 2000; Goodall 1986; Nishida 1990). They range more widely than do females, and cooperatively defend the community's territory, actively patrolling its borders (Boesch and Boesch-Achermann 2000; Nishida 1979; Williams et al. 2004). When males encounter individuals from neighboring communities, they react in concert with sometimes lethal aggression (Nishida et al. 1985; Watts et al. 2006; Wilson, Wallauer, and Pusey 2004; Wilson and Wrangham 2003).

Males are the more gregarious sex, forming parties with each other and with sexually receptive females (Boesch and Boesch-Achermann 2000; Goodall 1986; Halperin 1979; Wrangham, Clark, and Isabirye-Basuta 1992). Their cooperative defense of the community borders is reflected in friendly relationships within parties. Partners sit near one another, groom one another, share meat, and support one another in contests (Boesch and Boesch-Achermann 2000; Goodall 1986; Mitani et al. 2002b; Newton-Fisher 2002; Nishida 1979; Simpson 1973; Watts and Mitani 2002).

Adult male common chimpanzees dominate all females. Among themselves, aggressive males outrank less aggressive males. Social status is a major preoccupation of males. They spend considerable time and energy, often with support from allies, aggressively challenging the ranks of other males, and defending and reinforcing their own positions (Boesch and Boesch-

Achermann 2000; Muller 2002; Muller and Wrangham 2004; Nishida and Hosaka 1996).

Among eastern chimpanzees, adult females that are not sexually cycling are far less gregarious than are adult males (Goodall 1986; Wrangham 2000). For example, mothers at Gombe National Park, Tanzania, spend about 65% of their time with no other adult nearby, whereas males are alone for only 27% of the time (Wrangham and Smuts 1980). Estrous females, though, are constantly accompanied by adult males, usually several at a time (Boesch 1996; Stumpf 2007; Tutin and McGinnis 1981; Williams et al. 2002; Wrangham 2000).

In western chimpanzees, the sexes differ less in level of sociability. Thus in Taï Forest, mothers and non-estrous females associate and groom relatively frequently with adult males. Nevertheless, males are more frequently friendly with one another than they are with females, or than females are with one another (Boesch and Boesch-Achermann 2000).

Like gorillas, female chimpanzees usually leave their natal community at maturity and transfer to another community, although they are less likely than gorillas to transfer more than once (Boesch and Boesch-Achermann 2000; Nishida et al. 2003; Pusey et al. 1997; Wrangham et al. 1992).

The social system of bonobos differs from that of common chimpanzees in several notable ways. Female bonobos, whatever their reproductive state, associate relatively frequently with males or with each other (usually in mixed-sex parties) and are rarely found alone (Hohmann and Fruth 2002; Kano 1992; White 1988). Indeed, females cooperate with each other at high rates, sharing food and engaging in sociosexual interactions, which are thought to reduce tension and mediate competition between them (Fruth and Hohmann 2002; Furuichi 1989; Hohmann and Fruth 2002). They also form coalitions that enable them to dominate and, therefore, outcompete males (Furuichi 1997; Idani 1991; Parish 1994).

Male bonobos, on the other hand, are less bonded than are male chimpanzees. In fact, the most stable, cooperative relationships that adult males have appear to be those with their mothers (Hohmann and Fruth 2002; Hohmann et al. 1999). Bonobo males use aggressive coalitions with each other less than do common chimpanzees. Thus, bonobo males do not conduct border patrols and are not uniformly aggressive during interunit encounters, although males typically give displays during encounters (Hohmann and Fruth 2002; Kano 1992).

Despite these differences, bonobo society is similar to that of chimpanzees in being characterized by fission-fusion grouping, female emigration, and male philopatry (Furuichi et al. 1998; Gerloff et al. 1999; Hohmann and Fruth 2002; Stanford 1998b). The two *Pan* species also have similar mating systems, both of which, like other aspects of their societies, contrast dramatically with the gorilla's mating system. We consider this contrast in chapter 11.4 and what it can tell us about the nature of gorilla society.

4.2.3.2. Pongo

Orangutans are the least social of the great apes, and in fact, of most diurnal primates. At all sites, adults spend most of their time alone or, if they are females, with their immature offspring (Delgado and van Schaik 2000; Galdikas 1985b; Rodman 1988b). The most solitary individuals are those that can least afford the energetic costs of being with others. These animals are fully adult males, because of their large size and ponderous locomotion, and females with small infants because of the stresses of lactation (Knott 1998b; Rodman 1977; Sugardjito, te Boekhorst, and van Hooff 1987; van Schaik and van Hooff 1996). Even in Sumatra, where orangutans are considered more gregarious than in Borneo, average group size is less than two individuals (van Schaik 1999). The main contexts in which animals associate are sexual consortships, or when individuals happen to aggregate at the same fruit tree, where they interact little with each other, but instead focus on feeding (MacKinnon 1974; Mitani et al. 1991; Rijsken 1978; Schürmann and van Hooff 1986).

Both sexes disperse from their mother's home range at sexual maturity. But while it looks as if females might settle in an area next to their mother's, dispersing males disappear from the region and probably become nomads for a while (Galdikas 1985b; Rijsken 1978; Rodman 1973; van Schaik and van Hooff 1996). Adult females live with their immature offspring in overlapping home ranges, while males' ranges are much larger, incorporating those of several females (Delgado and van Schaik 2000). Since orangutans cannot move fast enough to be territorial (Mitani and Rodman 1979), adult males share their ranges with other males (Galdikas 1981; Rodman and Mitani 1987; te Boekhorst et al. 1990). However, these other males are primarily the unflanged males that have not developed full adult secondary sexual traits; indeed many of them do not look much different from females (ch. 3.2.7; box 3.2). Flanged males give "long calls" that carry long distances and announce their whereabouts in the forest. Dominant residents tend to call several times a day, possibly to repel other adult males or attract estrous females (Galdikas 1985b; Horr 1975; Knott and Kahlenberg 2007; Mitani 1985a).

While adults appear to have no social relationships at all, except for the agonistic relationships between adult males, orangutan society can still be described as a "loose community" in which the same adults live in the same general area for years at a time (Knott and Kahlenberg 2007; van Schaik and van Hooff 1996). These communities are centered around a locally dominant, resident male that is fully developed or flanged, but are regularly visited by itinerant unflanged and some flanged males as well (Delgado and van Schaik 2000; MacKinnon 1974).

Conclusion: Gorilla Society

Ecological differences among the great apes help to explain the dramatic contrasts in their societies, most obviously in the structure and cohesiveness of social groups. Contrasts among the apes in social processes such as dispersal and the nature of relationships within and between groups are linked to differences in the costs of being gregarious. In the following chapters we aim to show how the gorilla's social system (structure, processes, relationships) results from the interaction of male and female survival and reproductive strategies, which are shaped by life history variables and environmental factors, specifically, predator risk and the abundance and distribution of food. As we have said, we are presenting a case study in socioecology, illustrating the principles we outlined in chapter 2, and showing how these principles enable us to understand the diversity of social systems among the great apes, the primates, and across vertebrates.

Figure Details

Fig. 4.1. Number of fruit species in diet by altitude. Counting only a single (median) value each for Bwindi and Virungas, to ensure that data from sites are independent, $N = 8$, $r_s = 0.88$, $p < 0.01$. Data from western gorillas, in order of decreasing number of fruit species eaten: Ndoki, Rep. Congo (Nishihara 1995); Lopé, Gabon, (Tutin 1996); Bai Hokou, CAR (Remis 1997b); Mondika, CAR (Doran and McNeilage 2001); eastern lowland (Yamagiwa et al. 2003a); mountain, Bwindi (Robbins and McNeilage 2003); Virunga (McNeilage 2001; Watts 1984).

Fig. 4.2. Fruit in forest compared to fiber and fruit in gorillas' diet. Fruiting and non-fruiting seasons = upper and lower quartile fruit availability scores (omitting months on the quartile value). Fruit available (% trees in fruit/mo); fruit in diet (# species [judged by seeds] in dung/mo); fiber in diet (amount in dung/mo on 1–4 scale). Fruit available × fruit eaten, $N = 26$, Spearman $r_s = 0.4$, $p < 0.1$; Fruit available × fiber eaten, $N = 26$, $r_s = -0.6$, $p < 0.01$; Fruit eaten × fiber eaten, $N = 27$, $r_s = -0.6$, $p < 0.001$.

Fig. 4.3. Daily distances traveled by site and amount of fruit in diet. Data for western gorillas: Lopé, Gabon, *N* = 80 complete daily distances (Tutin 1996); Bai Hokou and Mondika, CAR, *N* = 95 and 94 complete daily distances, respectively (Doran and McNeilage 2001; Goldsmith 2003); Lossi, Rep. Congo, *N* = 63 daily distances; not included in the fruiting/non-fruiting comparison (Bermejo 1997); eastern lowland gorillas: Kahuzi and Itebero in DRC, *N* = 225 and 8 daily distances, respectively (Yamagiwa et al. 2003a; Yamagiwa and Mwanza 1994); mountain gorillas: Virunga Volcanoes, data from six different gorilla groups, *N* = 29–145 completed daily distances per group (Watts 1991a; Yamagiwa 1986); mountain gorillas, Bwindi, *N* = 77 daily distances (Goldsmith 2003). Note that in the comparison of daily distance traveled across populations (left three histograms), when the eastern lowland population (East) in the highland Kahuzi area (lower value) are lumped with mountain gorillas, there is no overlap with the daily distances from the western gorilla sites (West). Differences in daily distances according to amount of fruit in diet were significant at all sites for western and eastern lowland gorillas.

Fig. 4.5. Fruit part of diet shared between gorilla and chimpanzee. West 1: *N* = 12 months, range = 1.3–5; West 2: *N* = 48, range = 1.3–2.3; East: *N* = 23, range = 1.4–7; Mountain, *N* = 12; range = 1.5–16. Data from Ndoki, Rep. Congo (West 1) (Kuroda et al. 1996), and Lopé, Gabon (West 2) (Tutin and Fernandez 1993); Kahuzi-Biega, DRC (East) (Yamagiwa et al. 1996); and Bwindi, Uganda (Mountain) (Stanford and Nkurunungi 2003).

Fig. 4.6. Group sizes, median and maximum. *N* refers to number of groups from which median was obtained. Data from Lokoué, *N* = 37 (Gatti et al. 2004); Maya, *N* = 29 (Magliocca et al. 1999); Mbeli, *N* = 12 (Parnell 2002); Ndoki, *N* = 5 (Nishihara 1994); Lossi, *N* = 8 (Bermejo 1999); Equatorial Guinea, *N* = 13 (Jones and Sabater Pi 1971); Mikongo, *N* = 4 (de Mérode, Bermejo, and Illera 2001); Lopé, *N* = 4 (Tutin 1996); Itebero, *N* = 10 (Yamagiwa et al. 2003a); Kahuzi, *N* = 24 (Yamagiwa et al. 1993); Virunga '00, *N* = 15 (Kalpers et al. 2003); Bwindi, *N* = 28 (McNeilage et al. 2001).

Fig. 4.7. Number of one-male and multi-male groups. The numbers were calculated from table 4.2, assuming 100,000 western gorillas, 10,000 eastern lowland, and 1,000 mountain, and a median group size of eight weaned gorillas per group (sec. 4.2.1). While the figures for number of gorillas are not precise (see ch. 14.1.2), they roughly reflect the relative sizes of the three subspecific populations.

Statistical Details

Stats. 4.1. Density by body mass in African diurnal primates: $\log_{10}$ density = 2.1 − 1.0 $\log_{10}$ Mass (kg), $r^2 = 0.33$, $p < 0.001$. Predicted density for gorilla = $\log_{10}$ 0.08 (i.e., 1/km^2).

Stats. 4.2. Number weaned animals (excluding silverbacks, of course) per group by number of males in mountain gorilla groups. Virunga gorillas: $p < 0.05$ (Wilcoxon/Kruskal-Wallis, 1977, $N = 16, 11$; $z = 2.26$; 2000, $N = 6, 9$; $z = 2.16$); Bwindi gorillas: $p > 0.1$ ($N = 15, 13$, $z = 1.4$). Data from Kalpers et al. (2003), McNeilage et al. (2001), and Weber and Vedder (1983), comparing 1 versus >1 males per group.

Stats. 4.3. Number of adult males by number of females in mountain gorilla groups: $p > 0.1$ for all comparisons. Comparisons within Bwindi, Virunga; combined Bwindi, Virunga; separate census years in Virunga (1977, 2000); "adult females" classified as only known adult females, by half unsexed adults as females; by all unsexed adults as females; and comparisons by regression, correlation, ANOVA and non-parametric Wilcoxon/Kruskal-Wallis (equivalent to Mann-Whitney *U*) on one vs. more than one male. Sample sizes from 15 to 43 groups, depending on comparison, with roughly 50:50 split of one-male and multi-male groups. Data from Kalpers et al. (2003), McNeilage et al. (2001), and Weber and Vedder (1983).

PART 3

FEMALE STRATEGIES AND GORILLA SOCIETY

SUMMARY

We argue that a relative lack of competition in the foliage-eating gorilla allows females to be in groups, and that a relative lack of benefits of cooperation allows females to leave the group in which they were born. However, neither competition nor cooperation explains why they are in groups, nor why they leave them once in.

Grouping and emigration are instead explained by the unusually strong influence of the males. Females benefit by associating with a large, powerful male, twice their size, for protection against predation and for protection against infanticidal, non-father males. Breeding females emigrate if the male is not powerful enough and perhaps also if they have to share the male's protection with too many other females. Females born into the group are forced to emigrate to avoid inbreeding when their fathers manage to remain as the main breeding male in the group beyond the age of maturity of the female.

Thus, rather than the standard vertebrate scenario of females going to food and males mapping onto the distribution of females, we instead see in gorilla society a strong influence of males on the distribution of females. The distribution of food determines what gorilla females can do; the males then determine what the females actually do.

Female feeding on terrestrial herbaceous vegetation (THV), thistle. (Mountain gorilla.)
© A. H. Harcourt

CHAPTER 5

Female Strategies and Society: Food and Grouping

Summary

Gorilla females, unlike *Pan* or *Pongo* females, always live in more or less stable groups with a male. Group-living entails increased competition for resources, yet can also offer the benefit of cooperating with others, especially kin, in competition for those resources. If gorilla females group, do they compete less than do *Pan* or *Pongo* females, or cooperate more?

In many respects, gorilla females act much like other group-living primates. They compete over food to such an extent that dominance hierarchies are sometimes evident—in which case why do they live in groups? And they cooperate in that competition, doing so with kin more than with non-kin—in which case why do they emigrate?

The nature of gorillas' main food, foliage, is such (widespread and abundant) that competition is minimal. Consequently, any benefits from cooperation in competition are also minimal. Minimal competition means that gorilla females, unlike chimpanzee and orangutan females, can group without too much cost if there are other benefits to grouping. Minimal cooperation means that the females can emigrate without loss of benefit, if emigration is otherwise forced upon them.

But while minimal costs of competition and benefits of cooperation explain why they can group and can emigrate, the minimal costs do not on their own explain why they do either.

Introduction: Gorillas; Food, Competition, and Cooperation

Both sexes need to escape being food, that is, being preyed upon. After that, access to resources, especially food, is what differentiates the successful from the unsuccessful female, whereas access to mates is what differentiates males (ch. 2.1.2). Competition among gorilla females for access to food, and its relevance to the structure of their society, is the topic of this chapter.

With regard to competition for resources and group-living, it is an almost inescapable fact that group-living entails increased competition for resources by comparison to the alternative of living alone (ch. 2.2.3). And yet, gorilla females always live in more or less stable groups with other animals, almost always at least one male and several other females (ch. 4.2.1.1). Why do they not range largely alone, or in parties of changing size and composition that match the amount and distribution of food, as do *Pan* and *Pongo* females (ch. 4.1.8)?

To ask why female gorillas do not range on their own is not a purely hypothetical question. The females of many mammalian taxa range on their own (Wilson 1975): around 40% of 50 terrestrial mammalian families are largely solitary (Wilson 1975, ch. 23). While chimpanzees and bonobos are usually with others, nevertheless, they are not uncommonly on their own, especially when food is in short supply (ch. 4.1.8; 4.2.3.1). We are arguing, in other words, that it is unlikely that female gorillas are evolutionarily incapable of ranging on their own.

If gorilla females live in more or less stable groups, stable for months or even years at a time (ch. 4.2.1.1), why do they ever leave the groups, instead of remaining in them and taking advantage of the benefits of cooperation with kin in competition with others over resources (ch. 2.2.3)?

This chapter is about competition for food among gorilla females and its relation to their group-living and about cooperation among gorilla females and its relation to their emigration. Do gorillas compete? Do gorillas cooperate? How do the two balance to explain grouping and emigration by female gorillas?

Competition, cooperation, grouping, and emigration is a four-way interaction that needs to be simplified somehow. We separate the chapter into the topics of competition and grouping on the one hand, and cooperation and emigration on the other hand. We could have made the other split, into competition and emigration (too much competition and animals leave the area) and cooperation and staying. If those are the questions that interest you, think emigration while reading the competition section, and residence while reading the cooperation section.

5.1. Female Strategies: Food, Competition, and Grouping

5.1.1. Do gorilla females compete over food?

Do female gorillas compete over food? It has been implied that they do not (Isbell 2004). If they do compete, how seriously do they do so?

Scramble competition—getting to the food and eating it before another animal gets there—is difficult to demonstrate. Therefore, the tests for competition that we mostly talk about are those that indicate contest competition—aggression over food.

Gorilla females do indeed compete. We found in our two main study groups that more than three-quarters of all competitive incidents were over food. We defined competitive incidents as aggression (from mild threat to attack) and supplants, which are defined as an animal moving away at the non-threatening approach of another animal. All ten adult females were aggressive or supplanted other adult females more often in the context of food than in any other context (fig. 5.1A). Similarly, David Watts (1994b) found

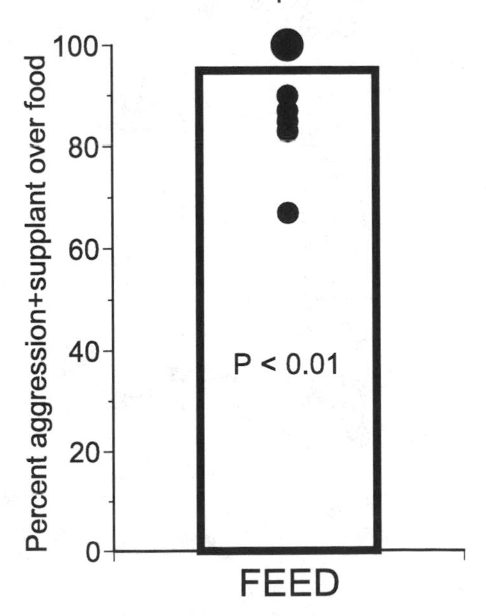

Fig. 5.1A. Gorilla females compete over food. Proportion of aggressive interactions and non-aggressive supplants among ten adult females in two groups that were over food. Supplant = one individual avoids non-aggressive approach of another. Circles are values for individual females; bar is median. (Mountain gorilla.) *Details at end of chapter.* Source: Data from Harcourt and Stewart (unpublished).

that food is by far the single most important context of aggression between gorilla females in a group: about half of all aggression in his four study groups was over food (fig. 5.1B). Moreover, the aggression can be fierce. In nearly half (44%) of 71 fights between females in two groups, at least one of the protagonists was wounded, and when one was wounded, usually (81%) both were (Watts 1994b).

The next most common context of aggression is unprovoked harassment (fig. 5.1B). No resource seems to be involved. One animal simply threatens another. Such unprovoked aggression is common in animal societies. The general interpretation is that dominants benefit by reminding subordinates of their competitive relationship: threatened subordinates remain subordinate. What is the benefit of such reminders, except easier access to a resource for the dominant the next time the animals compete?

In wild gorillas, a correlation between distribution of food and competition has not been experimentally demonstrated. However, a study in captivity indicates increase in aggressive competition among females when food

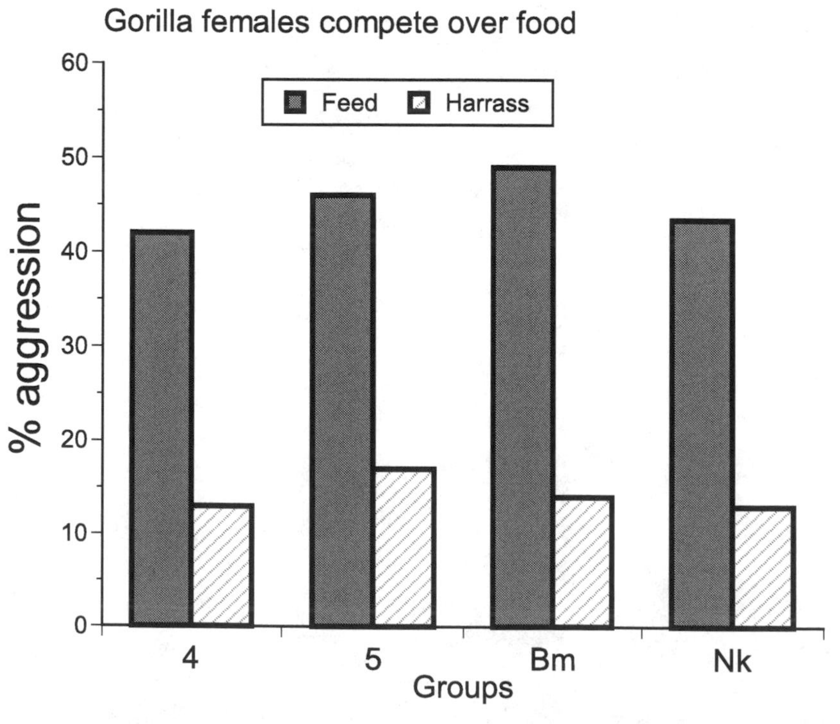

Fig. 5.1B. Gorilla females compete over food. Proportion of aggressive incidents among adult females that were over food, and in the next most common context, harassment (i.e., aggression with no obvious resource at stake). (Mountain gorilla.) *Details at end of chapter.* Source: Data from Watts (1994b).

is clumped, increased aggression from dominant to subordinates, increased submission by subordinates, and preferential access to the clumped food by the dominants (Scott and Lockard 2006).

If increased competition is a disadvantage of being in a group, then the larger the group, the more competition animals should face (ch. 2.2.3). And it is indeed the case that gorilla females experience more feeding competition in larger groups (fig. 5.2A, B). Figure 5.2A shows more feeding interruptions per hour of observation in larger groups. Figure 5.2B shows the same result, but far more precisely: the same female in a larger group is involved in more competition than when she was in a smaller group. Again the data come from Watts's long and detailed studies (Watts 1985).

Finally, if larger groups mean more competition, we can expect resident gorilla females to be particularly aggressive to newcomers, as happens in so many societies (Isbell and Van Vuren 1996). They are (Watts 1991c, 1994a). Those in large groups act as if they experience extra females as important competitors: residents and immigrants are nearly one-and-a-half times as frequently aggressive to one another as are unrelated residents within the

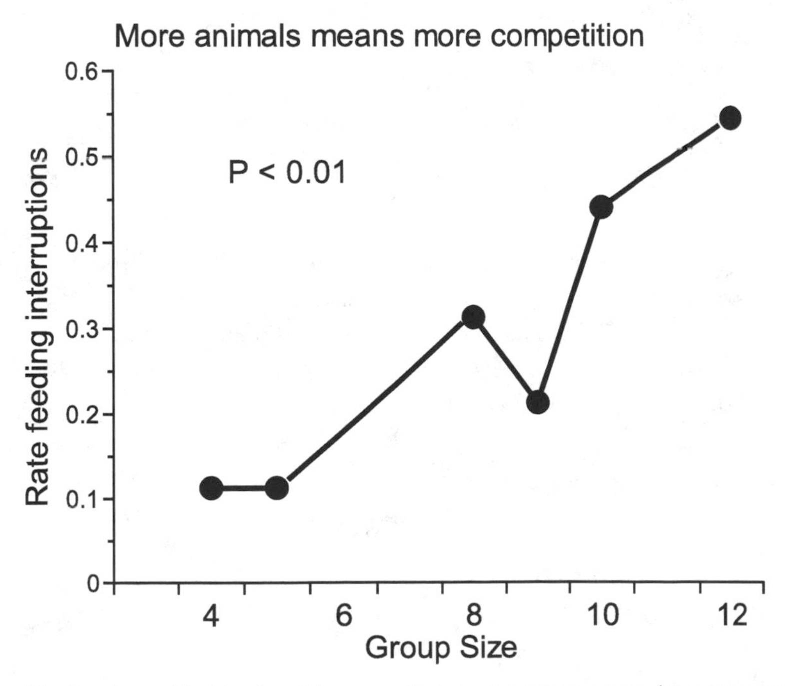

Fig. 5.2A. More frequent feeding competition among females occurs in larger groups. (Mountain gorilla.) *Details at end of chapter.* Source: Data from Watts (1985).

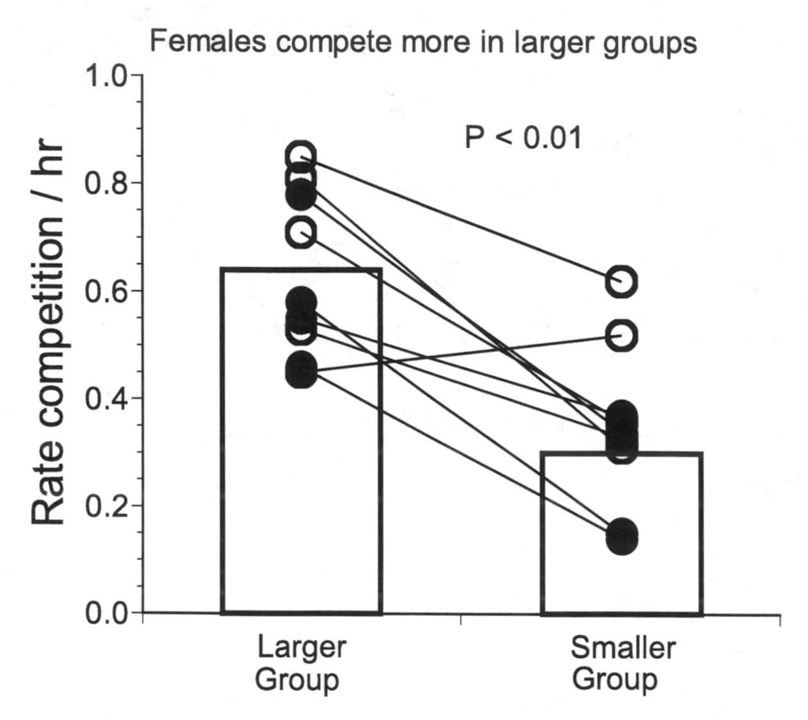

Fig. 5.2B. Individual females in different-sized groups experience more frequent competition (i.e., supplants given and received) in larger groups. Circles and lines are the same female in different sized groups. ● and ○ = two groups. Bars show the median value. (Mountain gorilla.) *Details at end of chapter.* Source: Data from Watts (1985).

group (fig. 5.3). (We compare to unrelated females, so as not to confound the resident-immigrant comparison with a kin–non-kin comparison).

So, gorilla females compete over food. But, is the competition important? In other words, does it affect females' ability to survive, mate, or rear healthy offspring? Whether it does or not will strongly influence the sort of hypotheses that will be necessary to explain why gorilla society is one of stable groups of females with a male.

5.1.2. Is the competition important?

When food is in short supply gorillas rely largely on foliage, not fruits (ch. 4.1.3). In other words, gorillas' fallback food is of relatively low quality, abundant, and evenly spread (fig. 5.4). Therefore, despite the fact that females compete over food, and despite the fact that animals appear to try to enforce their competitive status by aggression for no obvious reason

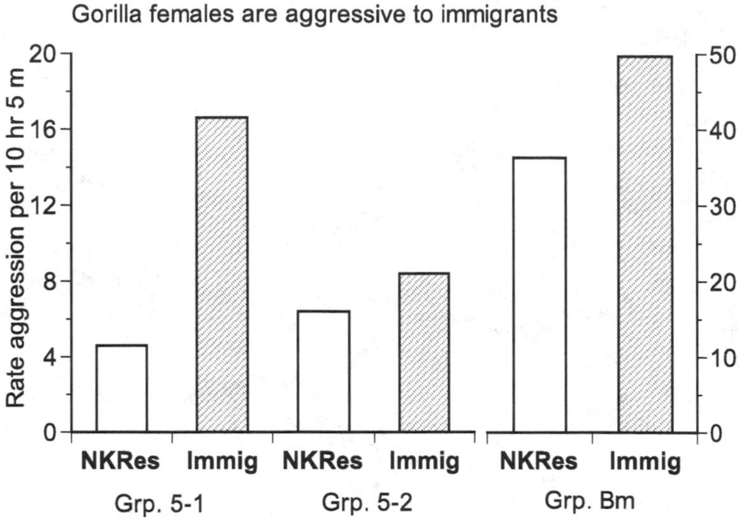

Fig. 5.3. Female gorillas are aggressive to immigrants. Data are rates of aggression per 10 hr females are within 5 m of one another among non-kin residents (NKRes, open bar) and between residents and immigrants (Immig, shaded bar). (Mountain gorilla.) *Details at end of chapter.* Source: Data from Watts (1994a).

("Harass" in fig. 5.1B), fighting for possession of some leaves when leaves are not very valuable nutrition, and abundant everywhere, is probably not very beneficial. It should be easier to simply go a few feet further away. But what quantitative evidence do we have to substantiate this arm-waving statement?

If competition could be shown to be rare in gorillas, its rarity might be evidence that it was unimportant. In our study groups, the average (median) pair of females supplanted one another less than once very 10 hours that we watched them (Stewart and Harcourt 1987), and were aggressive to one another about once every 2 hours (stats. 5.1). Whether the rates are high or low, we can tell only by comparison with other species. But the comparisons between species of rates of any behavior that are sometimes made are meaningless unless body size, among other factors, is accounted for, given that body size strongly affects rates of behavior (McMahon and Bonner 1983, ch. 5). We therefore use other measures than rate to judge the importance of competition.

First, what about the intensity of competition, as opposed to its frequency? It turns out that while total competition (aggressive and nonaggressive) over food increases with group size (fig. 5.2A, B), intense competition as indicated by aggression does not so obviously do so (stats. 5.2) (Watts 2001).

Fig. 5.4. A female gorilla, with her infant, feeding on the leaves of abundant herbaceous vegetation. (Mountain gorilla.) © A. H. Harcourt

Second, if competition were important to female gorillas, it should be the case that differences in competitive ability are obvious (Stewart and Harcourt 1987). The animals should often enough have competed that they have learned each other's competitive ability, and learned to react accordingly. Dominant animals should not give way to subordinates; subordinates should give way to dominants without argument. In other words, an obvious dominance hierarchy should be apparent. In other group-living primate species in which contest competition is readily apparent, so also is a more or less steep dominance hierarchy (de Waal 1989a, ch. 3, 4). Indeed the presence of an obvious hierarchy has been used to infer serious contest competition (Isbell 1991).

Dominance hierarchies, by which we mean linear hierarchies (i.e., A supplants B and C, B supplants C) are detectable in gorilla groups (ch. 4.2.2.2) (fig. 5.5). However, while hierarchies are certainly detectable in some gorilla groups, it is by no means the case that they are always obvious (ch. 4.2.2.2). Some clearly dominant individuals exist, and some clearly subordinate ones,

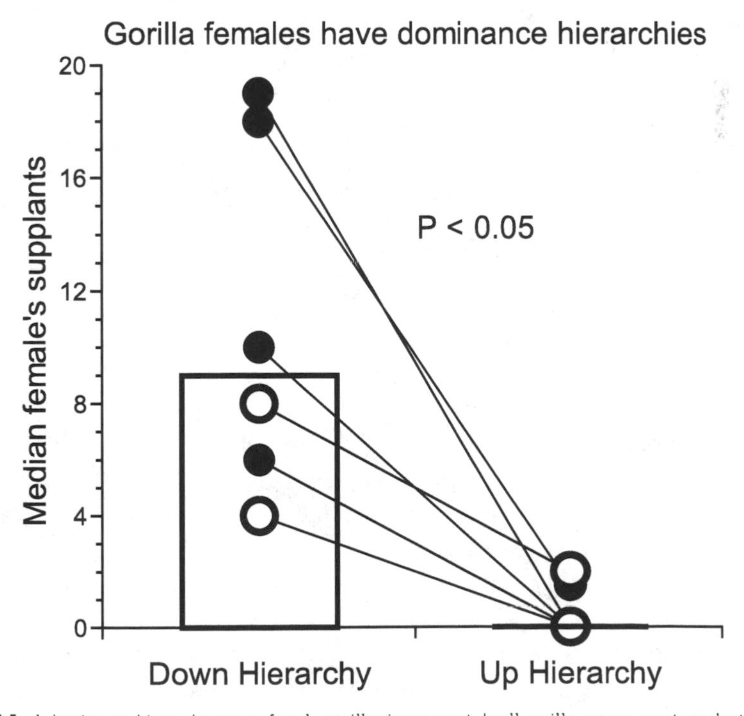

Fig. 5.5. A dominance hierarchy among female gorillas is apparent. In all gorilla groups, most supplants are down the hierarchy, that is, not often does a female supplant an individual that supplants it. ● and ○ = pairs of females; bar = median pair. (Mountain gorilla.) *Details at end of chapter.* Sources: Data from Harcourt and Stewart (1989) and Watts (1994b).

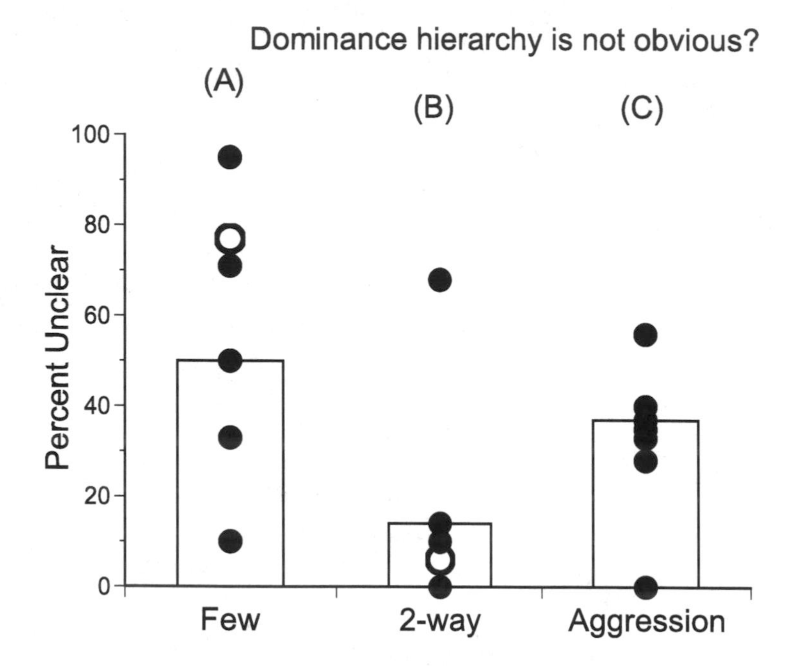

Fig. 5.6. A dominance hierarchy is sometimes not obvious. ● = data for individual groups; ○ = data summed across three decades. Among a high proportion of females, differences in competitive ability, that is, dominance rank, are not obvious, as judged by infrequent supplants: (a) Few, supplants in both directions; (b) 2-way, or frequent aggression up the hierarchy; (c) Aggression. (Mountain gorilla.) *Details at end of chapter.* Sources: Data from Harcourt and Stewart (1989), Robbins et al. (2005), and Watts (1994b).

with the relative competitive ability (i.e., difference in rank) stable over long periods (Robbins et al. 2005). But among several other individuals, no differences in ability are clear, in part because many females compete too infrequently for a difference to be seen (Harcourt 1979b; Robbins et al. 2005) (fig. 5.6[a]). Similarly, in a study in captivity, where female gorillas competed aggressively over clumped foods, nevertheless in none of the three groups observed did a linear hierarchy exist (Scott and Lockard 2006).

But just because we cannot detect a hierarchy does not mean that it does not exist. Studies could be of too short duration to detect differences among females that rarely interact (Robbins et al. 2005). Additionally, primatologists have known for decades that it is not unusual for the fewest supplants, or behavior of any sort, to be seen between the most distantly ranked animals (e.g., Alexander and Bowers 1967; Seyfarth 1977), perhaps in part because subordinate females stay well enough away from dominants that they are never in a position to be supplanted or attacked. However, as Watts points out, the lack of detected supplants between some gorilla females is not a result of lack of aggression, and hence lack of the chance for the females to learn contrasts in ability. One pair of the females that he studied (maternal sisters) was the most frequently aggressive in the group over one year, but neither ever submitted to the other in that year (Watts 2001).

Other measures, in addition to supplants and their frequency, can provide further information on the degree of difference among females in competitive ability.

If two females supplant one another, presumably little difference in competitive ability exists between them, or if it somehow does, then they do not recognize a difference: in gorilla groups, a somewhat substantial proportion of females supplant one another (fig. 5.6[b]).

If a supplanted female, that is, a subordinate, is obviously unafraid of the dominant animal that supplanted her, presumably little difference in competitive ability exists between them, or the subordinate does not recognize a difference: subordinate female gorillas often react with aggression to a dominant female's aggression (fig. 5.6[c]).

While these results come from only a few groups in one study site, they are consistent over many years. We started our studies of gorilla behavior in the early 1970s, and continued through to the early 1980s. Watts continued from the late 1970s to the early 1990s. And Martha Robbins and her colleagues (2005) have summarized results over three decades. Ever since gorillas have been studied, observers have seen competition and have identified obviously dominant and obviously subordinate females, but found it difficult to confirm obvious differences in competitive ability among others (Harcourt 1979b; Robbins et al. 2005; Watts 2001). Furthermore, among western gorillas, as among mountain, a female's most common reaction to aggression from another female can be to ignore it (73% of reactions to 30 threats in seven groups), as if competition is unimportant, and obvious rank differences do not exist (Stokes 2004).

Therefore, despite the fact that contrasts in competitive ability can be detected among gorilla females, and the fact that some of the differences last years, it seems unlikely to us that anyone will detect in gorilla populations the differences in body mass and reproductive output with rank that have been so nicely demonstrated for the Gombe Stream chimpanzee females (Pusey et al. 1997, 2005). At the same time, even in species in which competition and dominance hierarchies are obvious, dominant females do not necessarily do any better than subordinates, in part because of overriding effects of predation (Cheney et al. 1988; Cheney et al. 2004; Takahata et al. 1999).

Finally, if competition within the group were important, emigrating females should preferentially choose to join small groups (other things being equal). However, in mountain gorillas, emigration cannot be tied significantly to group size (Watts 1990a, 1996). Emma Stokes and her colleagues (2003) suggested that females in her western gorilla population preferentially immigrated into small groups and emigrated from large groups However, precisely this distribution of transfers would be expected if females emigrated and immigrated randomly with respect to group size.

5.1.3. Conclusion: Food, competition, and grouping

In sum, it seems that while in at least large groups (in which, as we have seen, more competition occurs), gorilla females compete often enough for contrasts in competitive ability to be apparent, nevertheless, those differences are unimportant for many females (Watts 2001). In other words, winning or losing in contest competition does not make much difference to females' abilities to survive, mate, or reproduce.

While that is not a surprising conclusion for a folivore, it is not inevitable. A premier folivore, the black-and-white colobus of Africa (*Colobus guereza* and relatives) is territorial (Oates 1994), as are some of the folivorous Asian colobines (Bennett and Davies 1994). In other words, they bother to contest abundant food that is of relatively poor quality, even if the defense of the food might be a strategy for defending mates (Fashing 2001). Why the difference from the gorilla? Perhaps it is because in the colobine's arboreal 3-D environment (compared to the gorillas' terrestrial 2-D environment), so much food can be packed into so small a space that even a poor-quality food becomes abundant enough to be worth defending (Dunbar 1988, ch. 12; Mitani and Rodman 1979; Wrangham 1979).

With gorillas, a further factor plays into the equation. Not only might competition among the females be unimportant, but any competition that a female experiences from another female is minimal by comparison to what she faces from a male. We will take up that part of the story later (ch. 6.1).

Nevertheless, even if competition is not so serious that it prevents grouping, a gorilla female in a group must face more competition than if she ranged on her own (Wrangham 1979). So why does she group? The answer cannot be that the costs of emigrating to a new site or group are too great (ch. 2.2.3.2), because female gorillas emigrate.

5.2. Female Strategies: Food, Cooperation, and Emigration

Given that female gorillas always live in groups—in other words accept the costs of competition that group-living entails—the default for a female should be to stay in her group, especially the group of her birth, her natal group (ch. 2.2.3.2). Emigration takes her to an area that she knows less well, maybe not at all; it places her with animals that she does not know, few if any of which will be cooperative kin, and many of which will initially be antagonistic (fig. 5.3). Some gorilla females do indeed stay and breed successfully (ch. 4.2.1.2). However, most leave (ch. 4.2.1.2). Why?

The benefits of cooperation with others in competition over resources has for over two decades been a strong component of explanations for grouping, and for friendly relationships within groups, especially among kin (ch. 2.2.3.2). Can a lack of cooperation, or lack of importance of cooperation, explain why gorilla females emigrate, leaving potentially cooperative kin?

We discuss the broader issue of cooperation in box 5.1, "Why cooperate?," where we emphasize that three ways exist for animals to benefit from cooperation, with one of them not depending on the partner being kin, or even of the same species.

5.2.1. Do gorilla females cooperate in competition over food?

Gorilla females do indeed cooperate in competition over food. The cooperation is most noticeable during contest competition (as opposed to scramble competition), that is, when animals threaten one another or even fight over food (as opposed to simply eating as fast as possible, like hungry students at a free barbecue). Scattered through the gorillas' home ranges are occasional small patches of food which the animals sometimes contest. Perhaps the oddest of these contested foods is dead trees. The gorillas cluster round them, pull off small chunks of the rotten wood, and nibble on it. And fights will break out over this dead wood (fig. 5.7). We used to have no idea why dead trees are treated as valuable by the gorillas. When we tasted the wood, it tasted of dead wood. But as we were doing the final copyediting of the manuscript before submission, it was discovered that the dead wood eaten by gorillas in the Bwindi Impenetrable National Park in Uganda had unusually high concentrations of sodium: dead wood was the source of over 90% of the gorillas' dietary sodium (Rothman, Van Soest, and Pell 2006).

The female gorillas cooperate in the conflicts over the dead wood. Animals help one another against other competitors (fig. 5.7). And kin help one another more than do non-kin. Mothers are particularly helpful toward their offspring, of course, but other kin help too. Gorillas, it seems, are acting like those female-resident species of macaques, vervets, and baboons, in which having helpful family present is so important to competitive success (Watts 1994a). And yet for these species, the help is used as an explanation of why they stay together, whereas gorilla females emigrate.

Box 5.1. Why cooperate?

"Cooperation (see also Kin selection and reciprocal altruism)." This index entry in a second edition textbook on primate behavioral ecology indicates perhaps the two most common explanations for cooperation. However, the entry misses a long-standing third explanation for cooperation, literally "co-working," and a more recent fourth explanation.

Evolutionary theory certainly predicts cooperation with kin or reciprocating partners. If you help kin, you are helping carriers of copies of your genes (with various provisos); and because if the help is reciprocated, and the benefit of the subsequent reciprocated help is at least equal to the cost of the original help, helping can result in overall benefit (again with various provisos) (e.g., Dugatkin 1997; Kappeler and van Schaik 2005).

But evolutionary theory has always also predicted cooperation with another set of partners—namely, every other individual out there, even of another species—if the cooperation brings immediate, mutual benefit (Clutton-Brock 2002; Dugatkin 1997; Harcourt 1991b; van Schaik and Kappeler 2005; Wrangham 1982). I help you, whoever you are, because I can benefit right now from the help. You might benefit also, but so do I. Two fieldworkers, lost in a bad area of the city at night, fortuitously meet, walk together for mutual protection, and part ways once they find their hotel. They are not relatives. No help is reciprocated later. Instead, both benefit, or potentially benefit, here and now from forming a group.

What the index entry means by "cooperation" when it refers to only kin selection and reciprocity as the causes of the behavior is not cooperation, but "altruism," namely, doing something for another individual at some personal cost now. An interesting relatively new field of hypotheses for such behavior has emerged. If my altruistic act is seen by a third party, I can gain status, respect, and trust, which attributes will help me in other interactions, including help from the third party (Milinski 2005; Nowak and Sigmund 2005; Silk 2005).

But back to kin selection, direct reciprocity, and mutualism. While kin often cooperate more than do non-kin, and preferences for kin as partners has been shown in both primates and non-primates, much cooperation that we might interpret as due to kin selection (preferential choice of kin with whom to cooperate because of the benefit of improved inclusive fitness) could simply be due to kin being more likely than non-kin to be together, and therefore to be available as partners (Baglione et al. 2003; Chapais 2005; Chapais and Bélise 2004; Kapsalis 2004; Silk 2002a). Nevertheless, kin are also probably more trustworthy partners, less likely to renege during a mutualistic act of cooperation, and so probably make better partners, other things being equal (Chapais 2005; Silk 2005; Wrangham 1982). None of which is to say, though, that kin do not compete (West, Pen, and Griffin 2002), as the most cursory reading of any country's history can attest.

With regard to reciprocity, it has been extraordinarily difficult to demonstrate conclusively among animals (Pusey and Packer 1997), not least because it is very difficult to demonstrate that an act of help now is in response to a previous act of help, though it is not impossible (Seyfarth and Cheney 1984). Helping now for help later is probably rare because it is risky, except between partners that can be trusted to reciprocate (Boyd 1992). Kin might be the most trustworthy partners—but the problem with proving reciprocity as the benefit of cooperation among kin is that the cooperation tends to be explained by only kin selection, not reciprocity.

By contrast to reciprocity, mutualism—cooperation for immediate personal benefit—is frequent and obvious (Clutton-Brock 2002; Wrangham 1982), and indeed, even when kin cooperate, the immediate personal benefit can often be greater than any benefit from inclusive fitness, in which case mutualism is the explanation for the cooperation (Chapais and Bélise 2004; Clutton-Brock 2002). A good example is male lions. As near as makes no odds, the only way for a male lion to obtain and keep access to a group of females is to form a partnership, because one male will almost always be defeated by a partnership of two, which will be defeated by a partnership of three (Bygott et al. 1979). And in the world's best studied wild lion population, 42% of 12 prides with males of known relatedness (judged by DNA analysis) contained unrelated males, which each fathered effectively equal numbers of offspring (Packer and Pusey 1982).

Among primates, one of the nicest examples of mutualism has been worked out in detail by Ronald Noë (1992). The alpha male does not cooperate to get access to estrous females (he does not need to). The lowest-ranking males do not cooperate with one another (even together they cannot defeat the alpha male). But two cooperating, middle-ranking, unrelated males can sometimes defeat an alpha male. Certainly, only one male will father the female's subsequent single offspring. However, the mutualistic cooperation gives the losing male a chance at fertilization that he would not have had without the cooperation.

In a wide variety of species (invertebrates, birds, mammals) large groups of cooperators can defeat small groups (Harcourt 1992; Harcourt and de Waal 1992b). Group members or partners do not have to be kin, nor do they have to be reciprocating partners. As long as all increase their chance of getting access to the resources by joining with others, join and cooperate they will, and do.

Fig. 5.7. Gorilla kin cooperate in competition over food. The adult female at the bottom left has just cough grunted and aimed a cuff at the young juvenile in front of the tree stump. She has made a mistake. Also present are that juvenile's mother (above the stump), grandmother (adult to right of stump), and aunt (young adult to right of stump). All are directing aggressive vocalizations (cough grunts) at the attacker, and the mother herself is aiming a cuff at the attacking adult female. In the face of this family response, the attacking female and her daughter (foreground at left) departed. (Mountain gorilla.) © K. J. Stewart

The fact that mothers help their immature offspring and that other adults also help young relatives is crucial evidence that the propensity to help in competition exists among gorillas. However, the fact of help to immature animals is not the crux of the question. The crux is whether adolescent and older females, the ones that might emigrate, benefit from the cooperation. If they do, the emigration is a puzzle.

Do adult females cooperate with one another? Indeed they do, and they do so most with kin (Harcourt and Stewart 1989; Stewart and Harcourt 1987; Watts 1997, 2001). Adult kin are generally friendlier to one another and less unfriendly with one another than are adult non-kin (fig. 5.8A, B). That friendliness includes several different sorts of behavior, with helping one another in contests being one of them (fig. 5.8B, C).

We show three sets of data to back up this claim of cooperation among kin. If separate people, studying different groups of gorillas, in different periods, with different methods and different measures come up with the same answer, then the answer is probably robust. We can be confident of the claim that adult female gorilla kin help one another more than they do non-kin, as do the females of many other Old World group-living, female-resident primate species. And yet gorilla females emigrate, leaving their friendly, helpful kin behind.

5.2.2. Is the cooperation important?

Kin are friendly to one another, for example helping one another in contests. But does the cooperation make any difference to females' ability to get access to resources? How do we tell?

If the help makes a difference, animals with help should be able to get resources that they would not otherwise have obtained. In other words, subordinate protagonists should be helped more than are dominant protagonists, and the help should enable subordinate protagonists to win contests.

We did not have enough records to analyze help of only adult females by adult females (no female had more than three records of such help). We thus analyze help to both adult and immature females. As help to imma-

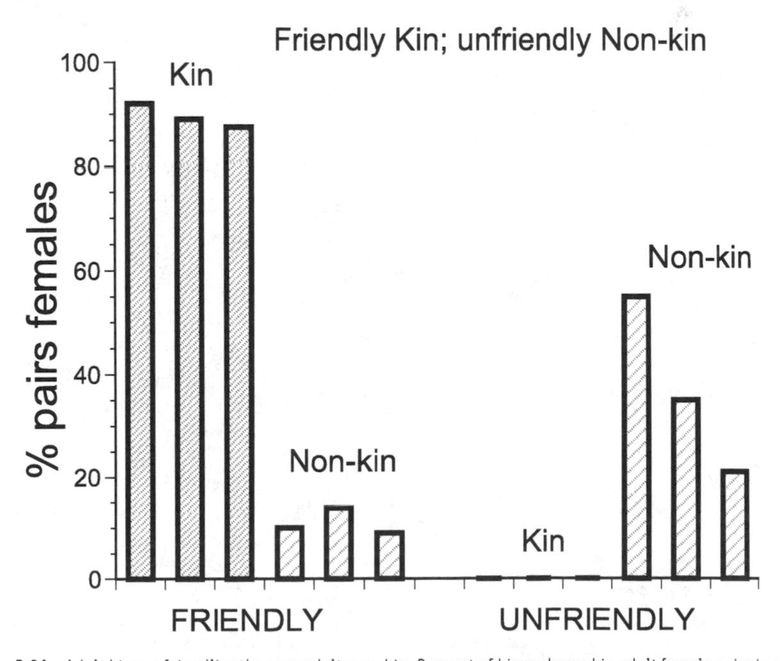

Fig. 5.8A. Adult kin are friendlier than are adult non-kin. Percent of kin and non-kin adult female pairs in one group over three periods of study across 8 years that showed above average friendly or unfriendly behavior within the pair. (Mountain gorilla.) *Details at end of chapter.* Source: Data from Watts (2001).

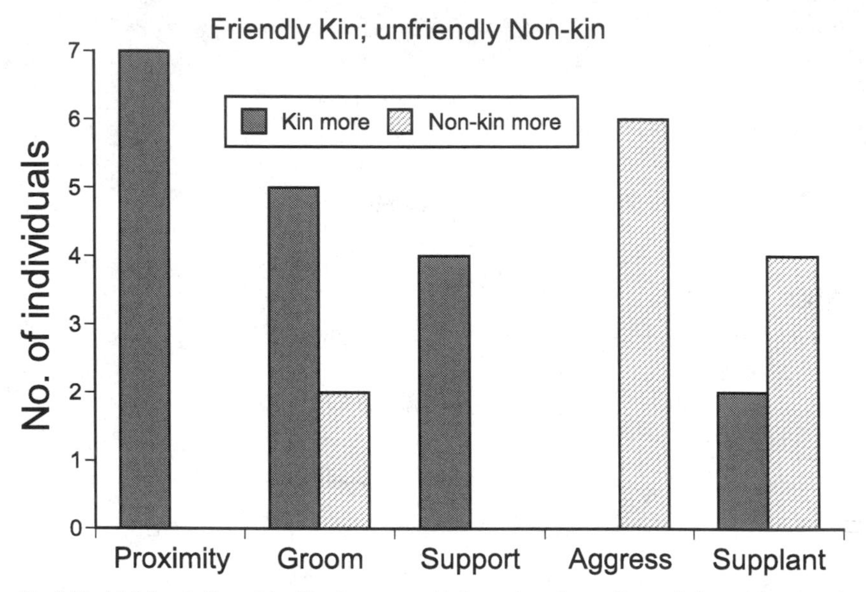

Fig. 5.8B. Adult female kin are friendlier than are non-kin in a variety of ways. The graph shows the number of adult females that had both adult kin and non-kin in the group (total $N = 7$), which showed more of the stated behaviors to their kin or to their non-kin. (Mountain gorilla.) *Details at end of chapter.* Source: Data from Stewart and Harcourt (1987).

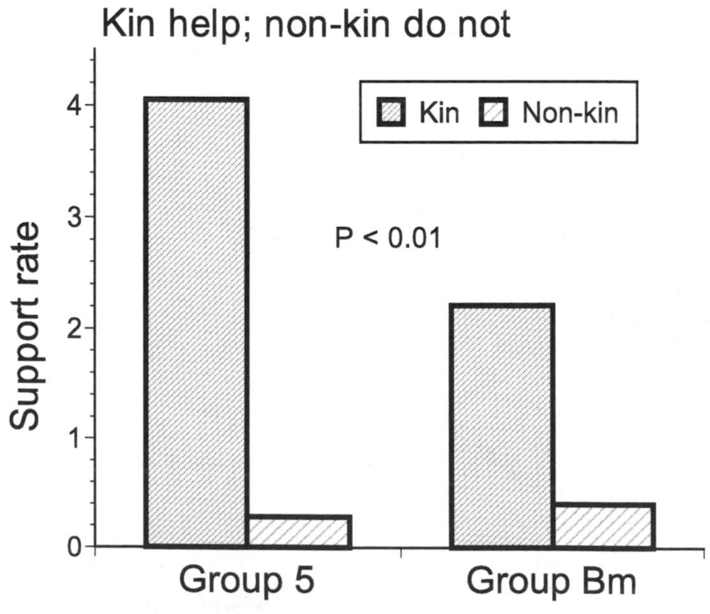

Fig. 5.8C. Adult female kin support one another in contests much more frequently than they do non-kin. (Mountain gorilla). *Details at end of chapter.* Source: Data from Watts (1997).

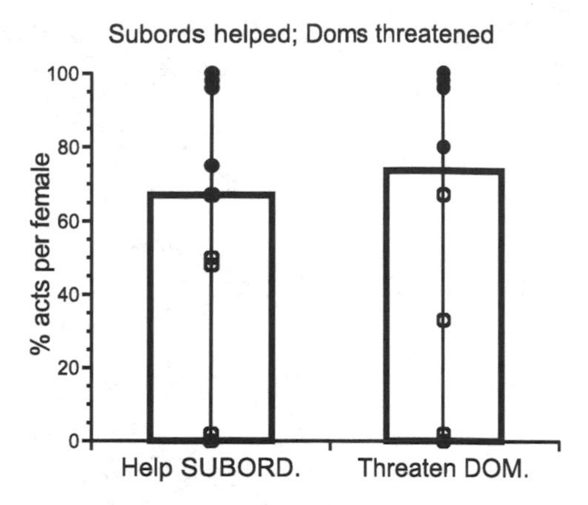

Fig. 5.9. Adult females help subordinate contestants against dominants. Helpers are adult females; contestants can be immature females. ● and ○ indicate females from two study groups. Bar is median individual. (Mountain gorilla.) *Details at end of chapter.* Source: Data from Harcourt and Stewart (unpublished).

tures can affect their rank as adults (see following discussion), help when immature might be relevant to decisions to stay or leave the natal group. In general, most females help subordinate contestants rather than dominant contestants, and do so by threatening the dominant contestants (fig. 5.9). In all cases of help to dominant contestants, the adult female helper was more closely related to the dominant, helped animal than she was to the subordinate opponent. Watts (1997) got somewhat different results, finding that non-kin did not distinguish rank of contestant in their help, whereas kin more often supported the aggressing kin, which will usually have been the dominant opponent.

Among adult females, help makes little to no difference to the outcome of contests (Watts 1997). However, where there potentially exists a lot of difference in competitive ability between the contestants, namely when an immature animal is the subordinate contestant, subordinates are helped frequently (Harcourt and Stewart 1989), and the help is useful, in the sense that the subordinate gains access to the contested resource (fig. 5.10). Watts (1997) found likewise that over two-thirds of support from adult females in contests between an immature and a dominant animal was to the subordinate contestant, and he found that the support was successful even more often than we found, about two-thirds of the time.

Help to immature animals is highly relevant to their competitive ability as adults. In those species or populations of primates in which daughters come to "inherit" the rank of their mother, that is, achieve a relative competitive ability close to that of their mother, the argument is that it is help in immaturity that determines the ultimate outcome. Immature offspring of

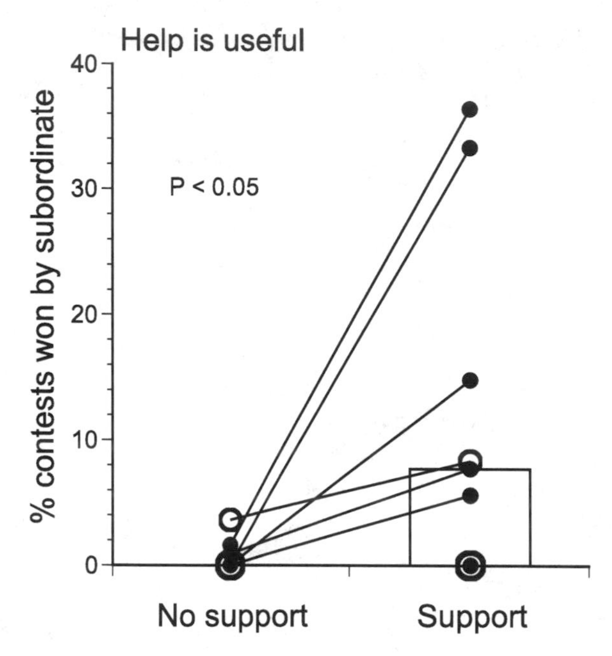

Fig. 5.10. Subordinate, immature animals are more likely to win contests (gain the resource) if they are helped than if they are not helped. Circles and lines connect same individual. ● and ○ indicate two study groups. Bar is median individual. (Mountain gorilla.) *Details at end of chapter.* Source: Redrawn from Harcourt (1992).

dominant mothers learn that they can dominate even adults subordinate to their mother, while the offspring of subordinate mothers learn that they can never win, even against smaller opponents if those opponents are offspring of females dominant to their mother. Consequently, offspring dominate all those which their mother dominates, and are subordinate to all those to which their mother is subordinate (ch. 2.2.3.2).

Does the same happen with gorillas? In other words, is help useful enough to counteract the contrasts in competitive ability provided by age and size? It is not. Rank among immatures and adult females in gorilla groups closely follows time in the group, which for immatures, of course, means age (ch. 4.2.2.4). We confirm that here by showing that far from adult female kin ranking next to one another, they have ranks more different than do non-kin: help as immatures (and as adults) does not translate to adult offspring achieving the rank of their mother (fig. 5.11).

Cooperation, then, is temporarily useful to gorilla females, especially to immature animals. However, it leads to no lasting effect on relative competitive ability. Why not, given that it apparently does so in some other group-living primate species (ch. 2.2.3.2)?

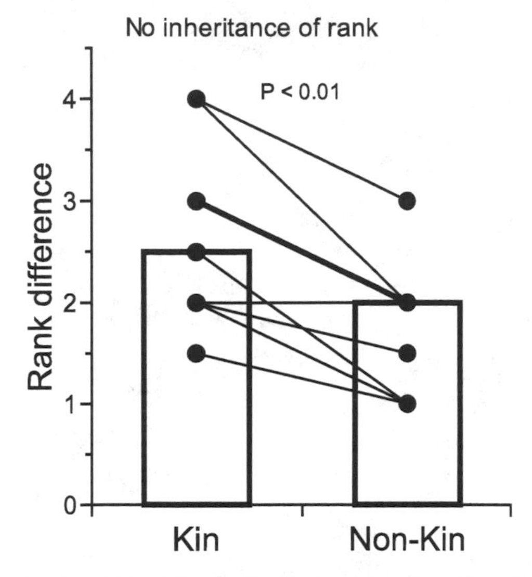

Fig. 5.11. Gorilla females do not inherit their mother's or family's rank: median differences in rank were greater between kin than between non-kin. Data for nine adult females in four groups with both kin and non-kin adults in the group. Thick line = two females. (Mountain gorilla.) *Details at end of chapter.* Sources: Data from Harcourt and Stewart (unpublished) and Watts (1994b).

One simple answer is that cooperation occurs too infrequently, and the spoils of victory are so lacking in value. We have previously suggested that gorillas support one another in contests as frequently as do several Old World female-resident species (Harcourt and Stewart 1989). However, our count for that claim included silverbacks, the animals that help by far the most frequently. Here we are concerned with females helping one another.

Among adult females, not only do not all kin help one another, but among those that do help, the help can be infrequent. Our data showed that adult females were helped by other females as subordinates (useful help) in less than 5% of their contests (stats. 5.3). Watts's (1997) results were similar: he found females to be supported in a median of 4% of all conflicts, though 10% of serious ones. The low value and abundance of the usual resource, foliage, is presumably the reason that useful cooperation is rare: why help another animal in a contest when she could obtain the reward so easily by shifting a few meters (Harcourt and Stewart 1989)?

Another reason for the lack of a long-term effect of help in fights concerns the general influence of demography and group size on society (Datta 1992; Dunbar 1979; Hausfater 1982; Hill 2004). In this case, with birth intervals of around four years (ch. 3.2.3.3) in the usually small groups of gorillas (ch. 4.2.1.1), the disparity in age and therefore size among contestants, especially immatures ones, is normally so great that no amount of help can reverse the consequent disparity in competitive ability (Harcourt and Stewart 1989). The disparity will be greatest among siblings (never closer in age than the minimum interbirth interval), whereas non-kin can obviously be more similarly aged, indeed they can be the same age.

Additionally, support from adult females does not make much difference to the outcome of contests between adult females, largely because the support rarely changes the outcome (Watts 1997). So, while the support temporarily helps immature animals (fig. 5.10), it does not even temporarily help the adults.

5.2.3. Conclusion: Food, cooperation, and emigration

Gorilla females benefit little enough from cooperation that if other factors favor emigration, foregoing opportunities to cooperate is not enough of a cost to prevent them from emigrating (Watts 1990a). Thus, lack of benefit from cooperation allows emigration (Watts 1990a). However, it does not explain it. Nor does competition (see previous discussion).

5.3. Comparison with *Pan* and *Pongo*

Chimpanzees (including bonobos) and orangutans are large-bodied animals that survive mainly on ripe fruit (ch. 4.1.8). Competition for food can therefore be intense enough for animals to have to range alone (ch. 4.1.8; 4.2.3), unlike for folivorous female gorillas. Indeed, competition among female chimpanzees in Gombe Stream National Park, Tanzania, seems to be important enough for higher-ranking females to produce more surviving offspring than do lower-ranking females (Pusey et al. 1997). Mild competition between folivorous gorilla females, compared to the potential for intense competition between frugivorous chimpanzee and orangutan females, means that gorilla females can group, whereas the option is less often available to chimpanzees and orangutans.

Because it is ripe fruit that the chimpanzees and orangutans survive on, more than unripe fruit, and because the animals are large-bodied, their food often comes in small packages that are often not sufficient for more than one individual, especially in the non-fruiting season (ch. 4.1.8). Therefore, cooperation in competition is not advantageous when most needed. Consequently, chimpanzee and orangutan females do not suffer greatly from lack of cooperative female kin, and hence can emigrate at maturity if other factors make it advantageous to do so.

Conclusion

One reason gorillas do what they do could simply be that they are gorillas. A pigeon can fly to escape a cat; a pig cannot fly to escape a leopard. Female gorillas live in groups because they have evolved to live in groups, and cannot now do anything different.

That is not the case. Gorillas compete and cooperate like other species, and experience the benefits and costs of grouping like other species (Dugatkin 1997; Harcourt and de Waal 1992a; Kapsalis 2004; Krause and Ruxton 2002; Stewart and Harcourt 1987; Watts 1994a). Not only that, but different individuals compete and cooperate in different ways: some adult females are clearly, even aggressively, dominant over others, some are not; some adult females help their adult kin; some do not.

In other words, we are not looking at an endpoint of evolution; we are looking at evolution in action. Gorilla society cannot be explained by features of gorillas unique to gorillas. Gorilla society needs to be explained by an interaction of gorilla features (large-bodied folivore) with aspects of the current social and physical environment in which individuals find themselves, which aspects differ among individuals. Different individuals find themselves in different environments, and make different decisions (or behave as if they make decisions) about what to do in those different situations, with different consequences. These themes of variation among individuals, and therefore evolution in action now will recur throughout the book.

Gorilla society can consist of groups of females because while gorilla females compete for food, they do not do so seriously enough that the costs of grouping are high, unlike for chimpanzee and orangutan females (Wrangham 1979). Competition for food neither limits grouping, nor limits group size in this large-bodied folivore (see ch. 2.2.3.1). The minimum competition between gorilla females allows grouping. And while gorilla females cooperate, the rate and rewards are low enough that the females do not suffer large costs of lost opportunity if they are forced to emigrate (Watts 1997).

But just because females can group, just because they can emigrate, does not mean that they have to do either. Given that there is competition, and that the competition would be avoided if females that emigrated traveled alone (and almost all females do emigrate), why do gorilla females group? Why do they never travel alone as do chimpanzee and orangutan females? Given that cooperation occurs, and can, if minimally, be useful, why do females leave kin, effectively the only animals besides males from which they will receive help?

Thus, feeding competition and cooperation among females are not sufficient to explain gorilla grouping. Female strategies in relation to food alone are insufficient to explain the society of gorilla females. Other explanations are needed. The influence of males needs to be understood in order to fully understand females' grouping and emigration strategies (ch. 6; ch. 7; ch. 8) (Wrangham 1979).

Figure Details

Fig. 5.1A. Females compete over food. Data are combined aggression and supplants. Binomial test, $p = 0.002$. (Virunga Volcanoes National Parks, Rwanda, Zaire.)

Fig. 5.1B. Aggression (including aggressive supplants) among females in four groups over a total of 40 months of field observation across seven years. (Group 5 data shown are from middle of 3 periods of observation; other groups watched for one period.) $N =$ adult females/aggressive incidents: Group 4: 3/172; Group 5: 11/485; Group Bm: 7/266; Group Nk : 6/141. (Virunga Volcanoes National Parks, Rwanda, Zaire.)

The contrast in values in figures 5.1A and 5.1B in proportion of competitive interactions that were over food is probably largely due to the fact that if we could not identify a specific resource, we omitted the incident from analysis, and thus omitted, for example, Watts's (1994b) incidents of harassment, intervention, and progression.

Fig. 5.2A. More competition in larger groups. Data are mean rate of feeding interruptions (i.e., supplants) per focal observation hour per group by group size: Spearman $r_s = 0.93$, $p < 0.01$. (Virunga Volcanoes National Parks, Rwanda, Zaire.)

Fig. 5.2B. Wilcoxon matched-pairs, signed ranks test: $N = 9$, $T = 21.5$, $p = 0.008$. (Virunga Volcanoes National Parks, Rwanda, Zaire.)

Fig. 5.3. Gorilla females are aggressive to immigrants. Rates of aggression among pairs of non-kin residents (so as not to confound residence with kinship) compared to rates of aggression between residents and immigrants (which are unrelated) in two groups with both non-kin residents and immigrants. Non-kin Residents vs. Resident-Immigrant—Group 5 in two periods a year apart: (ranges across pairs, 9–919, 8–175 vs. 0–919, 23–175); Group Bm in one period: (ranges, 53–881 vs. 115–182). Contrasts were similar when total aggression was counted. No statistical results for these particular comparisons are available. In western gorillas, it looks as if residents are more frequently aggressive to immigrants than they are to other residents, but there are too few data for statistical comparison (Stokes 2004).

Fig 5.5. Dominance hierarchy sometimes apparent. Supplant = approached individual moves more than 2 m within 1 sec of approach within 5 m by another (Harcourt and Stewart 1989); moves more than 2 m on approach within 2 m by another (Watts 1994b). Data based on median per group ($N = 6$ groups) of 4 females that supplanted or were supplanted, 6 pairs of individuals between which supplants were seen, and 81 supplants. (Per group, bottom to top in figure, total number of females-pairs-supplants: 7–20–32; 9–33–113; 6–14–48; 11–54–132; 4–6–81; 5–10–64. The second listed group is the same group in three different study periods, 1980–3, 1984–5, 1986–7, with 4, 9, and 11 females, respectively, per period.) In very small groups, with infrequent supplants, as is often the case in gorilla groups, the apparent linearity could be a sampling accident (Appleby 1983). One can test statistically whether the hierarchy is significantly unlikely to be an accident (Appleby 1983). • = groups in which hierarchy is significantly unlikely to be linear by chance ($p < 0.05$); ∘ = groups in which linearity is not significantly different from chance (these were the two smallest groups). Binomial test on six of six groups showing the same contrast in direction of supplants ($p < 0.05$). (Virunga Volcanoes National Parks, Rwanda, Zaire.)

Fig. 5.6. Dominance hierarchy sometimes not obvious. (a) Few = total percent of pairs per group with < 3 supplants per pair; (b) 2-way = total percent pairs per group in which both members of a pair supplanted one another. • = data for individual groups, $N = 5$ groups (one repeated by Watts four years after Harcourt and Stewart's study); median of 15 pairs per group (range, 6–55); ∘ = data summed across three decades, $N = 6$ groups (Virunga Volcanoes National Parks, Rwanda, Zaire). (c) Aggression = females' aggression to dominant partner expressed as percent total aggression between the two (excluding aggression during interventions in contests among others, which was quite often against dominant opponents [Harcourt and Stewart 1989]); median value per female is presented. $N = 3, 5$ females with dominant partners in the two study groups. Similarly, Watts (1994b) found that the median female reacted to aggression by ignoring it or by being aggressive back on 83% of occasions (range 76–89%, 4 group), that is, the median female was unafraid of those dominant to her, and unfeared by those subordinate.

Fig. 5.8A. Kin are friendly. Proportion of possible kin and non-kin pairs ($[N \times N - 1]/2$ individuals per group) in one group over three periods within which the females showed especially friendly (affiliative interaction rates above median and aggressive rates below median) or unfriendly behavior (affiliative rates below median, and aggressive rates above). (Affiliative behavior is proximity, grooming, and tolerated bodily contact, e.g., resting together.) Periods were 1984–5, 1986–7, 1991–2. $N = 13, 9, 8$ kin pairs; 42, 57, 70 non-kin pairs. If one assumes that the pairs are independent, the contrast between kin and non-kin within any year is highly significant. However, the same female can appear in several pairs. The values for different pairs are therefore not independent. We did not perform the necessary statistical test, repeated sample analysis, because the contrast is so obvious here, and the finding is replicated with the different sample shown in figure 5.8B, in which individuals are independent data points (Virunga Volcanoes National Parks, Rwanda, Zaire).

Fig. 5.8B. Individuals are seven adult females in two groups with both kin and non-kin adult females present in their group. Each individual appears only once. Binomial tests: Proximity, $p < 0.02$; Aggress., $p < 0.05$. (Virunga Volcanoes National Parks, Rwanda, Zaire.)

Fig. 5.8C. Support rate to adult female kin and non-kin is the proportion of total supports that were to kin or to non-kin divided by the number of pairs of adult female kin and non-kin (e.g., 80% supports to kin, but only 20% pairs as kin = support rate of 4.0). $N = 11$ adult females for group 5 (average over three periods of study), and seven in group Bm. Group 5 is the same as one of the groups whose data are in figure 5.7B, but studied at a later time. (Virunga Volcanoes National Parks, Rwanda, Zaire.)

Fig. 5.9. Help to subordinates. $N = 9$ adult females in two groups recorded to help; acts of help given per female = 1–12, median = 4. Contestants could include immature animals. (Virunga Volcanoes National Parks, Rwanda, Zaire.)

Fig. 5.10. Percent contests won (resource gained) by subordinate immature animals in relation to whether they were supported in the contest (another animal intervened on their behalf) or not supported. $N = 9$ animals (3 never won contests against dominants), Wilcoxon matched pairs signed ranks, $T = 2.9$, $p < 0.05$. Number contests per individual, supported contests, 4–21, median = 12; unsupported, 28–148, median = 110. (Virunga Volcanoes National Park, Rwanda, Zaire.)

Fig. 5.11. No inheritance of rank. Binomial test, $p < 0.01$. (Virunga Volcanoes National Parks, Rwanda, Zaire). Data for seven adult females from Harcourt and Stewart (unpublished) and two additional females from Watts (1994b).

Statistical Details

Stats. 5.1. Aggression rate among females, Group 5: median, 1.2 acts per female per hour within 5 m of another female (range, 0.4–2.7); Group Nk: 0.5 (0.3–2.6).

Stats. 5.2. Aggression × group size – N = 6 groups, 5–13 females per group; r_s = 0.7; *ns;* (Watts 2001).

Stats. 5.3. Incidents of support received by adult females from adult females in contests among adult females: N = 10 adult females in two groups; support received in 0%–3.5% contests (median, 2.2%); 10–148 contests per female (median, 36) (Harcourt and Stewart, unpublished).

Male feeding near non-feeding females and offspring. (Mountain gorilla). © A. H. Harcourt

CHAPTER 6

Female Strategies: Male Influences on Females' Competition, Cooperation, and Grouping

> The entire daily routine—the time of rising, the direction and distance of travel, the location and duration of rest periods, and finally the time of nest building—is largely determined by the leader [the dominant silverback male]. . . . The complete adherence of gorillas to one leader must be considered in all discussion of groups. . . . **George Schaller (1963, p. 237)**

Summary

A major facet of gorilla society is the absolute dominance of the male over females, which correlates with his being twice their size. For instance, he supplants the average female nearly ten times more often than does the average other female in the group. Not only does his dominant presence make competition and cooperation among females themselves almost irrelevant, but in addition, the male actively diminishes the effects of the females' competition and cooperation by preventing dominant females from exerting their competitive advantage through his usual help of subordinates against dominants. The effect of the male is so strong that group size of gorillas is unlikely to be limited by competition among females, because an extra female is going to make so little difference by comparison to the competition already imposed by the one male.

The fact that female chimpanzees and orangutans are probably prevented from traveling with males by the costs of competition, whereas bonobo females are dominant to males and usually travel with them, supports the hypothesis that gorilla males impose costs on females. Reasons for why female gorillas group with a male and each other therefore have to be sought elsewhere than in females' feeding competitive and cooperative strategies. They and the environment in which they operate allow grouping, but do not cause it.

6.1. The Male Is a Major Competitor

Ever since Schaller's (1963) groundbreaking studies of gorillas in the early 1960s, one of the most obvious aspects of gorilla society has been the absolute dominance of the silverback over the females in the group (ch. 4.2.2.1). That dominance extends beyond the leadership role noticed by Schaller, which we highlighted with this chapter's opening quote. The dominance extends also to competition over food.

The silverback is twice the size of females (ch. 3.2.2). Females receive far more competition from the silverback than they do from any other female. The silverback is readier to compete than is a female, and the females are less able to ignore a silverback's intentions to compete than they are to ignore each other's intentions. In our studies, no female was supplanted more often from food by any female than she was by a silverback (fig. 6.1). The average (median) female is supplanted from food nearly ten times as often by the silverback as she is by the average other female in her group (fig. 6.1).

Put another way, only in groups of ten or more females, will any female receive more competition from all females in total, than she will from a single silverback. Groups of ten or more females are very rare: the median number of females per group in Africa is only three or four (ch. 4.2.1.1), and less than 5% of over 150 counted groups have more than ten females.

This strong influence of the silverback could be another reason (in addition to the nature of the food and its distribution) why transferring females apparently do not take into account group size, that is, do not prefer to join smaller groups (ch. 5.1.2).

Further evidence of the potentially adverse effect of the silverback on females' access to food comes from where females feed. In a population of west-central African gorillas studied by Melissa Remis (1995; 1999), females fed on smaller branches and fed more often in the periphery of trees than did males (fig. 6.2). Feeding on small supports can be dangerous. Remis (1995) describes a young female falling 15 m to the ground after being forced out onto terminal branches by an older female.

Of course, the question immediately arises of whether instead of being forced out by competition with the male, the females are taking advantage of their own smaller size to feed higher in the canopy, or on the periphery

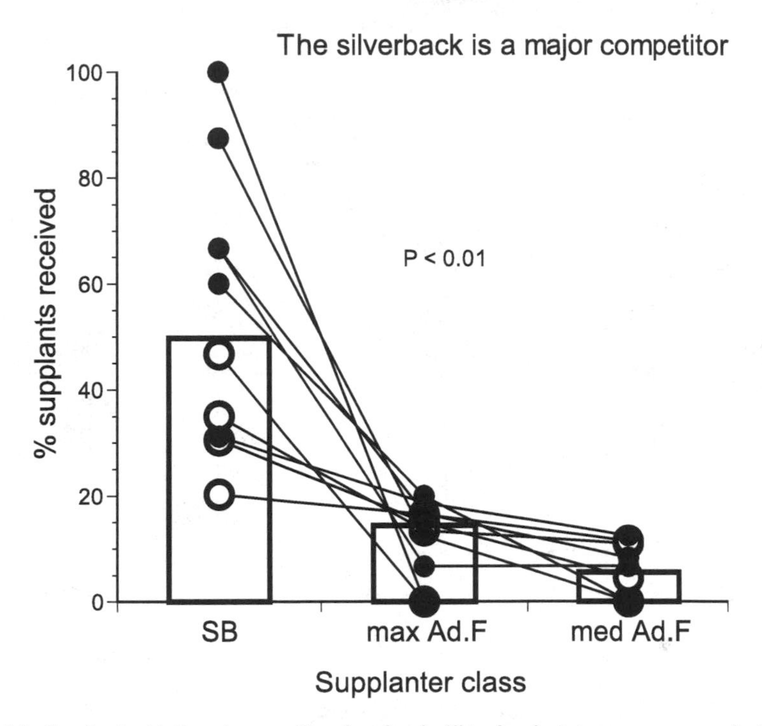

Fig. 6.1. The silverback is the major competitor of any female. All ten females in two groups were supplanted* far more frequently by a silverback (SB) than by either their most serious female competitor (Max Ad.F), or by their median female competitor (Med Ad.F). ● and ○ indicate individual females in two groups; bars indicate median values. (Mountain gorilla). *Details at end of chapter.* *(Supplant = non-aggressive approach that causes approached animal to depart.) Source: Data from Harcourt and Stewart (unpublished).

of tree crowns where food might be more abundant or of better quality, depending on what quality is being sought (Niinemets 1995; Schaefer and Schmidt 2002; Young 1995). Or alternatively, are the females simply feeding where predators are less likely to reach them (ch. 7.1.2)?

The idea that smaller animals are escaping competition by feeding in the parts of the environment that the larger animals cannot reach was one explanation for why immature Virunga gorillas fed up trees much more often than did adults (Fossey and Harcourt 1977). If this is what is happening with Remis's western gorillas, then absence of the males should result in the females moving to larger branches or lower down near the core of the tree. If predation is causing the smaller animals to be farther out and up, absence of their main protector, the silverback male, should cause the opposite to occur (ch. 7.1.3). Remis found that in the absence of the silverback in the

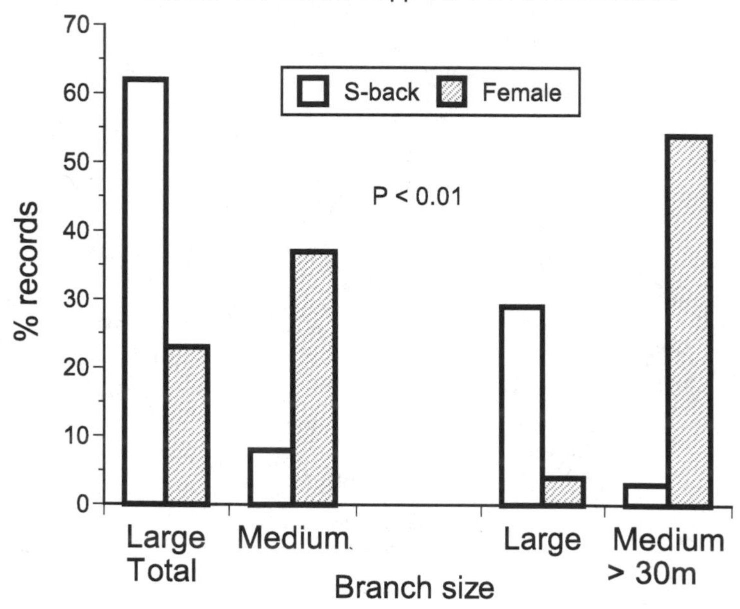

Fig. 6.2A. Females can escape competition from silverbacks by using smaller branches. Data show total records, and records when gorillas were more than 30 m up trees. (Western gorilla). *Details at end of chapter.* Source: Data from Remis (1995).

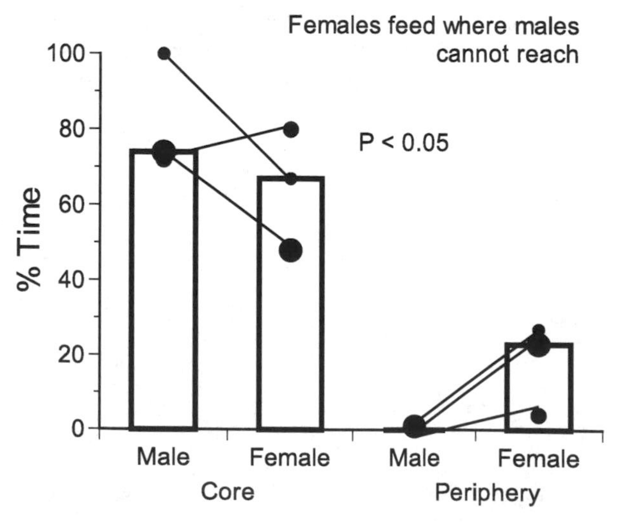

Fig. 6.2B. Females can escape competition by feeding in the periphery of tree crowns (of three sizes). Circles are three categories of increasing size of tree. (Western gorilla.) *Details at end of chapter.* Source: Data from Remis (1999).

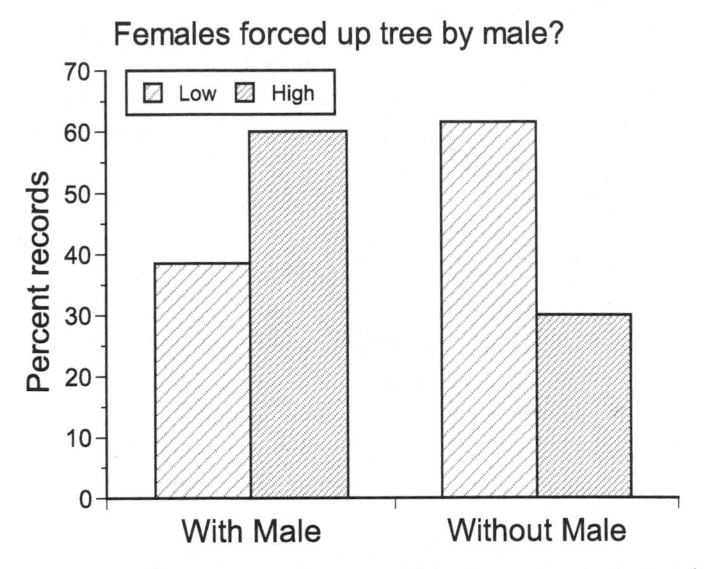

Fig. 6.3. Females feed higher in tree crowns in the presence of their main competitor, the silverback. (Western gorilla.) *Details at end of chapter.* Source: Data from Remis (1995).

tree in which the females were feeding, the females moved lower down in the crowns of the tree (fig. 6.3). Unless females are lower in trees in the absence of the silverback in the tree in order to stay closer to him wherever else he is, it seems that males force females up and out.

In sum, not only are males by far the most frequent competitors that females facc, they are, of course, an unbeatable competitor, unlike other females.

6.2. The Male Mitigates Competition Among Females

If the male is a major competitor of females, he also mitigates the effects of contest competition among females themselves. He intervenes far more often in contests among females than do the females. Thus, David Watts (1997), correcting in various ways for opportunity to intervene, found that males were twice as likely to intervene in contests among females as was the average female. We found males to intervene in contests that involved females over five times as often as did the females (Harcourt and Stewart, unpublished).

The relevant question here, though, is not who intervened, but who the females experienced interventions from. Do females experience more intervention from males than from females? The answer depends on the number of female kin in the group, because as we have already shown, kin intervene for each other far more often than do non-kin (ch. 5.2.1). Females with no kin in the group, the usual situation (ch. 4.2.1), receive more support from the male than from all other females combined (Harcourt and Stewart, unpublished.; Watts 1997).

While over three-quarters of interventions by a female were on behalf of one of the contestants (the female's aggression is directed at one of the contestants) (Harcourt and Stewart, unpublished), half of a silverback's interventions are undirected, with neither contestant obviously supported—what David Watts terms "control" interventions. The contest is simply stopped. A stopped contest benefits the subordinate contestant. Furthermore, if the male intervenes on behalf of one of the contestants (i.e., he directs his aggression at the other), he almost always intervenes on behalf of the subordinate female against the dominant female. The only time any of the three silverbacks that we observed did not help the subordinate female was when he helped the dominant female with which he was sexually consorting at the time (stats. 6.1).

Of course, the silverback's interventions are far more successful than are the females' interventions—not surprisingly given his massive size. Thus, 80% of males' interventions stop the contest, or directly help the supported animal to win, compared to only about 30% of females' interventions (Watts 1997).

The consequence must be that males reduce the cost of grouping for females of low competitive ability. The male's aiding strategy therefore encourages grouping, in the sense of lowering the costs of being a subordinate female in the group, and hence preventing subordinate females from leaving. The benefits to him are obvious but will be considered again in chapter 11.1.6.2 on male competition for sole access to females.

6.3. The Male Mitigates the Benefits of Females' Cooperation

The male also intervenes in coalitions among females, in addition to intervening in contests among them. And he intervenes for the losing party, as he does when intervening in contests, whether the losers are single females against a coalition of more than one, or a losing coalition against a more powerful coalition. The male is so powerful that even coalitions among females are ineffective in the face of the male's intervention on behalf of the subordinate animals (Watts 1997). Females that normally won when cooperating with another female against a third are far less likely to win when the male intervenes in the contest (fig. 6.4). Conversely, females that usually lose against coalitions of other females are far less likely to lose (fig. 6.4).

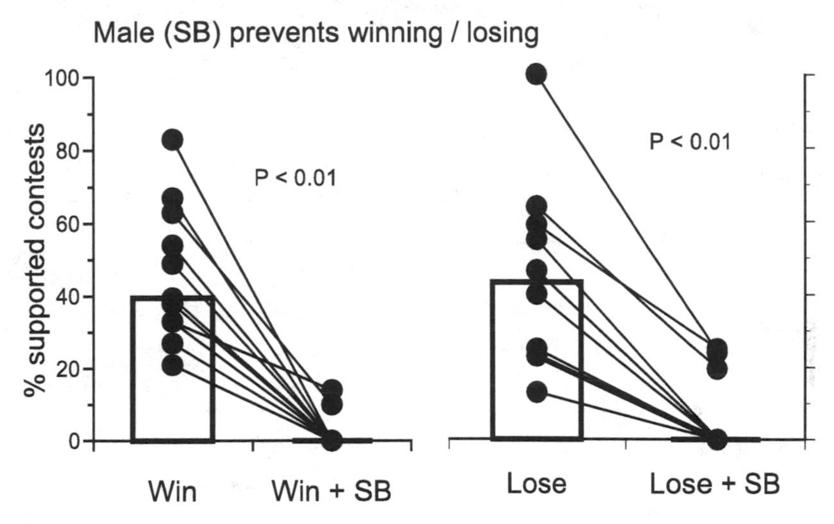

Fig. 6.4. The silverback prevents useful cooperation among females. When the silverback intervenes in contests in which a female is being supported by another female, females that win when he does not intervene (Win) are less likely to win when he does intervene (Win + SB); and females that lose against coalitions of other females when he does not intervene (Lose) are less likely to lose when he does intervene (Lose + SB). (Mountain gorilla.) *Details at end of chapter.* Source: Data from Watts (1997).

6.4. Comparison with *Pan* and *Pongo*

If chimpanzee and orangutan females impose so much competitive cost on each other that they travel in only small parties or alone when fruit, their main food, is in short supply (ch. 4.1.8; 4.2.3), then the larger males, especially the larger orangutan male, would presumably be impossibly competitive foraging partners for females at that time of shortage. Hence the orangutan sexes usually do not travel together, except when competition over food is not serious, such as at generously fruiting trees (ch. 4.2.3.2), or when other benefits to females are so large that they outweigh the costs of competition, such as when females are in estrus (Wrangham 1979). Even then, estrous female chimpanzees sometimes resist associating with a male, leaving the male to resort to force to coerce the association (Mitani et al. 2002b; Wrangham 2002).

It is not impossible, though, that female orangutans could escape the competition, even while consorting with a male (Delgado and van Schaik 2000). While the heights in the forest trees at which orangutans feed do not obviously differ between the sexes (Rodman 1977), female orangutans, half the male's size, can presumably feed on smaller branches and smaller trees than can the males, just as gorilla females do. Concomitantly, if it were advantageous for a male to travel with her (ch. 10.1.3), and if the male's presence were costly for her, the orangutan female could probably escape more easily than could a gorilla female. Not only must the female orangutan be able to use smaller branches and trees than can the male, but she travels faster, and feeds for less time than does the male (Rodman 1977).

The bonobo chimpanzee becomes an interesting case here. Bonobo females can more easily dominate males than can common chimpanzee or orangutan females (ch. 4.2.3.1). They can therefore more afford to travel with males, and they do so more often than do either common chimpanzee or orangutan females (ch. 4.2.3.1). Additionally, the bonobo females are surely less able to avoid males than are the orangutan females, given that they are far more similar in size to the males than are orangutan females, meaning that even if it were disadvantageous to the females to travel with males, they might not be able to avoid males for which it was advantageous to travel with females (see ch. 10.1).

There is one feeding context, however, in which female apes would at first sight seem to benefit from males, and that is when chimpanzee males provide females with meat after a hunt (Boesch and Boesch-Achermann 2000, ch. 8; Mitani and Watts 2001; Stanford 1998a; Watts and Mitani 2002). The excitement over meat and the effort expended in obtaining it suggests an important nutritional benefit. Nevertheless, females might receive so little meat that no nutritional benefit in fact accrues (Boesch and Boesch-Achermann 2000, ch. 8), especially as hunting is most common in the fruiting season (Mitani and Watts 2001; Watts and Mitani 2002).

Conclusion

The vertebrate socioecology framework has food influencing females, and females influencing males. However, even with respect to just food, it looks as though female gorillas are influenced as much by the male as by food.

The gorilla male is such a frequent and powerful competitor by comparison to females that any competition among females becomes almost irrelevant. Exactly how much of a cost the silverback imposes on females, nobody knows. The point here is that compared to another female, the male must be a major cost. Females receive so much more competition from the silverback than from other females that competition among females must make little difference to most females' abilities to survive, or rear their offspring. Grouping with each other does not impose any substantial cost on females by comparison to the costs imposed by the male.

And because of the male's frequent and powerful intervention in females' contests and in their coalitions, their cooperation with one another in contests over food is similarly irrelevant (Watts 1997). Not only can the male prevent individual females from defeating another, but he easily prevents even coalitions of females from winning against another female. We can illustrate with one example the relative lack of importance of competition among females and of cooperation in that competition. The noise of the altercation between the two families of females pictured in figure 5.7, in which a coalition of one family drove away the other, brought the dominant male galloping to the scene—where he then monopolized for the next quarter of an hour the resource over which the females had just been competing (fig. 6.5).

We suggested that female gorillas did not need to compete intensively, nor cooperate often, because food was so easily obtainable without contest competition (ch. 5). Why then does the male compete so frequently? We suspect it is because he can. Competition with a female is effectively without cost to him, because he is so likely to win and highly unlikely to suffer in the winning.

Rob Barton and colleagues suggested that baboon males can have a strong influence when "bonds" between females are weak, that is, they cooperate infrequently (Barton 2000b; Barton et al. 1996). We suggest that group size might be another factor that affects the male's influence (Stewart and Harcourt 1987). Because gorilla groups are usually small, of the order

Fig. 6.5. The silverback supplants all females from the contested resource. (Mountain gorilla.) © K. J. Stewart

of three or four females per group (ch. 4.2.1.1), each female frequently experiences competition from the male, and the effect of his interventions in their contests. By contrast, in species with large groups, or in large groups of gorillas, we suspect that the influence of the male would be less.

Baboon males, like gorilla males, are twice the size of females, and can easily supplant them (Altmann and Altmann 1970; Seyfarth 1978b). However, as baboon groups are often large, the males' influence could be diluted by comparison to gorillas, and hence cooperation among females is an important component of female competitive relationships in baboon groups (Barton 2000b; Barton et al. 1996; Seyfarth 1976). Indeed, in the very nice comparison of competition and cooperation that led to Barton and colleagues' suggestion that female bonding affects the males' influence on females, the males' influence was weaker in the larger baboon groups, in the sense of decreased ability to break up alliances of females (Barton et al. 1996).

We argued that the costs of competition and the benefits of cooperation among gorilla females were mild enough to allow grouping and not to prevent emigration (ch. 5). The male's influence on competition and cooperation among females makes it even less costly for females to group with one another (he helps losers), and less costly to leave behind on emigration the benefits of cooperation with kin (his help reduces the effectiveness of cooperation). While the male makes competition and cooperation among females irrelevant, the competition that they experience from him means that if access to food is the only determinant of male-female associations, the last thing a female should do is associate with a male (Wrangham 1979). Yet they do. In the next chapter we consider why.

Figure Details

Fig. 6.1. Silverback is major competitor. Ten of 10 females (Binomial test, $p = 0.002$). Median number supplants (including supplants as a result of aggression) received per female, Group Nk, • = 13 (range, 5–16); Group 5, ∘ = 42.5 (range, 30–50). (Virunga Volcanoes National Parks, Rwanda, Zaire.) We have already shown that most competition among females is over food (ch. 5.1.1, fig. 5.1A, B). Similarly, most supplants from silverbacks were over food: Group Nk, 79% of 29 supplants; Group 5 in two years, 95%, 75% of 20, 9 supplants. (Virunga Volcanoes National Parks, Rwanda, Zaire.)

Fig. 6.2A. Females can escape competition. Percentages apply to the total sample of seven possible types of support. Statistics apply to only the two-support sample shown. However, note that the data are almost certainly not independent, based as they are on a one-minute sampling regime (Remis 1995), and therefore the statistical results should be treated with extreme caution, even though the p-value presented in the figure assumes a sample size one-tenth the actual used by Remis (Total: $G = 127, p < 0.0001, N = 237$ observations of males, 196 of females; >30 m: $G = 43, p < 0.0001, N = 20$ observations of males, 39 of females. (Dzanga Ndoki National Park, CAR.)

Fig. 6.2B. Stated statistics are log linear model, $G(w) = 9$, $df = 2$, $p < 0.025$ (Remis 1999). Data shown come from tree sizes 1, 2, 5+6 in table 7. Size 4 ignored, because no males recorded; types 5 and 6 combined because same size. (Dzanga Ndoki National Park, CAR.)

Fig. 6.3. Females outcompeted. Females with male in same tree, $N = 161$; without male, $N = 150$. Statistical procedures and warnings as for fig. 6.2A. Full sample, $G = 23, p < 0.001$; sample size divided by five, $p > 0.05$. (Dzanga Ndoki National Park, CAR.)

Fig. 6.4. Silverback prevents cooperation. Counted only females with at least five contests involving intervention by females. $N = 11$, 10 females in two groups. Binomial tests, $p < 0.01$. (Virunga Volcanoes National Parks, Rwanda, Zaire.)

Statistical Details

Stats. 6.1. Silverbacks help subordinates against dominants. $N = 3$ silverbacks; help to subordinates against dominants, 86%–100% occasions, median = 100%. $N = 6$ per silverback.

Unhabituated gorillas cluster round the leading male in the presence of potential danger, a strange observer. (Mountain gorilla.) © A. H. Harcourt

CHAPTER 7

Female Strategies: Male Influences; Joining a Protective Male

Summary

Gorillas are preyed upon, and in some populations, infanticide has at times been common. We argue that in the face of the obvious feeding costs that a male imposes upon females (ch. 6.1), the only reason for females to group with males is to obtain protection against predators or infanticidal non-father males, or both.

Several lines of evidence indicate that females join males, rather than groups, and that groups result from independent choice by several females of the same male.

Observational evidence supports the antipredation argument for grouping: the male is almost the only protector of a group, and the females act as if he protects them.

Observational and modeling evidence support the anti-infanticide hypothesis also. Females that lose their main protective male lose their infants to other males. Also, modeled females that range alone suffer far higher risk of infanticide than do those that range with a male.

To decide which is the more important protective benefit of grouping, predation or infanticide, we need a prediction that will separate the hypotheses.

7.1 Protection from Predation

> I believe that the gorilla is normally a perfectly amiable and decent creature. I believe that if he attacks man it is because he is being attacked or thinks he is being attacked. **CARL AKELEY (1923, p. 196)**

> I have seen the native hunters, having dispatched the Old Man, surround the females and beat them over the head with sticks. They don't even try to get away, and it is most pitiful to see them putting their arms over their heads to ward off the blows, making no attempt at retaliation. **FRED MERFIELD AND HARRY MILLER (1956, p. 55)**
>
> . . . a lone black-backed male was surrounded by natives. One man fled but slipped, and the gorilla grabbed him by the knee and ankle and bit the outside of the calf, stripping off a large piece of muscle 7 to 8 inches long. **GEORGE SCHALLER (1963, p. 307)**
>
> The dominant males tended to face me when I first arrived, and perhaps roar a few times as the females and youngsters gathered near him to watch me intently. **GEORGE SCHALLER (1963, p. 309)**

Summary: Protection from Predation

We argue that gorilla females associate with a male (as opposed to join a group of conspecifics) to benefit from being with a powerful animal that will mob predators on their behalf. Evidence for antipredation benefits for a female of being with a male is (a) gorillas are preyed upon, despite their large size; (b) they clearly react to that risk of predation—for example, being more alarmed when on the ground than when up trees; (c) the adult male is the main attacker of predators; and (d) the females react as if he is the main protector against predators—for example, nesting at night higher up trees in the absence of a male than in his presence.

7.1.1. Female gorillas associate with a male for defense against predators

Why is it necessary to argue benefits to gorilla females of joining a male? In the first place, the socioecological framework that we are using has males mapping themselves onto females, not the other way round. Second, females should remain in the relative safety of the group of their birth, in the company of their cooperative kin, other things being equal (ch. 2.4.2.6), and yet gorilla females do not (ch. 4.2.1.2; ch. 8.2). Third, having emigrated, females should range on their own, and so avoid the competitive costs of association with others (ch. 2.2.3.1). And yet gorilla females not only join others, which are sometimes aggressive to them as immigrants, but they always join a male, which is twice their size and the equivalent of several females as far as competition for food is concerned (ch. 4.2.1.2; ch. 5.1.1; ch. 6.1). With all those costs to gorilla females of traveling with others, and especially with a male, some obvious other benefit must exist to offset the feeding costs. We suggest that a main benefit is protection from predation.

The influence of predation on grouping usually comes under the heading of influence of ecological effects on social system (e.g., Nunn and van

Schaik 2000). Here we consider it under the heading of the influence of the male on female strategies, because we argue that we see groups of gorillas as a result of females associating with a male, not with each other, for protection against predation.

We write here of the females associating with the male, rather than the male associating with females. In fact, we argue later that males do associate with females because of the benefit to the males of the association (ch. 10). However, in this chapter, it is the benefit to females that is our concern.

Food does not explain female gorillas grouping with one another (ch. 5) and especially does not explain why they always group with a male (ch. 6.1). Instead, in a species in which males are twice the size of females, an obvious reason for females to associate with a male is to gain his protection against predators (ch. 2.3.3.3). We then see groups of gorilla females with a male, the standard harem social system of gorillas, because several females choose the same protective male (Wrangham 1979). That females associate with males for protection against predation is an old hypothesis (DeVore and Hall 1965), and one that continues to receive support (Anderson 1986; Hill and Dunbar 1998; Hill and Lee 1998), especially for species in which the male is much larger than the females, as is the gorilla.

Nevertheless, current opinion appears to be that gorilla females associate with a male for protection, not against predators, but against other infanticidal males (sec. 7.2), even if antipredator benefits also accrue (Watts 1996). Richard Wrangham first suggested protection from infanticide (indeed protection from harassment in general), rather than protection from predation, as a reason for females to associate with a male. He largely did so for heuristic reasons, to see how far a consideration of just the main components of the socioecological framework, female competition for food, male competition for females, and the interaction of the two competitive strategies, could explain observations (Wrangham 1979). Also, he was beginning to think that antipredator strategies could not explain the variation in primate societies (Wrangham 1980). His experiment was very successful. It led to a flourishing of ideas on the importance of infanticide in the nature of primate and other animals' societies (ch. 2.4.2.4; 2.4.3).

However, what is infanticide but, from the female's and infant's point of view, merely a form of predation (Treves 2000; van Schaik 1996)? Does it matter to the infant or its mother whether a gorilla male's canines or a jackal or leopard's canines go through the infant's skull? If not, and if death from each source is equally prevented by the presence of a male, the antipredation hypothesis for gorilla females grouping with a male is difficult to distinguish from the anti-infanticide hypothesis. Indeed, twenty years after Wrangham explicitly stated that the data to distinguish the hypotheses are not available

(Wrangham 1986), we go a stage further and suggest that for gorillas (as opposed to gibbons [van Schaik and Dunbar 1990] and Thomas langur [Sterck et al. 2005]), nobody has yet shown that they can be distinguished, whether we have the data or not (sec. 7.3) (Harcourt and Greenberg 2001). Nevertheless, as said, the anti-infanticide hypothesis for gorilla female-male association appears to hold sway.

To redress the balance, we will spend more time than we might have done otherwise arguing the case for the antipredator hypothesis for female gorillas associating with males. In pulling the antipredation hypothesis's foot from the grave, we are not arguing that the anti-infanticide hypothesis is wrong. Quite the opposite (sec. 7.2). Rather, we suggest that the antipredation hypothesis for female-male association in gorilla populations should not be buried yet. It should not be buried until predictions that separate it from the anti-infanticide hypothesis are produced, or predation can be shown to be an irrelevant danger for gorilla females by comparison to infanticide.

There is good reason to think that protection from predation might be a particularly important influence on gorilla society, however important protection from infanticide might be in explaining other primate species' societies. That good reason concerns the gorilla's large body size. The gorilla, especially the male, is by far the largest-bodied primate. The gorilla male is, as we have quite often said, twice as large as a female gorilla, which itself is twice as large as females of the next largest species (the other great apes), which themselves are more than twice as large as females of the next largest species (baboon). The gorilla's large body size has been taken to mean that the gorilla is more or less immune from predation (Palombit 2000; Wrangham 1979).

On the contrary, however, the large size means, first, that the gorilla is largely confined to the ground—and terrestrial primates are assumed by many to be more at risk from predation than are arboreal primates (e.g., Nunn and van Schaik 2000). Second, large body size correlates with standing and fighting, not fleeing, as an antipredator strategy (ch. 2.3.2). That being the case, predation could play an unusually large part in gorilla society. This not to say that gorillas are preyed upon more than are other primates. They are surely not. The point is the relative effect of predation as against other influences on the nature of gorilla society.

Most arguments about antipredation benefits of being with others concern benefits of being with several others, such as better vigilance, safety in numbers, confusion of the predator, and communal mobbing (ch. 2.3.3.1). The larger the group, the more these benefits accrue, although with some provisos concerning membership of the group (Krause and Ruxton 2002, ch. 6).

We are making a different argument, although a subset of the communal mobbing hypothesis. The female gorilla benefits from being with a male that will mob on her behalf. She is not joining a group, especially not joining a group of females. She is not joining a male merely for safety in numbers. Her goal is to join an actively protective male. Of course, if several females join the same male, each then must also benefit from the normal antipredator benefits of grouping.

7.1.2. Females join males

The argument is that gorilla females join a male for protection against predators, infanticidal males, or both, and are found in groups with other females because several females make the same choice of male. If there are other benefits of being in a group, they are secondary benefits to the prime one of association with a powerful protector. What is the evidence that gorilla females seek association with a male?

1. Emigrating females in both eastern and west-central Africa almost always immediately join a male, even a lone male (ch. 4.2.1.2). They do not travel on their own, or join other females in the absence of a male.
2. Even if two females emigrate together, they almost always immediately join a male, even a lone male, and do not wander on their own (ch. 4.2.1.2).
3. On the death of the male of a one-male group, females disperse and go their separate ways to join other males (ch. 4.2.1.2).
4. After the dispersal, even females with young infants join another male, despite the fact that the association will probably result in infanticide by the new male (ch. 4.2.1.2; ch. 4.2.1.3; sec. 7.2). In other words, the benefits to a female of association with a male are greater than the costs of loss of a current infant.
5. Unhabituated females spend more time near the male in the absence of their favorite female partner than they do near their favorite female partner in the absence of the male, and are obviously responsible for maintaining proximity with him (Harcourt 1979b). By contrast, females completely habituated to the observer do not obviously distinguish the silverback from their favorite female, and are as content to let the silverback keep near them as to try to stay near the silverback (Harcourt and Stewart, unpublished; Sicotte 1994; Watts 1992, 1994a).
6. Females within large groups seem to compete to be near the silverback (ch. 4.2.2.2) (Watts 1994b), whereas competition to be near other females, even kin, has apparently not been seen.

7. While we do not have data to show it, everyone who has studied gorillas has remarked on the way in which females assiduously follow the timing, direction, and speed of the males' movements; he far more rarely follows theirs—see the opening quote of ch. 6, taken from George Schaller, doyen of gorillaologists.

Evidence against the hypothesis that females join males rather than groups is the persistence for several months after the death of the leading male of all-female groups in the Kahuzi-Biega population of eastern gorillas (Yamagiwa and Kahekwa 2001).

A general test for whether indeed females are joining a male or joining a group might at first glance be data on the size of the group that transferring females join. Joining a large group could indicate a preference for groups over males; joining a lone male or a small group would indicate a preference for males. However, the two hypotheses cannot be separated this way. A general preference for a powerful male would lead to powerful males having large groups; and females joining a small group could result from avoidance of competition in large groups, despite a preference for large groups (ch. 2.2.3.1).

In sum, then, the bulk of the evidence is that females join a male, not a group. Of course, if other females also join the male, their presence will provide all the other benefits of grouping—more eyes, safety in numbers, and confusion—which is an explanation for the all-female groups in the Kahuzi-Biega population (Yamagiwa and Kahekwa 2001).

7.1.3. Gorillas are preyed upon

What is the evidence for predation on gorillas, the largest living primate? After all, one argument for the evolution of large body size is that it reduces the variety and number of predators that can kill the organism in question (ch. 2.3.2), and as we have said (sec. 7.1.1), the argument has been made that the gorilla is so large that it is safe from predation.

With respect to gorillas specifically, part of the argument for why infanticide rather than predation explains female gorillas associating with males is that gorillas are too large for predators to be a serious threat. However, large as they are, especially the male, gorillas are chased by leopards, and are occasionally caught and killed by them in both eastern and west-central Africa (Fay et al. 1995; Goldsmith 1999; Watson 2000). Thus, Schaller recounts a resting, maybe sleeping, mountain gorilla male killed by a leopard in the night in the Virunga Volcanoes (Schaller 1963, Ch, 7, p. 303). And, of course, gorillas are hunted by humans. We see no reason why they would not have been hunted by humans for millennia, and there is some evidence that they might have been: 5,000-year-old gorilla (and chimpanzee) bones have been found in a cave in Cameroon, a perhaps unlikely place for the animals to have entered voluntarily (Tutin and Oslisly 1995), and for at least five times that long, people have lived in Africa's forests (Mercador 2002).

In the twentieth century, the "gentle giant" reputation replaced Du Chaillu's "fierce giant" reputation (box 7.1) as a result of the work of Carl Akeley, George Schaller, and Dian Fossey—in the mostly unhunted Virunga Volcano region of Rwanda, Uganda, and Zaire (Akeley 1923; Fossey 1983; Schaller 1963). However, go to where gorillas are hunted in west-central Africa, and the first reaction of a gorilla frightened at close quarters by a potential hunter can be to attack. We have spoken with African hunters who describe coming unawares upon a silverback, and being attacked without provocation (excepting, of course, the provocation of being previously hunted). These are not mock attacks. The gorillas bite hard with very large canines: we have seen in southeast Nigeria the severe wounds that they cause.

Not only are gorillas, large as they are, caught and occasionally killed by predators, but they are caught sufficiently often, and have been caught over a sufficiently long time, for the species to have evolved antipredator strategies. One is the males' roaring, screaming charge (Schaller 1963, ch. 7). Better than waiting until the predator appears, though, is to avoid it to start with. Terrestrial primates spend the night in places safe from predators, often up trees (ch. 2.3.1). So do gorillas. And the smaller gorillas, the more vulnerable ones, nest higher up trees than do the larger ones (fig. 7.1).

Additionally, all gorillas act as if they perceive trees as being safe. Their main response when up a tree is either to ignore the observer, or be curious, whereas on the ground they are usually frightened (fig. 7.2).

Finally, what other explanation is there for the existence of the all-male groups in both eastern and west-central Africa, which all have at least one silverback (ch. 4.2.1.1; table 4.2) than that gorillas do not escape predation, despite their size, and that grouping is advantageous to them, especially grouping with a large male (ch. 2.3.3)? The explanation is not that a takeover of a breeding group is easier if males cooperate, for takeovers appear never to occur (ch. 4.2.1.2; ch. 11.1.3.2).

7.1.4. Males protect females

The argument that female gorillas join a male for his protection against predation needs evidence that he so protects, and that the females treat him as a protector (see van Schaik and Dunbar 1990). Some of the evidence applies

Box 7.1. Paul Belloni Du Chaillu, 1831–1903

The name of Du Chaillu is indelibly associated with the hunting of gorillas, and with lurid descriptions of the ferocity and power of hunted male gorillas. Du Chaillu's father was a trader in west-central Africa and, as is often the case then and now, volunteered or was chosen as a more or less prominent member of the local business community to be his country's consul. The son as a young man explored equatorial west-central Africa, in what is now Gabon, though it is just possible that the first gorilla he killed was in the very southeast corner of Equatorial Guinea. He became famous, even notorious, through several rather sensational accounts of his travels and his game hunts (Du Chaillu 1861). He can be blamed for the fearsome image of a hunted gorilla: ". . . I knew that we were about to pit ourselves against an animal which even the tiger of these mountain fears . . . " (p. 85), ". . . hellish expression of face, which seemed to me like some nightmare vision" (p. 98), ". . . his powerful fangs were shown as he again sent forth a thunderous roar . . . truly he reminded me of nothing but some hellish dream creature. . . " (p. 101). And so Du Chaillu shot the gorilla.

His books, though, are full of admiration for Africa and the Africans, full of ethnological and anthropological detail. Oddly enough, the other part of the world he traveled through later in life was Scandinavia, where his accounts of the people, their customs (different skis for different skiing conditions), the land, and the wildlife followed much the format of his African accounts, although with hunting less of a theme. Perhaps Du Chaillu's fame was helped by the fact that he converted accounts of both travels into books for children.

to protection against only predators; some that the male protects against danger, which could include infanticidal males.

7.1.4.1. The male protects females (and their offspring) against predators

The silverback is not only twice a female's body size, but has canines to match (ch. 3). Anyone approaching a group of unhabituated gorillas will quickly realize the male's readiness and ability to threaten and attack predators (fig. 7.3A–D). He is particularly likely to threaten, even attack, under conditions of increased danger for his group, in other words, when animals are on the ground in the presence of a strange human who surprises them at close quarters (fig. 7.4A, D).

If these threats or attacks deter the predator, then the male protects the females. Observers have certainly been deterred from approaching groups by the presence of aggressive males. The silverback's roaring, screaming

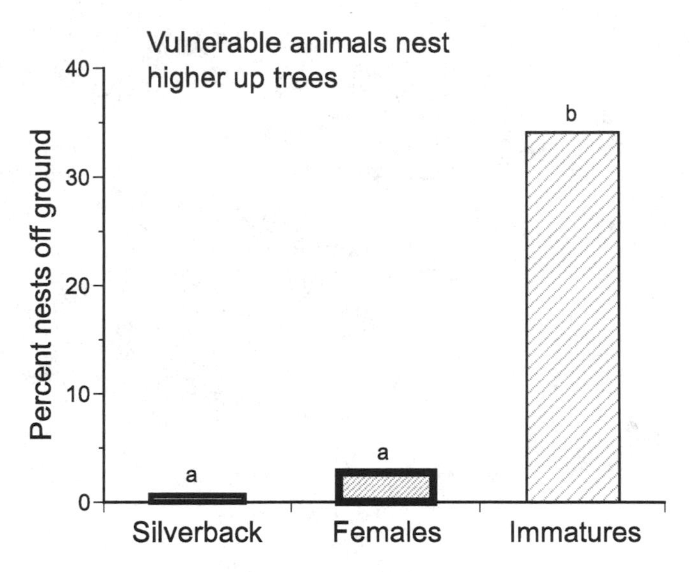

Fig. 7.1. Smaller, more vulnerable gorillas (immature animals) are more likely to nest in trees. (Eastern lowland gorilla.) *Details at end of chapter.* Source: Data from Yamagiwa (2000).

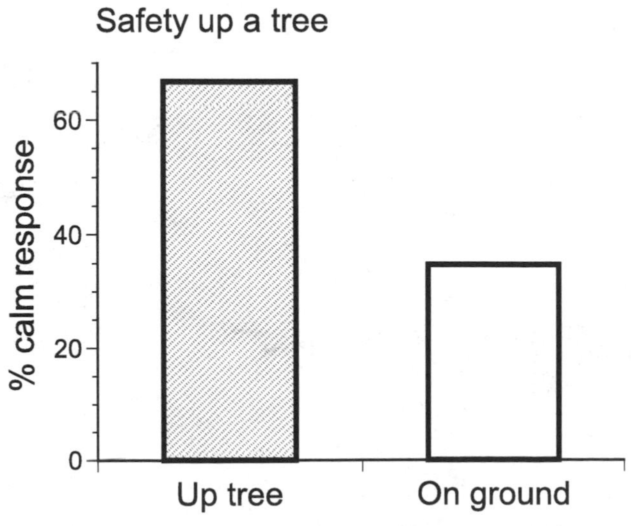

Fig. 7.2. Gorillas are calmer (feel safer?) up trees. (Western lowland gorilla.) *Details at end of chapter.* Source: Data from Blom et al. (2004).

Fig. 7.3A. Silverbacks are the main defenders. Two silverbacks interpose themselves in a threatening stance (strut with compressed lips) between the observer and the rest of the group. (Mountain gorilla.) *Details at end of chapter.* © A. H. Harcourt.

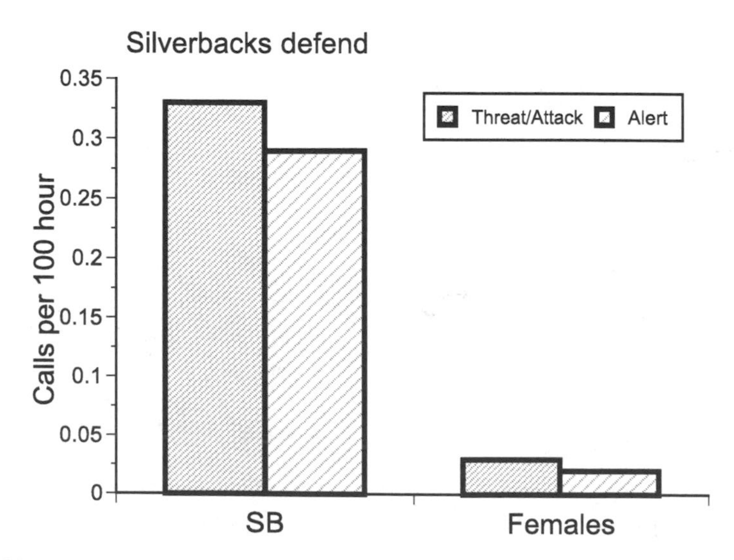

Fig. 7.3B. Silverbacks are the main defenders against potential predators. Data are average vocal responses of different sorts per individual to observers by unhabituated animals. (Mountain gorilla.) *Details at end of chapter.* Source: Data from Fossey (1972).

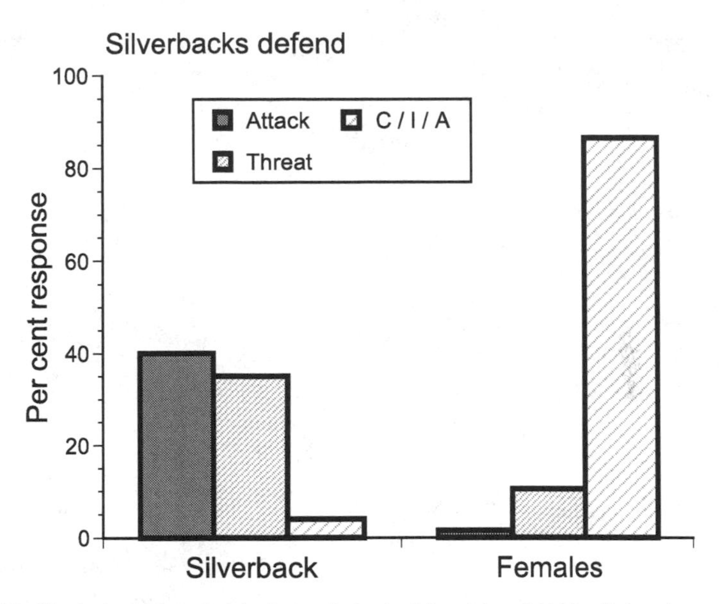

Fig. 7.3C. Silverbacks are the main defenders against potential predators. C / I / A = Curious, Ignore, Avoid. (Western lowland gorilla.) *Details at end of chapter.* Source: Data from Tutin and Fernandez (1990).

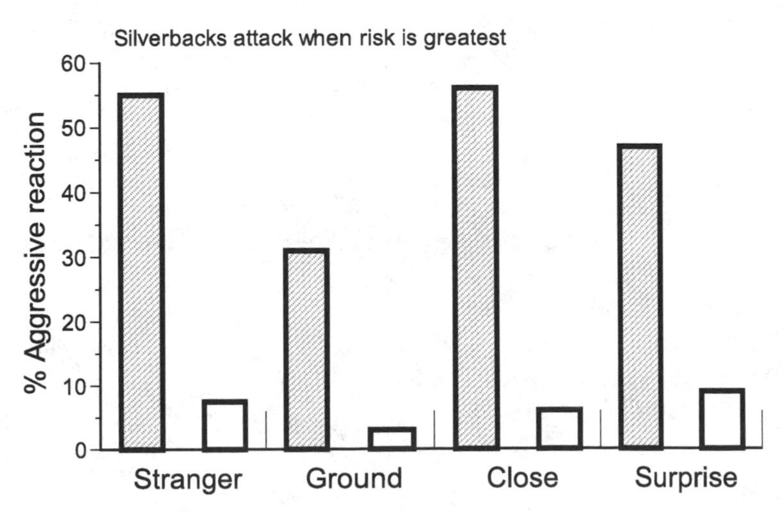

Fig. 7.3D. Silverbacks defend most against potential predators in the riskiest conditions. The data are groups' aggressive reactions (mostly silverbacks') on detecting observers– in relation to estimates of risk posed by the observers: observers strange to gorillas (shaded bar) or familiar (open bar); gorillas on ground or not; observers close to gorillas or not; gorillas surprised by observers or not. (Western lowland gorilla.) *Details at end of chapter.* Source: Data from Blom et al. (2004).

charge—specific to predators, and not used against other males—is very frightening. A group of gorillas in the Volcanoes National Park of Rwanda was never visited by observers or tourists, except by mistake, so famously aggressive was its silverback, the aptly named Brutus.

7.1.4.2. Females (and immature animals) act as if they perceive that the male is a protector against predators

Females act as if they perceive that the male is a protector. In moments of danger (either from a predator or another male), the females' frequently move close to the male (see ch. frontispiece, p. 186). Indeed, might it be that the silverback's silver back evolved precisely to make him more visible to his females, setting him apart from other animals in the group, and making it easier for females to find him in moments of danger?

While we have found no published quantitative evidence for this movement toward the male in times of danger, we can present quantitative evidence that vulnerable females spend more time near the male. The observations do not separate protection against predators from protection against infanticide. They simply illustrate that females act as if the male is a protector. For instance, Virunga mountain gorilla females that have just given birth sometimes spend more time near the male than they did previously, and more time than do other females (ch. 4.2.2.1). If the female loses her infant, then she moves away from the male again (Harcourt 1979a). In general, females with infants spend more time near the male than do females without; and among females with young infants, if they differ in time near the male, then those with the youngest infants spend the most time near him (ch. 4.2.2.1). All of these behaviors indicate to us that females perceive the male as a protector, and concomitantly, that his protection explains these aspects of male-female behavior.

These findings came from one-male groups of females without adult kin. Once more than one male was present, or adult kin were present, it was not so obvious that a change in vulnerability coincided with a change in proximity to the male (ch. 4.2.2.1). However, we can think of several reasons for the difference. For instance, the multi-male Virunga groups formed after many years of close study: could it be that the gorillas in the study groups learned that they were safe in the presence of an observer, and so did not go near the male while observations were being made (cf. Isbell and Young 1993)?

When females emigrate, they in effect always do so only when a male that they can join is nearby (ch. 4.2.1.2). In the one case that we saw of a female leaving a male without another anywhere near, she was clearly frightened. Previously completely habituated to us, she acted terrified on detecting us.

She left a silverback to which she had transferred three weeks previously when the group to which she was moving, her former group, was perhaps 2 km away. Perhaps she made a mistake as to how close her former group was. We followed her for the next two days, as she desperately searched for her group, rarely eating, seemingly tracking their trail as we were doing hers (Fossey 1983, ch. 12; Harcourt and Stewart, unpublished).

It is at night that many terrestrial primates most obviously choose sites safe from predation, spending the night in tall trees or on cliff sides. For gorillas, the presence of an adult male makes an obvious difference to where other group members spend the night. Juichi Yamagiwa (2000; also Yamagiwa and Kahekwa 2001) very nicely showed that when the leading male of a group died, the females and immature animals were twice as likely to nest up trees as when he was alive (fig. 7.4). And most importantly for the argument, when a new male joined the group, in other words, when the females once more had a protector, the females returned to near their previous frequency of sleeping up trees (fig. 7.4). The females clearly acted as if more afraid when a silverback was not there, and safer when one was present. If height of sleeping up a tree is indicative of fear, the immature animals continued fearful after the arrival of the new male.

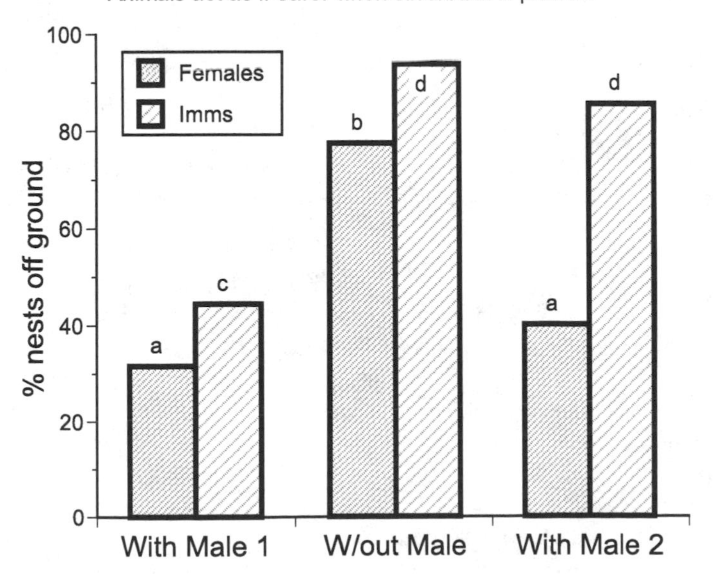

Fig. 7.4. Gorilla females and immatures act as if they are safer when a silverback male is present. (Eastern lowland gorilla.) *Details at end of chapter.* Source: Data from Yamagiwa (2000).

7.1.5. Comparison with *Pan* and *Pongo*

Chimpanzees (including bonobos) and orangutan females are smaller bodied than are gorilla females (ch. 1.3). They could presumably benefit even more from being in a group. Quite often they are, especially the bonobo females (ch. 4.2.3.1). However, often they are not, especially female orangutans and eastern chimpanzees (ch. 4.2.3.2). What do these contrasts tell us about gorillas and our argument for why gorilla females group with males?

Chimpanzees are certainly preyed upon. In Christophe and Hedwige Boesch's study site in Taï Forest in Côte d'Ivoire, leopards injured six chimpanzees and killed four over five years (Boesch 1991). Lions probably killed and certainly ate at least four chimpanzees in the Mahale National Park of western Tanzania (Tsukahara 1993).

Furthermore, chimpanzees of both sexes obviously recognize and react violently toward large cats, especially leopards, which are the commonest large predator in chimpanzee habitat. A frequent reaction of chimpanzees faced with a leopard is to mob it, sometimes flailing at it with sticks (Boesch 1991; Goodall 1986, ch. 18, p. 555; Hiraiwa-Hasegawa et al. 1986; Kortlandt 1962). With attack as part of the chimpanzees' form of defense, the potential exists for grouping to be beneficial as an antipredator tactic or strategy of individuals in this species, and therefore a component of the society (Hiraiwa-Hasegawa et al. 1986).

Nevertheless, preferential grouping with a male might not be the best solution for a female common or bonobo chimpanzee. In the first place, common and bonobo chimpanzee females, being half the size of gorilla females, can more easily flee into trees, an antipredator strategy that does not come with the competitive cost of association with a male. Also, a *Pan* male is less than 30% larger than a female (ch. 3.2.2; fig. 3.3B), and so presumably not nearly as efficacious a protector against the same predator as is the far larger gorilla male (ch. 3.2.2; fig. 3.3A). Certainly, the unhabituated common chimpanzee male's first reaction to observers is very different from the gorilla's: the common chimpanzee male's usual reaction is flight or retreat, as it is the females' (fig. 7.5, compare with fig. 7.3A–C).

The situation is different with orangutans. The male could protect the female. He is twice her size (ch. 3.2.2), extremely powerful, and ready to attack when attacked, as the very first picture in Alfred Wallace's marvelous *The Malay Archipelago* shows (Wallace 1869). And yet female orangutans do not associate with males (ch. 4.2.3.2). Our argument for gorillas seems to be in trouble. However, in so large and so arboreal an animal as the orangutan, protection from predation might not be important to females, either by grouping with a male or by grouping with each other (Dunbar 1988,

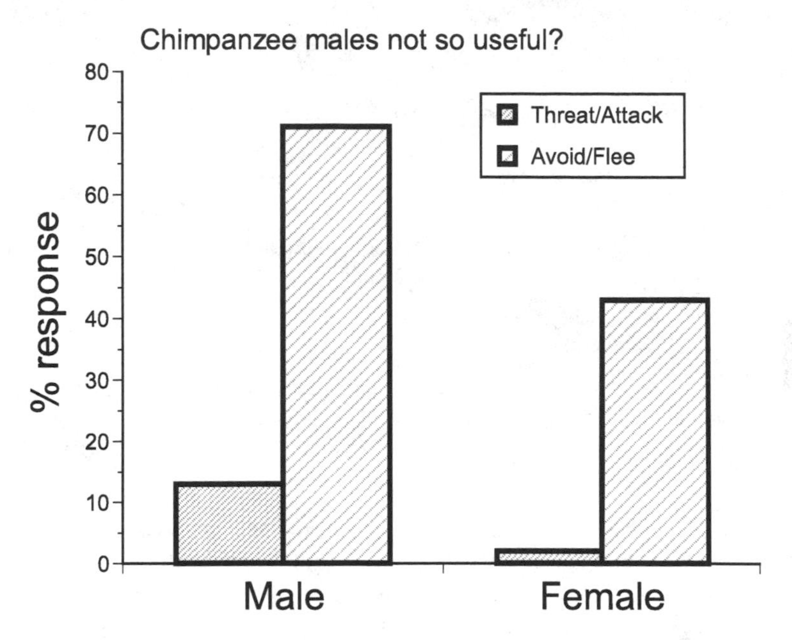

Fig. 7.5. Chimpanzee males often flee in response to danger. (Eastern chimpanzee.) *Details at end of chapter.* Source: Data from Tutin and Fernandez (1990).

ch. 13, p. 320; Janson 2003). While orangutans are preyed upon on the ground (Rijksen 1978, ch. 8), they can easily escape up trees, where no arboreal predator or raptor could take an adult. Humans with weapons can take orangutans up trees: Wallace killed fifteen of them (Wallace 1869, ch. 4). However, unless male orangutans were prepared to come to the ground to attack hunting humans, it is not clear that his protection would be beneficial.

Nevertheless, predation is very final. Any protection is better than none, other things being equal. But the fact is that even if it were advantageous for female orangutans to travel in a group or with a male, the sparse distribution of their food in poor periods, and their slow arboreal travel, especially of males, surely make it impossible (ch. 4.1.8). Chimpanzees too seem to be prevented from association by the sparse distribution of food in poor periods (ch. 4.1.8).

Whether *Pan* and *Pongo* solitariness is in fact due to more feeding competition than gorillas face, or to lower risk of predation (because *Pan* and *Pongo* can more easily escape up trees), we cannot know without measures of risk of predation and escape behavior and mechanics. We are sure, however, that relative lack of predation on larger primates cannot explain *Pan* and *Pongo* solitariness; if it could, the gorilla would surely also be solitary (Wrangham 1979).

Conclusion: Escaping Predation as the Cause of Females' Association with a Male

We have argued so far in this chapter that a fair amount of evidence indicates that gorilla females associate with a male in order to benefit from his protection. Predation, we argue, causes females to not only join a male, but to remain with him—as long as he continues to demonstrate his prowess. Here might be the place to repeat that we consider that contest competition between gorilla males for mating access to females is what led evolutionarily to the gorilla male's large size (ch. 2.4.1); the ability of the male to protect females from predators is secondary. The claim might be more an article of faith than a tested proposition, but the high degree of sexual dimorphism in the orangutan, a species in which males compete for mates, but do not protect females, supports the argument.

Animals of all sorts can reduce their risk of being eaten by grouping with others (Ch 2.3.3). For many primate species, it appears that females associate with males (as opposed to simply grouping with others) for protection against infanticide. However, we suggest that the gorilla, because of its unusually large size, might be unusually susceptible to predation, and unusually likely to attack as a means of defense against predators. Therefore, it might be a species whose society is unusually likely to be one in which females associate with males for protection.

Of course, evidence to support the antipredator hypothesis for female-male association is not the same as evidence to show that the antipredator hypothesis is better than another. Is the antipredation hypothesis better than the anti-infanticide hypothesis for explaining why female gorillas associate with a male? We will consider this question after we have presented the evidence that females group with a male for protection against infanticide.

7.2. Protection from Infanticide

> Twenty-seven days after Rafiki's [the father's] death, eleven-month old Thor was killed during a violent physical interaction between the two groups. Uncle Bert [leading silverback of another group] bit the infant fatally in the skull and groin, both typical infanticide wounds causing almost instant death. **DIAN FOSSEY (1983, Ch. 7, p. 150)**

Summary: Protection from Infanticide

Gorilla males will kill infants that they have not fathered. It appears that females that associate with a powerful male are protected from infanticide. But is protection from infanticidal males the primary reason why females associate with a male?

We can answer the question by comparing infanticide rates suffered by a female that associates with a male with the rates suffered by a female that adopts the alternative great ape strategy. That strategy is the one adopted by female chimpanzees, namely, roaming effectively alone and mating promiscuously with a large enough proportion of the males within her community that she is unlikely once with an infant to meet a male with which she has not mated, and that will therefore be infanticidal. To make the comparison, we have to quantitatively model the roaming, promiscuous strategy, because lone female gorillas do not exist in reality.

The model's answer is that the anti-infanticide hypothesis for females associating with males could be correct: the virtual, roaming, promiscuous gorilla female cannot under most scenarios mate with a sufficient number of males in the population to reduce subsequent infanticide to levels below those seen in reality. However, some gorilla populations are at high enough density and the females in those populations travel far enough each day for them to adopt the lone, promiscuous strategy. And yet they never do.

7.2.1. Infanticide in gorillas

The first time we saw infanticide in gorillas, in 1973, it took us a long time to realize that a gorilla, a leaf-eating primate, had killed the infant. We found the infant's body on the ground, with what later turned out to be the classic signs of infanticide by gorillas, serious wounds to the head and, in this case, especially the abdomen (fig. 7.6). Our initial reaction was that the gorilla group had been attacked by poachers or feral dogs. The vegetation was flattened, blood-bespattered, diarrhea bespattered. Eventually, we (by which we mean here everyone at the Karisoke Research Center then (Dian Fossey, us, field assistants) worked out from the trail evidence that a lone male had done the deed (more details in Fossey [1984]). Infanticide was far from our mind, not only because we did not know of it before in gorillas but also because this and other observations from gorillas were some of the earliest to indicate the wide taxonomic and geographic extent of the phenomenon, and hence to substantiate its likely adaptive basis (ch. 2.4.3; box 2.4; ch. 11.1.2).

Infanticide occurs in probably all populations of gorillas (ch. 4.2.1.3). It seems likely, therefore, that gorilla females could associate with a male for protection against infanticidal, non-father males (ch. 2.4.2.4). That being the case, we then see groups of females with a male, the standard harem social system of gorillas, because several females choose the same protective male, as we have argued for protection from predation as the cause of the association.

Fig. 7.6. One of the Karisoke Research Center's earliest cases of infanticide. (Mountain gorilla.) © A. H. Harcourt

7.2.2. Female gorillas associate with a powerful male for protection against infanticide by other males

The anti-infanticide hypothesis now seems to be almost embedded in the literature as the single reason for female-male association in gorilla populations: "Female need for male protection against infanticide probably led to male-female association in its present form" (Watts 1996, p. 23); ". . . the mountain gorilla is probably the best-documented example of a primate that has evolved male-female bonds as an anti-infanticide strategy . . . " (Palombit 1999, p. 118); ". . . the increasingly compelling explanation for groups of gorillas is that females benefit from grouping around a single male, limiting infanticide by other males . . . " (Rodman 1999, p. 322); "Infanticide has long been considered the adaptive reason for group living in polygynous mountain gorillas" (Palombit 2000, p. 252).

The anti-infanticide hypothesis is sometimes argued with little attempt to distinguish predictions of the anti-infanticide hypothesis from those of the antipredation hypothesis (sec. 7.1), in part because the anti-infanticide hypothesis is so well supported in so many species (ch. 2.4.2.4). In fact, protection from predators is difficult to distinguish from protection from infanticidal males as a reason for females to associate with a male because, as we said when discussing predation, what is an infanticidal male the equivalent of, but a mild predator (Treves 2000; van Schaik 1996; Wrangham 1986)?

The argument that female gorillas associate with a male for protection against infanticidal non-father males requires several lines of evidence, as did the alternative hypothesis that female gorillas associate for protection against predation (sec. 7.1). We need evidence that females seek association with a male or males. We need to show that males do protect females from infanticide. And then we need to show that females join males for that protection. It is not sufficient to show that females that associate are protected, because they might have joined the male for other reasons (protection from predation), and only subsequently or secondarily benefit from protection from infanticide.

We have already presented evidence that females seek association with a male or males (sec. 7.1.2). The two other steps remain.

7.2.2.1. Males protect against infanticide

Males will fight vigorously to keep other males away from their group (ch. 11.1). They are surely both guarding their own access to females when they do this, but also equally surely, protecting their own infants against infanticide.

How do we know that males are efficacious protectors? One answer is that in gorilla society, infanticide almost always occurs when the protective male dies (fig. 7.7). If his absence leads to infanticide, his presence is protective. In these cases, the infanticidal male is usually a strange male to which the females transfer after the death of the male with which they have bred.

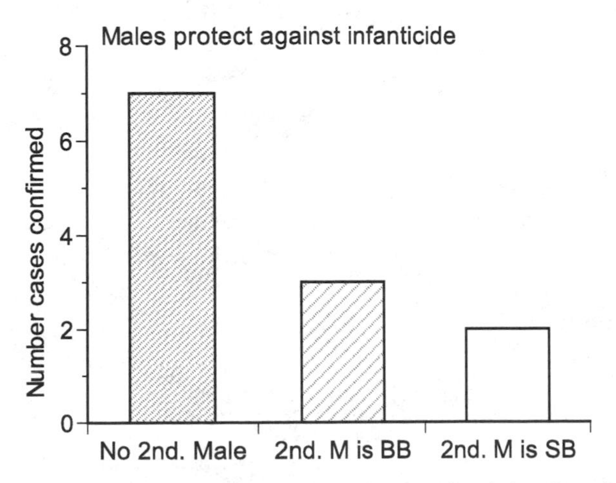

Fig. 7.7. Males protect against infanticide. Infanticide mostly occurs when the breeding male dies, and no other male is present (No 2nd. Male). 2nd. M is BB = only other male is a blackback; 2nd. M is SB = another mature male (silverback) is present. (Mountain gorilla.) *Details at end of chapter.* Source: Data from Watts (1989a).

On one instructive occasion, however, the infanticidal male was one that had earlier immigrated into the group as a young blackback male (similar to the "follower" strategy of hamadryas and gelada baboons [Dunbar 1984; Kummer 1968; Sigg et al. 1982]), and that had not mated with the female whose infant he killed (Fossey 1984, Case 2).

Furthermore, in cases where a protective male dies but another is in the group to take his place (and therefore females do not need to emigrate, and therefore do not have to associate with a non-father male), infanticides are rarer, especially if the second male is a silverback (fig. 7.7). However, as the incident described in the previous paragraph illustrates, it looks as if males are protective only if they are related to the other males in the group, or have mated with the females. We will turn next to within-group promiscuity by gorilla females (sec. 7.2.2.2).

7.2.2.2. Females associate with a male for protection against infanticide

If infanticide occurs, and association with a powerful male prevents it, what more evidence is needed that females join males for protection against infanticide? After all, this was how we argued that females associate with males for protection against predation. The argument is not proven, however. For instance, species exist in which females do not permanently associate with males, and yet infanticide is seemingly no more common than in one-male gorilla groups. The gorilla's closest ape relative, the chimpanzee, seems to be such a one (ch. 4.2.3.1) (Arcadi and Wrangham 1999; Hamai et al. 1992; Nishida et al. 2003).

Also, females that associate for protection against infanticide could suffer the cost of increased risk of infanticide. This perhaps anomalous consequence occurs because once a female associates with a particular male all other males in the population are then free to act as if any infant of that female is not theirs—and kill it. Compare that with the opposite strategy, the chimpanzee female's strategy, of ranging without males and mating promiscuously. Any male with which the female mated might be the father, and it should pay them to act accordingly. By comparison, the faithful, monogamous strategy changes a population of N potential fathers, potential non-infanticidal males (under the roaming, promiscuous strategy) into a population of $N - 1$ infanticidal males. This cost of association does not exist for protection from predation as the benefit of association.

Of course, costs of the polyandry exist, such as aggression, disruption of feeding, and transmission of disease (Matsubara and Sprague 2004; Matsumoto-Oda 2002; Nunn, Gittleman, and Antonovics 2000; Wrangham 2002). However, evidence that this $N - 1$ amplification effect of association on infanticide rates could work is on the face of it already available.

Infanticide appears to be more prevalent among one-male than multi-male primate species (Palombit et al. 2000), and to be more prevalent in one-male than multi-male populations within species. The savanna baboon of Africa is an example (Collins, Busse, and Goodall 1984; Palombit 2003; Palombit et al. 2000). Part of the reason for the effect is surely that with more males in a group, there are more protectors. But could another part (among others also) be that there are fewer infanticidal attempts because multi-male mating by the females has reduced the number of local infanticidal males?

Therefore, to answer whether female gorillas associate with a male for protection against infanticide, we have to compare that strategy with its opposite, the roaming, promiscuous alternative. In the first strategy, $N - 1$ males are infanticidal. In the second, mated males are converted into non-infanticidal potential fathers, with the consequence that the number of infanticidal males is potentially far smaller.

This roaming, polyandrous strategy should not be impossible for gorilla females. A close relative, the chimpanzee, uses the strategy (ch. 11.4; sec. 7.2.3), implying that gorillas could, given that closely related species have similar behaviors. Also, breeding gorilla females sometimes leave their current male to join and mate with a new male (ch. 4.2.1.2; ch. 8.1), and so can be serially promiscuous. Moreover, within groups, females will mate with more than one male in the group (Harcourt et al. 1981b; Robbins 1999; Watts 1990b), that is, they are polyandrous, perhaps indeed adopting the anti-infanticide multiple-mating tactic (ch. 11.1.5). Gorilla females could be chimpanzee females. Why are they not?

To answer that question we need to compare the rates of infanticide experienced by gorilla females that faithfully associate with a male to the rates experienced by lone, polyandrous females. But how, given that lone, polyandrous gorilla females do not exist? The answer is that we have to create them in a computer. We have to model them. Does the virtual lone, polyandrous gorilla suffer more or less infanticide than either the virtual or real gorilla that associates with a male?

We need to be very clear here that we are asking about the causes of females joining a male. Clearly, once she has joined him, both hypotheses, the anti-infanticide hypothesis and the antipredation hypothesis, would have her remain with him.

7.2.2.3. Females associate with a male for protection against infanticide: A model

The model. The quantitative model compares how often a female that is nursing an infant meets either a male with which she has mated (and which is consequently a non-infanticidal potential father), or a male with which she has not previously mated (and which is consequently infanticidal). The relative frequency is the risk of infanticide—low if she has managed to mate with most of the males, high if she has not mated with many of the males she meets. Note that this model is different from Broom and colleagues' findings (2004). They investigated infanticide by immigrant males, which is the normal context for infanticide in primates, but not the normal one for gorillas.

Our model builds on the work of Peter Waser (1976), who used the James Maxwell gas equation for frequency of encounters between moving molecules to ask questions about the frequency of expected encounters between primate groups. The model assumes that the chance of two bodies meeting (be they particles or primates) is determined by the density of the bodies in space and their speed of movement. Collisions are more likely between many fast-moving bodies than between a few slow-moving ones. Because we are dealing with primates, a few additional factors are added than Maxwell used, such as the distance at which bodies can detect one another.

And in the case of the question we are asking here, a few more factors are added too (Harcourt and Greenberg 2001). One is the *duration of estrus* (longer estrus means a greater proportion of males mated by the female and hence made non-infanticidal) (Harcourt and Greenberg 2001). For simplicity's sake, we assume in the model that any male with which the female has mated will not commit infanticide; any unmated male will do so when met. *Duration of nursing* is in the model also, on the assumption that the longer the duration of nursing, the more chance there is that a female will meet an infanticidal male.

The model's answers. Results of the model are in figure 7.8, and the details behind it are in the details for figure 7.8 at the end of the chapter. The model indicates that if a Virunga gorilla female traveled on her own, she would experience a 58% risk of infanticide, instead of the observed 14% of infants dying from infanticide in groups with one male (fig. 7.8[a]). Over half her infants would be killed by one of the many males with which she had not managed to mate in her brief estrous period (Harcourt and Greenberg 2001). We did not run simulations with variation of values in the formula to test the difference statistically, but its magnitude suggests that it should pay a female to choose a protective male and stay with him. In sum, the anti-infanticide hypothesis for female-male association in gorilla society looks as if it is correct.

That answer is for the Virunga mountain gorillas only, though. What about other gorilla populations? The model indicates that over quite a wide range of values for the parameters in the model, a lone, promiscuous gorilla female would face unacceptably high infanticide risk (fig. 7.8[b]). For instance, with densities and distances traveled per day twice the African average, infanticide risk is 30%, over twice the observed high rate for one-male groups in the Virungas. It seems that the anti-infanticide hypothesis could explain most gorilla populations.

The fact that the model produced a rate of infanticide of females that associate with a male that was near the observed value (Mod-M compared to w/ M in fig. 7.8[a]) substantiates the validity of the model. We can test its validity further. The model is not specific to gorillas. Any species' data can be fed into the model. Among Gombe chimpanzees and orangutans, rates of infanticide appear to be lower than in the Virunga gorilla population, indeed zero so far in orangutan studies (Arcadi and Wrangham 1999; Delgado and van Schaik 2000; van Schaik and van Hooff 1996). If values for orangutans and chimpanzees are entered in the model, the results again match reality; in this case the reality of very low rates on infanticide in these two species (fig. 7.8[c]).

Thus, far from association with a male increasing a female's risk of infanticide by converting all other males in the population into infanticidal males, as we had thought, a female seems to benefit from the start by the association. Without association, the chances of infanticide are so high that association is the way to go. Infanticide has the same effect, then, as predation is argued to have. They both cause females to associate with a protective male.

However, some gorilla populations not only exist at densities of more than four times the average but also travel at least twice as far per day as do the Virunga gorillas (ch. 4.1.4; 4.1.7). The model indicates that females in such populations could travel on their own with low risks of infanticide (fig. 7.8[a]). If so, and given that in such populations all females group with a male, we now have contradictions to the anti-infanticide hypothesis for male-female association in gorillas. We will return to them.

In the meantime, what other factors than density and daily distance traveled might influence whether females can use the lone, promiscuous strategy for minimizing infanticide, or whether they need to group with a protective male? We can answer this question by altering one at a time the values for the parameters from the gorilla values to the chimpanzee or orangutan values. If changing a gorilla value to a chimpanzee or orangutan value reduces the risk of infanticide to the low chimpanzee and orangutan levels, then we can argue that the tested parameter is crucial in explaining the high rate of infanticide in gorilla populations and the low rate in chimpanzee and orangutan populations.

It turns out that no single chimpanzee value (i.e., daily distance traveled or duration of estrus) reduces infanticide risk to the observed very low level (fig. 7.8[d]). However, combine the chimpanzee values for the two parameters that most obviously differentiate the gorilla and chimpanzee (daily distance traveled [2 km for the chimp, instead of 1 km], and duration of estrus [more than a week per month for the chimpanzee compared to the gorilla's 2–3 days]), and infanticide risk of the virtual lone gorilla

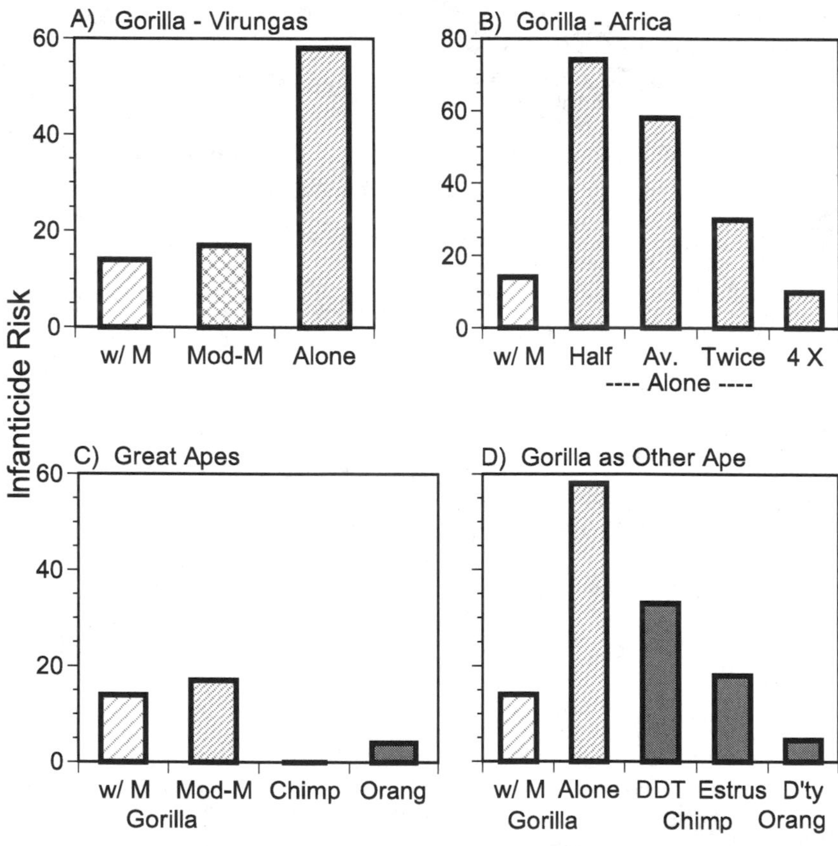

Fig. 7.8. Infanticide scenarios as determined by a quantitative computer model (Harcourt and Greenberg 2001). *Details at end of chapter.* (a) Females who travel without males and mate promiscuously (Alone) experience higher rates of infanticide than do females who travel with males, whether virtually in the model (Mod-M), or in fact in the Virunga population of mountain gorillas (w/ M). (b) In most populations in Africa, if females traveled on their own and mated promiscuously, they would (according to the model) experience higher rates of infanticide than observed in the Virunga population (w/ M). Half, Av., Twice, 4 × = values for density and day journey length varied within realistic range of half to four times the average African population (Av.). (c) The model gives realistic answers, not only for the gorilla (Mod-M in part [a]) but also for the other great apes, the chimpanzee and orangutan, in which observed infanticide rates in the wild are nearly zero. (d) If roaming gorilla females traveled as far each day as do chimpanzee females, or had as long an estrous period as they did (DDT, Estrus, Chimp), or lived at the density of orangutan females (D'ty, Orang), they would meet more males while estrous, and have lower infanticide rates.

drops to 4.5%, very close to the real chimpanzee value of about 5% (Arcadi and Wrangham 1999). In other words, gorillas do not have a long enough estrous period, nor travel far enough each day to use the roaming, promiscuous anti-infanticide strategy.

If the value for the orangutan's high average density is put into a model, with gorilla values for all the other parameters, the result is a realistic near-zero rate of infanticide (fig. 7.8[d]). Otherwise, the orangutan values are very similar to the gorilla's (fig. 7.8 details at end of chapter), and thus do not change the results. The implication is that in gorilla populations at high density, female gorillas should be able to use the lone, promiscuous strategy. Nevertheless, not all orangutan populations are at such high density, and the model predicts infanticide rates of over 10% for lone orangutan females at population densities of 2/km^2 and below, well within the densities recorded for some areas (Rijksen et al. 1995; Rodman and Mitani 1987; van Schaik et al. 1995). We thus predict that infanticide will be recorded in these low-density orangutan populations, unless one or other of the sexes of orangutans there are traveling unusually long distances when females are in estrus and mating, or the nursing females are able to avoid being detected or caught by males.

All the great apes have roughly the same duration of nursing/weaning periods of several years, so the length of that period does not explain contrasts in infanticide rate among them. Indeed, the model indicates that after a time (three months) that is short relative to the duration of nursing (more than a year), the female with an infant encounters all males in the population (Harcourt and Greenberg 2001). The absence of an effect of nursing duration was initially a surprising finding to us, because Carola Borries and Andreas Koenig (2000) got, at first glance, the opposite finding. However, among other differences between our analyses, they were asking about frequency of infanticide in a population, not as here, per female.

7.2.2.4. The model's conclusions

In sum, then, the model largely supports the anti-infanticide hypothesis for females associating with a male. We have tested the results in a variety of ways, and they all produce the same answer. However, finding evidence to fit a hypothesis is only the first stage in testing of a hypothesis. Just because the hypothesis works does not mean that it is the best hypothesis. Does the anti-infanticide hypothesis work better than the antipredation hypothesis is the question that has to be answered (sec. 7.3).

We might add that the model is, we think, the first quantitative demonstration with a model of the long-standing verbal argument that by being promiscuous, females can decrease male aggression against their offspring

(Hrdy 1979), and furthermore that they can do so both by working hard to find males (long daily travel distance) and by being receptive over a long time (long estrus) (Hrdy 1979; van Noordwijk and van Schaik 2000).

7.2.2.5. Criticism of the model

Our use of a model to investigate whether female gorillas might join a male for protection against predation has been criticized. As the criticism illustrates some fundamental misunderstanding of the use of models, we will present it and our response as a means of discussing the use of models. (For more discussion of the aims and use of models, see box 10.1.)

One criticism of the model is that lone ranging is not an option for females because of predation. We entirely agree that predation will prevent lone ranging. But the fact that one influence prevents a behavior (in this case, lone ranging) is no argument to stop an attempt to see if another influence on its own might also prevent it. Also, chimpanzee females sometimes roam on their own despite predation, so why could not gorillas? If chimpanzees can do it, so can gorillas. And finally, it is not always going to be the case that predators are present. An important use of models is to predict what will happen when conditions change. Predators are likely to be the first to go when habitat is fragmented, because they need more land than does their prey (see Brashares, Arcese, and Sam 2001; McNab 1963; Terborgh et al. 2001). The model allows us to predict the nature of gorilla society in the absence of predators, information that could be very useful in managing conservation of the species (ch. 14.2.4; 14.2.7).

A second criticism of the model is that if gorilla females roamed alone, a male would gain nothing by killing the infant because he would not reliably meet the female again by the time she is ready to conceive. That is probably true, although its veracity would need testing. However, in fact, in such a situation, the infanticidal male would presumably remain with the female until she came into estrus again, maybe even coercing her to stay with him, rather than killing the infant and then leaving it to chance whether he met her again. This kill-and-stay-to-next-estrus tactic is definitely a detail that needs to be incorporated in future models.

Finally, it has been suggested that the fact that males can give long calls that allow estrous females to approach the males—and nursing females to avoid the males—makes the gas model inappropriate. We do not agree. Part of the reason for the 500 m detection distance that we used (by comparison to Dunbar's latest models, which had detection distances of only 50 m [Dunbar 2000]) was the males' long calls. Calling does not make the model inappropriate: one simply modifies the parameters in the model. In reality, we wonder how hidden a nursing female could afford to remain. Could she really afford to, for instance, avoid good food sources—where there are likely to be males?

Of course, the model is by no means complete. Several other behaviors, constraints, and payoffs need to be considered to test its robustness (sec. 7.3). It remains to be seen whether their incorporation will negate or strengthen the anti-infanticide hypothesis for male-female association. And some statistical testing of the observed and modeled results needs to be done.

7.2.3. Comparison with *Pan* and *Pongo*

The chimpanzee and orangutan have already been compared in detail in the main part of the argument, because their differences from the gorilla in the nature of their mating system, combined with low infanticide rates, made them a perfect test of the model, and a perfect means to explore why the gorilla did not, or could not, use the roaming, promiscuous strategy to escape infanticide. Chimpanzee males, for the most part, commit infanticide in the standard situation of males meeting females that are carrying infants not fathered by the males, a situation that occurs when females are at the edge of the community range, or newly immigrated into a community (Hamai et al. 1992; Hiraiwa-Hasegawa and Hasegawa 1994; Nishida et al. 2003). Tellingly, these are females that have not yet had the opportunity to mate promiscuously with most of the community's males. Nevertheless, in the Mahale Mountain community of Tanzania, males have killed infants that could be their offspring (Hamai et al. 1992). The suggestion offered is that the male might be using infanticide to force the female to mate solely with him, in other words, as a coercive mating strategy (Hamai et al. 1992). Infanticide is then equivalent to a demonstration of power, a function that our model was not designed to test.

An obvious difference between chimpanzee and gorilla infanticide, however, is that cannibalism after infanticide appears to be common in chimpanzees (Arcadi and Wrangham 1999; Hamai et al. 1992; Hiraiwa-Hasegawa and Hasegawa 1994; Nishida et al. 2003). At one time people thought that chimpanzees killed more male than female infants, but it turns out that the bias is spurious. It seems that the infants of young, immigrant mothers are preferentially killed (their infants were probably fathered by other males), and young mothers are more likely to give birth to sons than to daughters (Nishida et al. 2003).

Conclusion: Escaping Infanticide as the Cause of Females' Association with a Male

Observational evidence indicates that males protect females from infanticide; quantitative modeling evidence indicates that females that joined a powerful, protective male would be at less risk of infanticide than those that ranged alone. We ran the quantitative model to investigate infanticide rates suffered by lone, promiscuous females, in large part because we thought we had detected a fatal flaw in the anti-infanticide hypothesis. Association seemed, at first glance, to necessarily increase, not decrease, the risk of infanticide. The reason was that as soon as a female associated with one male, every other male in the population could act infanticidal. As it turned out, the virtual, lone, promiscuous female gorilla suffers unacceptably high rates of infanticide in most realistic populations.

At the same time, in some populations, densities are high enough and daily distances traveled long enough for the lone ranging, promiscuous anti-infanticide strategy to work (according to the model). Indeed, with only slight increases in the duration of estrus and length of daily distance traveled, females in the Virunga population could, according to the model, travel safely on their own (Harcourt and Greenberg 2001). They effectively never do.

As we were writing this book, the gorilla population of Kahuzi-Biega National Park in DRC was an anomaly: no infanticide had been seen, and groups of females existed without males for months at a time (Yamagiwa and Kahekwa 2001). One explanation was that neighboring males were relatives, and therefore non-infanticidal (Yamagiwa and Kahekwa 2001). Alternatively, the lack of infanticide and the all-female groups indicated that something was wrong with hypotheses about selective advantages to males of infanticide, and with the anti-infanticide hypothesis for females associating with a male. In a previous draft, we wrote, "However, our opinion is that if there are no infanticidal males in a study site, then it is a temporary situation, and the females must be acting as if infanticide is a risk. Gorillas too completely fit the profile of a species in which infanticide would be a beneficial competitive strategy for males for there to be no risk of infanticide."

In the last few years, a culling "experiment" in Kahuzi-Biega was performed that would have broken up any neighborhoods of male relatives—lawless soldiers and rebels slaughtered a tragically large proportion of the population (Yamagiwa 1999b). And now infanticide has been recorded in the Kahuzi-Biega gorilla population (Yamagiwa and Kahekwa 2004).

High density of males inhibited infanticide in the model. Zebras and hamadryas and gelada baboons all congregate in large, dense herds. Yet females of those species associate in harems with a powerful male or two. Why are the females not promiscuous? One answer could be that there are so many males in total in the herds that the females could never mate with a great enough proportion of the males to prevent meeting after she had given birth a male with which she had not mated. The model could tell us quite easily. Alternatively, it seems likely that females would suffer active or passive harassment during estrus as a result of the close proximity of so many males if the females did not associate with a protective male (Clutton-Brock and Parker 1995; Matsubara and Sprague 2004; Smuts and Smuts 1993; Wrangham 2002).

7.3. Conclusion: Predation or Infanticide?

As we have repeatedly said, we suspect that escaping infanticide (or reducing its likelihood) as a reason for gorilla females to associate with males is going to be difficult to separate from escaping predation: the infant is dead, whether it was a male gorilla's canine or a leopard's canine that pierced its skull (see Treves 2000; van Schaik 1996; van Schaik and Dunbar 1990). Also, given that males keep away both predators and competing males, protection from either is protection from the other: whether escape from infanticide or escape from predation causes a female to join a male, she escapes from both once she has joined him. How do we separate the two as the cause of the initial joining?

So far we have presented arguments in favor of each hypothesis separately, as if the other did not exist. That is how most arguments for the anti-infanticide hypothesis have so far been presented. The first step in testing of a new hypothesis is simply to see if the data fit the hypothesis. If they do not, there is no point in continuing. Once the hypothesis has survived that first step, the next step is to see if the new hypothesis is better than alternative hypotheses. That is what we now do here.

7.3.1. Predation is the primary cause of association

1. The quantitative model for infanticide (sec. 7.2) showed that certain populations of gorillas are at high enough density, with animals traveling far enough each day, for females to range on their own and yet suffer only low rates of infanticide. Given the feeding costs of association with a male (ch. 6.1), females should range on their own if they could. If infanticide is not keeping them with a male, what else other than predation is? In other words, predation is the main cause of female gorillas associating with males.

However, the gorilla populations at very high density could be at a temporary demographic peak. If so, it would be evolutionarily disadvantageous for the females to respond to such a temporary situation by ranging alone. Females that responded to perception of a high density of males by traveling alone would not leave more descendants than females that retained the evolutionary predisposition to range with a male. Evidence for such a population peak is the crash in numbers caused by disease that is now apparently occurring in the high-density western gorilla populations (Huijbregts et al. 2003; Walsh et al. 2003).

2. Danger from infanticide cannot explain why males gather into all-male groups, almost always with a fully adult male in them. And yet they do (Elliott 1976; Fossey 1983; Gatti et al. 2003; Levréro et al. 2006). The counter to this argument is that just because males, which cannot suffer from infanticide, react to predation by grouping with a powerful male does not mean that protection from infanticide is not a more important reason for females to associate with a male.
3. Females group with males in apparently non-infanticidal or nearly non-infanticidal populations (Stokes et al. 2003; Yamagiwa and Kahekwa 2001). However, this observation suggests only that predation might be one cause of association; it does not negate infanticide as another cause or the main cause (see no. 2 in this list). In the first place, infanticide rates might be low precisely because the association is working. Secondly, the lack of infanticide might be an anomaly (sec. 7.2 Conclusion).
4. van Schaik and Dunbar (1990) suggested that the antipredation hypothesis would predict that all females would move to a new male on loss of the current male, whereas the anti-infanticide hypothesis predicts that only females without vulnerable young infants would do so. The former is the case in gorillas (ch. 4.2.1.2). However, our infanticide model for grouping shows that it would pay any female to join a male, even if she did not have an infant (sec. 7.2).
5. The strongest argument for predation as the main cause of grouping, rather than infanticide as the main cause, could be that a predator that takes a female will end her reproductive career forever, whereas infanticide merely pauses the career (Harcourt and Greenberg 2001). Infants are easily replaceable. One year sees the replacement of the lost infant (females give birth about a year after losing an infant); and most infants are killed at less than two years of age (ch. 11.1), so making three years the time needed to replace the lost infant. Females will surely respond more strongly to situations that reduce

the likelihood of permanent cessation of reproduction—that is, predation—than they will to situations that merely delay it—that is, infanticide.

7.3.2. Infanticide is the primary cause of association

van Schaik and Peter Kappeler (1997) have argued that escaping infanticide explains better than does escaping predation the variation across primates (and other mammals) in female-male association. However, one of the more detailed quantitative analyses of the competing hypotheses indicates that once strict statistical control of confounding factors is used (especially phylogenetic relatedness of species, and therefore non-independence of data), infanticide does not explain better than does predation the differences between species in a number of aspects of the structure of primate societies, such as the number of males or females in a group (Nunn and van Schaik 2000).

In which case, why does the only quantitative model of the consequences for infanticide of association compared to lone travel produce a result so strongly in favor of the anti-infanticide hypothesis (Harcourt and Greenberg 2001)? One answer is that the model takes no account whatsoever of any other payoffs than infanticide (Harcourt and Greenberg 2001).

For example, it does not consider the feeding costs to females of associating with a male (ch. 2.2.3; ch. 6.1) (Janson and Goldsmith 1995; Matsubara and Sprague 2004; Williams et al. 2002). Accounting for that cost would reduce the benefit of association. On the other hand, the model does not account for the multiple costs of promiscuity, for example, the simple, immediate ones of time and energy in travel searching for males, or the longer-term cost involved if a substandard male achieved fertilization. Added to these must be the cost of increased likelihood of disease transmission (Nunn et al. 2000).

Also, there will be a variety of costs associated with harassment by males in a polyandrous society, including aggression from males (Clutton-Brock and Parker 1995; Matsubara and Sprague 2004; Matsumoto-Oda 2002; Smuts and Smuts 1993; Wrangham 2002). In fact, infanticide is merely an extreme form of harassment. If all harassment by males were counted, and association decreased all forms of harassment, and maybe induced the opposite, positive care (ch. 2.4.4.1), association could be even more advantageous than the model indicates. Alternatively, a male could harass the female to coerce association (Clutton-Brock and Parker 1995; Matsubara and Sprague 2004; Matsumoto-Oda 2002; Smuts and Smuts 1993; Wrangham 2002).

Adding yet more parameters to a model begins to obviate the main advantage of a model—simplification of the real situation. However, all the forms of male harassment could be subsumed in the infanticide model with a single parameter of decrease in reproductive output, as opposed to loss of infants by infanticide. A problem could be that realistic valuation might require immense amounts of data.

7.3.3. The hypotheses cannot yet be separated?

The two hypotheses for why female gorillas associate with a male cannot yet be distinguished, we suggest, although both have been around for a long time (Harcourt and Greenberg 2001; Stewart and Harcourt 1987; Wrangham 1986). Quantitative modeling of predictions of relative effects of predation and infanticide are needed, or more complete accounting of costs and benefits of faithful monogyny versus polyandry. For instance, one solution for separating the two hypotheses would be to quantitatively model the relative costs to females of infanticide (minor but frequent) and predation (devastating but rare) in order to see whether the observed low rate of predation on gorillas is anywhere near frequent enough by comparison to rates of infanticide for predation to be a useful hypothesis to explain why females associate with males. Some immediate complications would need to be included. As just one example, given that an infant can be predated independently of the mother, predation rates of each would have to be separate in the model. And the model would need a parameter that incorporated age of the female, so that total future offspring lost when she was taken by a predator could be calculated. Nevertheless, the modeling is possible.

Alternatively, we need a clean qualitative prediction that separates the two possibilities. van Schaik and Dunbar (1990) suggested some means of distinguishing them, but the ones that work are mainly applicable only to the gibbons that they were talking about. For instance, they suggested that the anti-infanticide hypothesis would predict high levels of vigilance only when young infants were in the group—but young infants will usually be present in the average gorilla group.

Elizabeth Sterck and her colleagues (2005) argued that infanticide, not predation, determines the choices of dispersing Thomas langur females. The reason was that while the females did not join groups of more females (as expected if group size protected from predation), but joined younger, presumably more powerful males (as expected if males protected against infanticide). However, female Thomas langur might be using the male, rather than the group, as the means of protection from predators (see Wich and Sterck 2003).

Additionally, not only can group size and male quality be confounded, as suggested in section 7.1.2, but a small group with a young, powerful male might be as good an antipredator choice as a large group that has an old male. Sterck and her colleagues (2005) showed that emigrating females prefer young males, which if they are starting their breeding career are presumably likely to have smaller groups than males just past their prime.

Finally, Nunn and van Schaik (2000) suggested that ecology (food and predation) better explain differences between species in the nature of their societies, while social factors (including infanticide) better explain differences within species. If so, primatologists will need to think carefully about whether their predictions are better tested with comparisons across species, or comparisons within species.

Figure Details

Fig. 7.1. Age-sex class differences in tree nesting. Data for secondary vegetation (gorillas' most-used habitat) from > 100 nests per age-sex class. Letters indicate statistical differences: same letter = no difference; different letter = significantly different. b: $p < 0.001$, $N = 166, 1253, 652$ nests per age-sex class. Note that the data are, strictly, not independent, for they involve multiple counts of the same individuals. (Kahuzi-Biega National Park, DRC.)

Fig. 7.2. Gorillas calmer up trees. Percent of responses to observers that are calm or ignore when gorillas are relatively safe (up tree) compared to unsafe (on ground). Comparison based on 874 ground encounters, 233 tree encounters. Problems of independence (sometimes same gorilla group encountered more than once a day, for instance) make statistical analysis invalid. (Dzanga-Ndoki National Park, CAR.)

Fig. 7.3A. Silverbacks in aggressive stance. The picture is of the extremely well habituated Group 5 of the Karisoke Research Center, which nevertheless was always nervous when crossing open spaces, as here. (Virunga Volcanoes National Park, DRC, Rwanda, Uganda.)

Fig. 7.3B. Rate of responses to presence of observer by silverback and females. Records from 4–14 calling silverbacks per nature of call, 1–8 females. No statistics possible. Threat/attack calls are roar, scream, "wraagh"; Alert calls are barks. (Virunga Volcanoes National Park, DRC, Rwanda, Uganda.)

Fig. 7.3C. Percent silverback, female responses that were attack (roar and charge), threat (roar), or curiosity, ignore, or avoid. $N = 80$ responses by silverbacks, 67 by females. No statistics possible (Lopé Reserve, Gabon.)

Fig. 7.3D. Reactions to observers. *Stranger:* median of monthly counts of aggression toward observers on encountering gorillas ("threatening 'wraagh' vocalization without movement to a charge to direct physical contact"), comparing first four months (Aug.–Nov. 1998) and last four months (Sep.–Dec. 1999) of habituation ($p < 0.05$, $U = 0$, Mann-Whitney U Test). It was not until the seventh month of habituation that the aggression rate dropped below any of the last four months. *Ground:* comparison based on 874 ground encounters, 233 tree encounters. *Close:* comparison based on 109 encounters at < 10 m, 149 at > 30 m. *Surprise:* gorillas surprised by observer ($N = 470$), observers made selves known to gorillas ($N = 640$). Problems of independence (sometimes same gorilla group encountered more than once a day, for instance) make statistical analysis of these last three comparisons difficult. (Dzanga-Ndoki National Park, CAR.)

Fig. 7.4. Percent group members' nests on ground in relation to presence or absence of silverback. Letters indicate statistical differences (same letter = no difference; different letter = significant difference): females: b = $p < 0.02$; immatures: d = $p < 0.04$. $N = 7, 14, 5$ night-nests counts, respectively, for With Male 1, i.e., first male; W/out Male; With Male 2, i.e., after the new male joined the group. The statistical tests involve multiple counts of the same individuals. (Kahuzi Biega National Park, DRC.)

Fig. 7.5. Percent male, female chimpanzee responses that were threat/attacks (loud calls and charge) or avoid/flee (avoid is stealthier than flee). $N = 55$ responses by males, 60 by females. No statistics possible. (Lopé Reserve, Gabon.)

Fig. 7.7. Silverbacks protect against infanticide. Binomial test, $p < 0.05$. (Virunga Volcanoes, Rwanda, DRC). Infanticides known or strongly suspected from Watts (1989a, table 1).

Fig. 7.8. Infanticide rates modeled. *The model.* The equation for the model giving the infanticide risk (*IR*) looks horribly complicated. However, it boils down to a comparison of, on the one hand, the proportion of total males with which the female mated (and hence which were "safe," that is, non-infanticidal [the left-hand side of the equation, with c in it]) with, on the other hand, the proportion of total males with which the female met when she was nursing (the right-hand side, with n), all non-mated ones of which would be infanticidal:

$$IR = [(M_t - M_g) / M_t]^2 \left(e^{-(v_{c,m}^2 + v_{c,f}^2)^{1/2} a_c [(M_t - M_g) / H] c}\right) \left(1 - e^{-(v_{n,m}^2 + v_{n,f}^2)^{1/2} a_n [(M_t - M_g) / H] n}\right).$$

The parameters in the model are density of males (***M*/*H***); days available for mating with males per conception (***c***, i.e., duration of estrous period multiplied by number of estrous periods per conception); speed of travel per sex ($\boldsymbol{v_m}$, $\boldsymbol{v_f}$) (distance traveled per day) entered separately for conception (***c***) and nursing (***n***) periods; detection distance (***a***) entered separately for conception and nursing periods (which might be influenced by, for instance, estrous swellings, or silent ranging by nursing females); and duration of nursing (***n***), that is, the period for which the infant is susceptible.

This equation states that the risk of an infant being attacked by an extragroup male is the product of the proportion of the extragroup males that the female mates with during her conception period $[(M_t - M_g) / M_t](e^{-(v_{c,m}^2 + v_{c,f}^2)^{1/2} a_c [(M_t - M_g)/H] c})$, and the proportion of the extragroup males that are met during nursing $[(M_t - M_g) / M_t](1 - e^{-(v_{n,m}^2 + v_{n,f}^2)^{1/2} a_n [(M_t - M_g)/H] n})$. The first set of males will be non-infanticidal (because the female mated with them); the second set of males contains both non-infanticidal males and those infanticidal ones with which the female did not previously mate when potentially fertile. The relative proportions of the two sorts of males are described by ***e***.

In detail:

$\boldsymbol{M_t}$ is total number of males in the population; $\boldsymbol{M_g}$ is number males in a group with the female. In the case of lone females, there are no intragroup males, and therefore the expression $M_t - M_g$ reduces to M_t, and the expression $[(M_t - M_g) / M_t]$ thus reduces to 1. This number of males needs to be converted to a density, which is done by dividing by the size of the female's home range, **H**.

The variable ***e*** is a decay function, which gives the proportion of infanticidal males remaining in the population after mated males become safely non-infanticidal. Further, *e* is affected by the rate at which infanticidal males are converted to non-infanticidal males—that is, what everything after the *e* within each of the main brackets refers to. That rate is determined by the following:

The variable ***v*** is the velocity at which females ($\boldsymbol{v_f}$) and males ($\boldsymbol{v_m}$) travel (i.e., their daily travel distance). The faster they travel, that is, the greater the distance they cover, the greater their chances of running into one another. The details of those chances (e.g., that v is squared $[v_m^2, v_f^2]$) are based on an equation that described the chances of molecules of gas hitting one another. Waser was one of the first to apply the gas equation to primates, in his case modeling the chances of groups of primates running into one another (Waser 1982; Waser 1976). Others then developed it further, or used it with different goals, such as to calculate the chances of males of a variety species finding estrous females, or whether groupings of individuals were random or due to deliberate association (Barnes 1982; Barrett and Lowen 1998; Dunbar 1988, 2000; van Schaik and Dunbar 1990).

The subscripts ***c*** and ***n*** in the section of the equation affecting *e* refer to the conception and nursing periods, respectively. In the conception period *c*, encountered males become non-infanticidal through copulation. In the nursing period *n*, encountered males that have not already mated with the female commit infanticide.

Finally, ***a*** is the distance at which animals can detect one another.

We examined each parameter's effects on infanticide probability using Microsoft Excel 4.0 Names function. The Excel formula is

```
= ((M-G)^2*(EXP(-((vf^2+vm^2)^0.5)*ac*c*((M-G)/H)))*(1-EXP(-
   (((vf^2+vm^2)^0.5)*an*n*((M-G)/H))))))/M^2),
```

where M = the total number of males in the population, v_f and v_m are the male and female velocities, and a_c and a_n are the detection distances in the copulation and nursing periods.

Model parameters and their values. Virunga Gor = values for Virunga gorilla population; Africa Gor = values for mean African gorilla population; Chimp = values from Gombe Stream National Park and Mahale National Park populations; Orang = mean values from several sites. Details concerning the values entered and the sources for those values are in the original paper that described and ran the model (Harcourt and Greenberg 2001).

Parameter (symbol), measure	Virunga Gor	Africa Gor	Chimp	Orang
Male density (*D*), km ^{2}	0.25	0.125	0.25	1
Annual home range (*H*), km ^{2}	7.5	17.5	30	5
Time estrous per conception (*c*), days	6.25	6.25	17.25	6.25
Average speed (*v*), km/day	0.5	1.1	2	0.75
Detection distance (*a*), km	0.5	0.5	0.5	0.5
Time to weaning (*n*), days	1,095	1,095	1,095	1,095

A female—looking for a male? (Mountain gorilla.) © A. H. Harcourt

CHAPTER 8

Female Strategies: Male Influences; Emigration and Choice of Males

Summary

Emigration by female gorillas, and especially their immediate transfer to another male, make gorilla society unusual among group-living mammals. The nature and distribution of food explains only why females *can* emigrate and travel with a new male, but not why they do so. Mating and rearing strategies, not feeding strategies, explain females' emigration.

Breeding females that leave one male for another do so, the circumstances suggest, to find a more powerful protector.

Females born into a group are forced to emigrate to avoid inbreeding if their father is successful enough to be still alive and still the main breeding male when they mature. In other words, the male's mating strategy of long-term monopoly of a group of females constrains the females' mating strategies.

However, costs of inbreeding are only relative. If either costs of emigration or benefits of staying outweigh the costs of inbreeding, daughters should stay. Thus, when more than one adult male is in a group, that is, when females can benefit from the extra protection afforded by more males and from the possibility of mating with a male other than their father, some females stay to have their first offspring in the group before then emigrating. In so doing they benefit also from bearing their first offspring in familiar surroundings, even with friendly kin still in the group, and they might benefit from delaying emigration until they are more mature.

Introduction

Most gorilla females eventually leave the group in which they were born, and some leave a group in which they have bred, either successfully or unsuccessfully (ch. 4.2.1.2). When they emigrate, they transfer immediately to another male (ch. 4.2.1.2). Gorilla females are, effectively, never on their own. We have two main questions. Why do females emigrate? And what determines the choice of male that they join? While we have followed the classic socioecological framework of females-to-food, males-to-females, and presented females as competing over mainly food and males as competing over mainly mates, we continuously emphasize the two-way influence of the sexes on one another. Males certainly do compete over mates, but the females' choice of mates, the fact that females exercise choice of mates, is a major component of gorilla society. That choice is perhaps nowhere so obviously apparent as when females emigrate and transfer (Harcourt 1978a; Sicotte 1993, 2001).

A vital background to the question of why females emigrate must be the fact that emigration imposes costs (ch. 2.2.3.2; ch. 2.4.2.6). The gorilla is no exception, even though competition for food appears to be relatively unimportant in this species (ch. 5.1.2). For instance, resident gorilla females are more aggressive to immigrants than they are to one another, and especially more aggressive to immigrants than they are to residents to which they are related (fig. 8.1; see also fig. 5.3). Yet gorilla females leave. And when they leave, they join a male. Gorilla society seems not to fit the females-to-food, males-to-females framework (ch. 2.1.2).

It is unlikely that competition forces the females out (ch. 5.1). Another explanation is needed. The explanations differ depending on whether one is talking about the emigration of breeding females or of natal females (i.e., those born into the group) (Harcourt 1978a; see also Packer 1979; Pusey and Packer 1987; Sicotte 2001).

8.1. Female Emigration and Mate Choice: Finding a Better Protector

When a leading silverback of a one-male group dies, the group disintegrates, and the breeding and prebreeding females almost always disperse to join other males (ch. 4.2.1.2). This forced dispersal is an extreme of emigration to find a better protector (ch. 7)—going from no protector to any protector.

But breeding female gorillas also will voluntarily leave one male for another, some of them several times in their life (ch. 4.2.1.2). Why? It looks as if they are seeking a better mate, a better protector (Sicotte 2001; Wrangham 1979). This explanation for their emigration and transfer is most obvi-

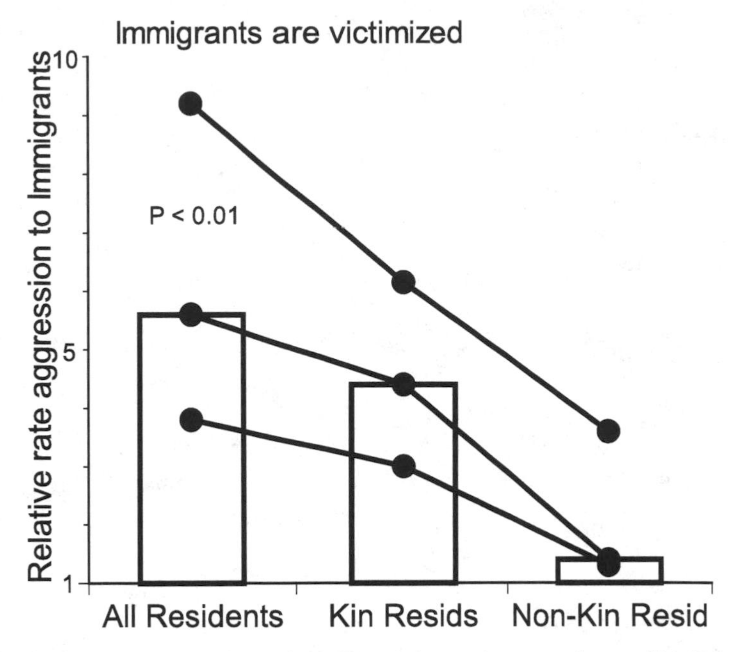

Fig. 8.1. Immigrants are victimized by residents. Data are the rate of aggression (corrected for time in proximity) directed by residents at immigrants compared to the rate residents direct at all other residents, to kin residents and to non-kin residents, that is, residents aggress immigrants over five times more often than they do other residents. Data are per group; median shown by bar. (Mountain gorilla.) *Details at end of chapter.* Source: Data from Watts (1994a).

ous when they leave after the killing of an infant by an invading non-father male (Fossey 1984; Watts 1989a). In at least two cases, the bereft mother joined the male that had killed her offspring, presumably a powerful male as indicated by the fact that he had defeated the father's attempts to stop the killing.

Other stimuli for the breeding females to emigrate might be more subtle. Do the females leave old males for younger males, even without the immediate stimulus of an infanticide (Wrangham 1979), as do Thomas langur females (Sterck et al. 2005)? Given that a common result of death of a resident male is infanticide by the male that the mothers join, it might be advantageous for females to leave an old male for a young, prime one once her current offspring are independent. Certainly, some female gorillas have emigrated, leaving behind weaned offspring. It might also be the case that if females find themselves with a male that is ranging in a region of poor-quality food (or some other aspect of poor quality, such as dangerously close to human habitation), she would leave him (Harcourt 1978a).

However, we suspect that power of the male might be the most important trait on which breeding females base their decisions to leave or stay. In addition to using loss of an infant, or age of the males to judge relative power, females could judge males' power by the intensity and duration of their threat displays (Sicotte 2001), the hooting and chest-beating displays in the case of the gorilla (ch. 4.2.1.4; ch. 11.1.3.1). These must be directed as much at females in a female-emigrant society as at rival males. We know of no good evidence from gorillas for this idea, and Sterck and colleagues (2005) could not identify traits of Thomas langur male loud-calling that correlated with choice. Nevertheless, we are sure that females must discriminate among males using their displays.

Finally, if power of a male is the crucial influence, what better than one male as a protector, but two or more males? One problem is that if one male can impose costs on females (ch. 6.1), two males presumably impose even more costs. Indeed the costs might be exacerbated if aggression between the males themselves results in aggression being redirected at the females, and longer distances traveled on days of aggression between the group males (Sicotte 1995). Nevertheless, in the Virunga Volcano region of the mountain gorilla's range, where multi-male groups are common (ch. 4.2.1.1), emigrating females appear to actively prefer multi-male groups (fig. 8.2).

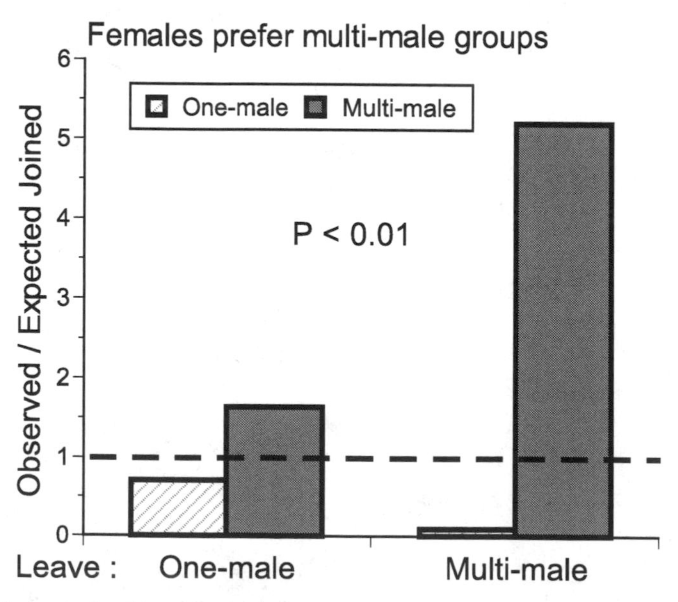

Fig. 8.2. Emigrating females prefer multi-male groups. The x-axis shows the number of males in the group *from* which the females emigrated ($N = 23$ one-male, 14 multi-male); the differently shaded bars show the number of males in the groups *into* which the females immigrated. The expected ratio of one-male versus multi-male groups joined is 1. (Mountain gorilla.) *Details at end of chapter.* Source: Data from Watts (2000a).

If females prefer multi-male groups, and multi-male groups are better protected, such groups should have more adult females (and immatures) than do single-male groups. They do not have more adult females (ch. 4.2.1.1). Given the obvious preference of Virunga females for multi-male groups, the fact that such groups do not have more females than do single-male groups is a puzzle. Perhaps a female's best choice is a multi-male group with few females, because she then gets the best of both worlds, both good protection as well as little competition (Sicotte 2001; Wrangham 1979)—except that emigrating females appear to pay no attention to the size of the group that they join (ch. 5.1.2).

However, any test of transferring females' preferences needs to be done with only females that have already left their natal group. That is because natal animals might pay no attention to the quality of the group that they join, because the goal is merely to leave the natal group (see Packer 1979). If so, natal emigration would hide any discrimination that emigrant breeding females might be exercising. None of the published analyses has yet distinguished preferences of breeding and natal female gorillas.

While multi-male Virunga mountain gorilla groups do not contain more adult females than do single-male groups, they do contain more weaned animals (ch. 4.2.1.1) (unlike Bwindi mountain gorilla groups), implying a benefit of membership in the Virungas of multi-male groups. While better protection should improve survival, some evidence exists also of easier reproductive opportunity, for females in multi-male groups might breed a little over a year earlier than do females in single-male groups, but there are problems with the data (Gerald Steklis and Steklis, in press). In addition to daily improved protection, a main benefit of membership of a multi-male group is the protection afforded at the death of one male by the immediate presence of a second male that is potentially related to the females' offspring (ch. 7.2.2.1; fig. 7.7).

8.2. Female Emigration and Mate Choice: Avoiding Inbreeding

As we have pointed out, emigration by females is unusual in group-living Old World simians (ch. 2.4.2.6). Yet gorilla daughters usually emigrate (ch. 4.2.1.2). What is special about gorillas that leads to habitual emigration by daughters? We have argued that it cannot be food. At most, food allows emigration; it does not cause it. If competition with other females for food were the only determinant of a female's strategies, a female should stay in the demonstrably safe group of her birth (if it were not safe, she would not have survived to be faced with the decision).

And if the male is powerful enough to have brought a daughter to maturity and the age of emigration, it is presumably not that he is a failure as a protector that leads her to leave.

Instead, the long-standing and well-substantiated argument for emigration (by either sex) is that a main benefit is avoidance of inbreeding and the associated potential costs of reduction in survival, mating, or rearing ability of offspring (ch. 2.4.2.6) (box 8.1). Gorilla females are particularly at risk of inbreeding. They usually live in one-male groups (ch. 4.2.1.1), whose resident breeding males are not ousted—takeovers do not occur (ch. 4.2.1.2). Consequently, successful breeding males appear to be able to retain breeding tenure of groups for longer than the time it takes their daughters to mature (ch. 4.2.1.2). That being the case, the nubile female finds her father as the only male with which to breed. To avoid inbreeding, she has to emigrate (Harcourt 1978a; Harcourt et al. 1976; Watts 1996).

8.2.1. Evidence for avoidance of inbreeding

One way to test the idea that avoidance of inbreeding is a cause of emigration by gorilla females would be to see if the health of a female's offspring was worse when she breeds in her birth group with a relative than after emigration and breeding in another group with a non-relative. As some female gorillas do remain to breed in their natal group, the test could be done. All sorts of confounding factors would have to be assessed. Nevertheless, the test is logically a means of proving the inbreeding avoidance hypothesis, and the hypothesis has been substantiated with examination of the consequences of extra-pair copulations in birds (Foerster et al. 2003).

An alternative test is to compare the potential for inbreeding by females that remain with the potential by females that emigrate. Is the potential for females that remain less than it would have been for emigrant females if they had remained? We ask the question of gorilla females, primate taxa, and mammalian taxa.

The most crucial factor that distinguishes gorilla females that remain to have their first offspring in their natal group from those that emigrate before breeding is, as far as we can tell, the presence or not of an adult male besides their father (Watts 1990b). In the Virunga gorilla population, roughly half the groups have one breeding male and half more than one breeding male. When only one male is present, all females emigrate before giving birth (fig. 8.3). Similarly, in west-central Africa, where almost all gorilla groups are one-male, no female has yet been known to breed in her natal group (Stokes et al. 2003). But when any additional male is present that was not the dominant male when the female was an immature, as happens in the Virunga gorilla population, some females remain (fig. 8.3). It seems that

Box 8.1. How costly is inbreeding?

Across a variety of wild plants, butterflies, amphibians, reptiles, birds, and mammals, the offspring of inbred individuals survive and reproduce less well than do the offspring of outbred individuals (Keller and Waller 2002). It is not always clear why wild inbred individuals do poorly, but studies of wild sheep and sea lions have shown that inbred individuals are more susceptible to disease and parasites, and in the case of the sea lions, take longer to recover from disease (Acevedo-Whitehouse et al. 2003; Coltman et al. 1999).

But how much less well do related parents reproduce, how much less well do inbred offspring survive? An important analysis by Katherine Ralls and Jonathan Ballou (1983) showed that inbreeding in zoo-bred primates at any level of consanguinity is associated with a median 75% drop in number of offspring and 50% increase in mortality of juveniles (box fig. 1). The values are similar for zoo-bred ungulates and small mammals (Ralls and Ballou 1983).

If inbreeding has such a strong effect in captivity, how much more severe must its effects usually be in the wild: the costs of inbreeding can be high.

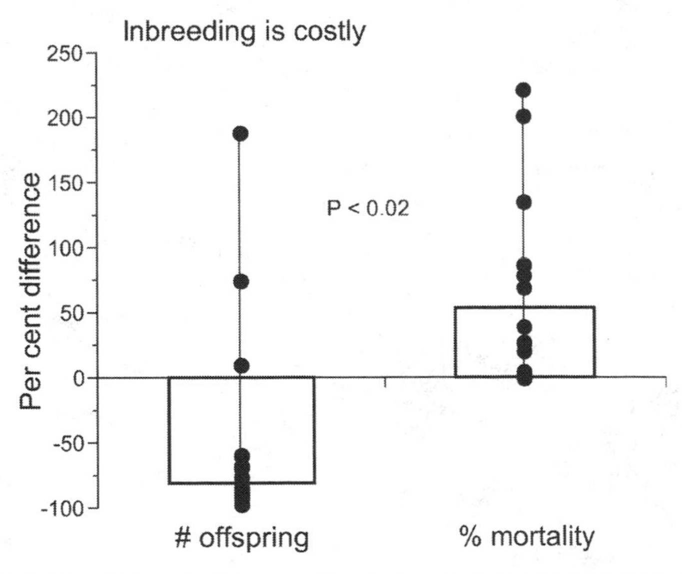

Box 8.1, Fig.1. Inbreeding is costly. Circles are captive primate species/subspecies; bar is median. *Details at end of chapter.* Source: Data from Ralls and Ballou (1983).

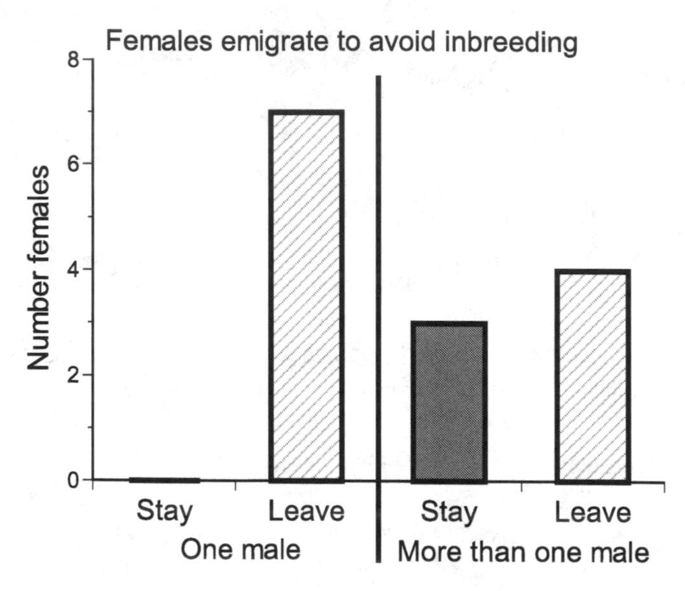

Fig. 8.3. Gorilla females emigrate to avoid inbreeding. When only one male, the father, is in the group, all daughters emigrate; with more than one male, some females have stayed stay to breed. (Mountain gorilla.) *Details at end of chapter.* Source: Data from Watts (1991b).

gorilla females remain if they have an alternative to their father to mate with. They seem to be following the tactic of remaining if they can. But if the alternative to emigration is to inbreed with their father, female gorillas are forced to emigrate, as with the Thomas langur (Sterck et al. 2005).

The data for gorillas are suggestive, but the difference in likelihood of emigration from one-male and multi-male groups is not statistically significant. Others have made similar comparisons across species.

Anne Pusey and Craig Packer asked whether female primates were more likely to emigrate, the fewer males immigrated into groups, and hence the more likely it was that the only males with which the female could mate were kin (Pusey 1987; Pusey and Packer 1987). Their analyses supported their contention (fig. 8.4).

Tim Clutton-Brock asked whether in species in which a successful male's breeding tenure was normally longer than the time it took a female to mature (and hence inbreeding was likely), females were more likely to emigrate than in species in which the male's tenure was shorter than the time to maturity of females (Clutton-Brock 1989a). His results, like Pusey and Packer's, indicate that females stay if they can (male tenure less than female's time to maturity), but emigrate to avoid inbreeding if they are forced to by the males' mating strategy of long tenure (fig. 8.5).

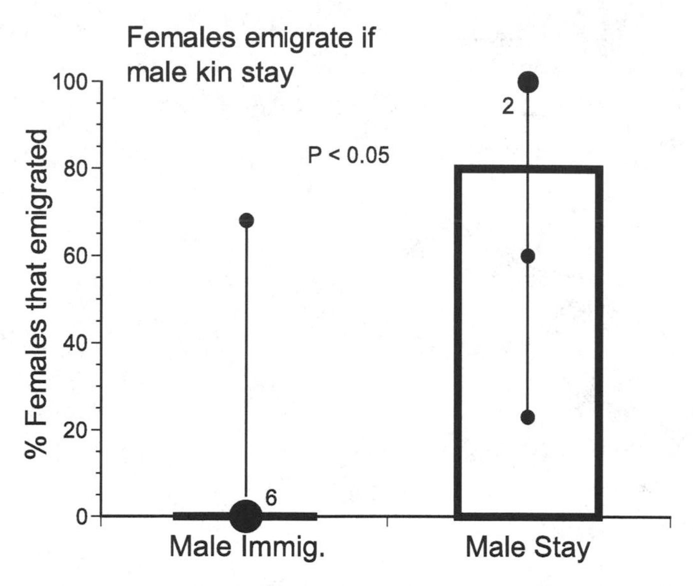

Fig. 8.4. Across primate genera, females remain in their natal group if most males immigrate (Male Immig.) but emigrate if most males stay in their natal group (Male Stay). Bars show medians; circles are values for individual genera; numbers by circles are number of genera if more than 1. *Details at end of chapter.* Source: Data from Pusey and Packer (1987).

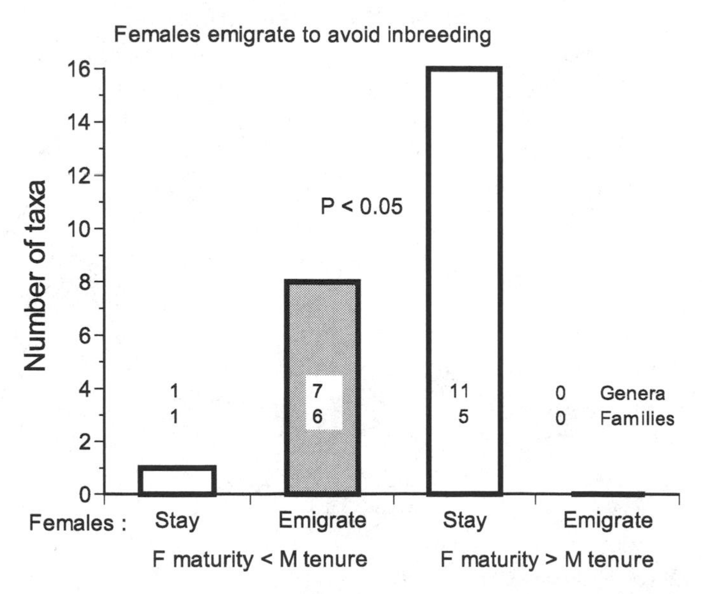

Fig. 8.5. Across mammalian taxa, females stay if they can, but emigrate to avoid inbreeding if male breeding tenure is longer than the time it takes a female to reach maturity, that is, if a female's father is likely to be still in the group when she reaches maturity. *Details at end of chapter.* Source: Data from Clutton-Brock (1989a).

Of course, both the gorilla data and the cross-species comparisons beg the question of why species differ in duration of male tenure. The breeding tenure of gorillas is considered in Part 4, on male strategies.

8.2.2. Incest in gorillas: Inbreeding is not always avoided

The story as told so far is incomplete. Female gorillas that stay in the group in which they were born when more than one male is in the group could still inbreed. That is because other males in a gorilla group are probably kin (ch. 4.2.1.2). They have not immigrated (ch. 4.2.1.2). They will be sons, brothers, or half brothers of the dominant male, and hence maybe even a brother, but more commonly half brother or half uncle of the maturing female in this society in which usually only one male mates (ch. 4.2.1.2; ch. 11.1).

The usually suggested mechanism by which females (and males) detect kin and avoid mating with them is that familiarity in immaturity leads to lack of sexual interest in adulthood (ch. 2.4.2.6). The female that stays is then in a dilemma. She either does not mate with the familiar males because familiarity has bred sexual contempt, or if she does not avoid them, she inbreeds, given that even in multi-male groups, the dominant male nearly monopolizes matings and conceptions (ch. 11.1).

Various other mechanisms for detecting kin than familiarity have been suggested (Rendall 2004). For instance, it is just possible that females in species that habitually live in multi-male groups, such as baboons (*Papio*), might have evolved the ability to recognize relatives by what they look or smell like, as opposed to as individuals that they happen to spend a lot of time near. Thus, in a long-studied population of baboons in Kenya's Amboseli National Park, the researchers have a hint that males might be able to distinguish their offspring from other males' offspring independently of the males' previous or current association with the mothers (Buchan et al. 2003).

However, we suspect that female gorillas have not evolved the mechanism. How could they, when gorilla society consists almost entirely of one-male groups? In any case, the records for the Virungas show that in this population of mountain gorillas, females have mated with potential half brothers, and one perhaps even with a full brother, as if they do not detect consanguinity (stats. 8.1) (Harcourt et al. 1981b; Watts 1990b). While these females might not have mated completely willingly, it tends to be the female rather than the male that initiates courtship with preferred males (Harcourt et al. 1980; Harcourt et al. 1981b), and we suspect that a female can prevent mating with unsuitable males. Yet, not only did one female mate with her full brother, she bred with him, producing a perfectly healthy offspring, which as we write is still doing fine at eight years of age. Inbreeding problems are probabilistic, as we have said.

Nevertheless, no female has bred with her father (Watts 1990b, 1992). Therefore, we continue to argue that, beneficial as continued residence in the group of birth might be, the male mating strategy of long-term breeding tenure forces emigration on females if they are to avoid the potential costs of inbreeding. In the cases where female gorillas mate with close relatives, they do so, we argue, because in this normally one-male species, the ability to detect male kin, other than by identification with the breeding male in the female's immaturity, has not evolved.

8.2.3. Minimal costs to inbreeding?

Even if females that remain in multi-male groups in the Virungas inevitably inbreed, the costs need to be balanced against the costs of emigrating (ch. 2.4.2.6). It is possible that a female could do better breeding with a relative in a safe area of good habitat than emigrating to breed with a non-relative elsewhere.

With gorillas, the strategy of giving birth to the first one or two offspring in the natal group, and then emigrating (ch. 4.2.1.2) could be especially advantageous. Females are not fully grown, not fully experienced by the time they first breed (ch. 3). Some indication exists that the infants of first-time mothers might be more likely to die than are infants of multiple-birth mothers (Harcourt 1987b; Nadler 1975b). If so, the costs of inbreeding are not as great as otherwise: if the infant is likely to die anyway, even with no inbreeding, inbreeding carries no costs. If birth itself, and the strain of lactation is greater for a first-time mother, is it the case that the mother that gives birth the first time in her home group is gaining at this time of increased strain the benefit of residence in a known environment, perhaps even with relatives to help?

8.3. Comparison with *Pan* and *Pongo*

Female common chimpanzees do not range with males all the time, and bonobo chimpanzee females are dominant to males (ch. 4.2). The influence of the efficacy of the male's antipredator protection on breeding females' decisions to emigrate or stay could therefore be irrelevant in these species. However, chimpanzee males defend their community territory, and therefore defend their females, from invasions by neighboring males (ch. 4.2.3.1). In the same way that the power of a single male might determine whether a female gorilla stays or moves, so the number of males in a chimpanzee community could determine whether breeding chimpanzee females stay or emigrate. Witness in Mahale National Park, Tanzania, the emigration of the females of one community of chimpanzees to an adjacent community on the death of almost all the males of their own community (Nishida et al. 1985).

As regards inbreeding avoidance, in both chimpanzee species, males remain in the area and community of their birth, while females emigrate (ch. 4.2.3.1). However, in the Gombe Stream population of chimpanzees, some females return to their natal group after emigration, and some do not emigrate (ch. 4.2.3.1).

If all male chimpanzees remain, why do not all natal females emigrate permanently to avoid inbreeding? The answer could be that the chance of a female mating a close relative is in fact slim, given that males and females range apart. Also, the chance of a female conceiving with a close relative could be slim, given the females' promiscuous mating behavior. Certainly, genetic evidence indicates that close inbreeding is rare (Constable et al. 2001). If inbreeding is unlikely, the costs of a new range that is poor quality, or has a high density of females, could outweigh the costs of inbreeding (ch. 2.4.2.6). In that case, returning to the natal community or staying in it could be the optimal strategy for a female.

The orangutan is a normal mammal in that both sons and daughters appear to disperse from the area of their birth (ch. 4.2.3.2). Both sexes could disperse to avoid inbreeding, if parents are still resident in the area. If so, the fact that they do the same as gorillas (and indeed many other mammalian and bird species [Greenwood 1980]) supports our inbreeding argument for female gorillas: stay if it is possible; emigrate to avoid inbreeding.

Alternatively, it could be argued that the sexes are emigrating to avoid competition with necessarily dominant individuals, their parents, which are larger than them, and to which they have learned to be subordinate. However, their solitary ranging in overlapping home ranges (ch. 4.2) might make it unlikely that parents impose much more of a cost than do other orangutans in the area. Whether or not female orangutans choose, like gorilla females, to go to the home ranges of particularly powerful males is completely unknown. If they do so, the benefit would be proximity to a desirable mate. It is unlikely that protection from the male plays any part, given that the sexes range separately and that apparently dominant males cannot prevent the small-bodied, male morph (see box 3.2) from more or less forcibly mating with females (Delgado and van Schaik 2000; Utami and van Hooff 2004; Utami et al. 2002).

Conclusion

The data indicate that breeding gorilla females that emigrate might be choosing more powerful males; and that young female gorillas that leave their natal group (and young females of other species) avoid the costs of inbreeding. Pascale Sicotte (2001) has suggested another reason for breeding females to emigrate and transfer, namely, as a means to increase her number of mates over her lifetime. Several benefits could result (Jennions and Petrie 2000; Keller and Reeve 1995; Madsen et al. 1992). Mating with more than one male can increase the female's chances of finding a male with better quality sperm. This sperm-quality benefit might be especially beneficial, given the poor quality of sperm of males in single-male mating systems, such as the gorilla's (Gomendio et al. 1998; Harcourt 1991a; Møller 1988). Multiple mates can also increase the chances of a mate with more compatible sperm, and more compatible genes. And finally, mating with more than one male can increase the genetic heterogeneity of the sum of the offspring that the female produces, and hence their performance, and therefore the female's reproductive success. While in other taxa there is plenty of evidence for these genetic advantages of multiple mating, they have certainly not been demonstrated in gorillas, and evidence is largely lacking in primates that any aspect of female choice of mates is associated with genetic advantages (Manson 2007).

Nevertheless, if any of these benefits of emigration operate, then the gorilla's is a society in which the males' mating strategy of long-term breeding tenure of a group of females strongly affects female mating strategies. The socioecological framework of females-to-food, males-to-females needs the addition of the direct influence of male mating strategies on female survival, mating, and rearing strategies.

We discuss further in Part 4 why males use the long-tenure strategy that forces emigration on their daughters, and how the females influences the males' use and success of that strategy.

Figure Details

Fig. 8.1. Immigrants victimized. Data are from two groups, with one counted twice in separate periods two years apart, producing the maximum and minimum values in the figure. Statistical data are available for only the comparison with all residents, because kin and non-kin residents were not separated in Watts's statistical analyses: aggression per time within 5 m, statistic $K_r = 28–277$ per group, $p < 0.01$ (Watts 1994a). (Mountain gorilla; Virunga Volcanoes, Rwanda, DRC.)

Fig. 8.2. Females join multi-male groups. Original group: One male: $N = 23$, $G = 2.36$, $p > 0.1$; Multi-male: $N = 14$, $G = 16.5$, $p < 0.001$; Both: $G = 12.7$, $p < 0.01$. (Mountain gorillas; Virunga Volcanoes, Rwanda, DRC.) The data are for both breeding females and females leaving the group of their birth.

Fig. 8.3. Emigration and inbreeding. Fisher exact probability test, *ns.* (Mountain gorillas; Virunga Volcanoes, Rwanda, DRC.)

Fig. 8.4. Avoidance of inbreeding. Percent females that emigrate per primate genus (median species' value) in male emigrant compared to male resident genera (separation at 50% emigrant/resident). Wilcoxon/Kruskal-Wallis test (equivalent to Mann-Whitney U test), $z = 2.17$, $p < 0.05$.

Fig. 8.5. Avoidance of inbreeding. Number mammalian taxa (species as bars; genera, families numbers shown) in which females usually remain in or emigrate from the group of their birth. As species are not independent data points (Harvey and Pagel 1991), we tested for families the likelihood of this result arising by chance ($p < 0.05$, Fisher exact probability test [Siegel 1956]). Families for Male tenure longer than female maturity—*Cebidae, Cervidae, Cercopithecinae, Colobinae, Felidae, Lemuridae, Sciuridae;* for Male tenure less than female maturity—*Alouattinae, Cercopithecinae, Colobinae, Equidae, Pongidae.*

Box 8.1, Fig. 1. Inbreeding is costly for zoo-bred primates. Data are for 16 species/subspecies, 10 genera, 7 families, including lemurs, New World monkeys, Old World monkeys, and the common chimpanzee. Of the 16 species/subspecies, 15 suffered more when inbred; all 10 genera and 7 families suffered more when inbred than when not (7/7, $p < 0.02$, Binomial test).

Statistical Details

Stats. 8.1. The full sibling mating is identified by the fact that the brother was the dominant breeding male when the sister conceived, that the siblings were themselves born when there was the same obviously dominant silverback in the group, and that anatomical, behavioral, and genetic paternity analyses show near monopolization of conceptions by dominant males (ch. 11.1).

Adult female gorilla that left one male to join another, leaving behind a juvenile daughter.
(Mountain gorilla.) © A. H. Harcourt

CHAPTER 9

Female Strategies:
Conflict, Compromise, and Cooperation Between the Sexes

Summary

Gorilla female society can be described as one of groups of mostly unrelated females that transfer between males, especially at maturity. The grouping is allowed by the distribution of the gorilla's main food (widespread, dense vegetation), which results in a lack of contest competition and therefore of cooperation in competition. Lack of competition allows grouping; lack of benefits from cooperation allows emigration. The grouping, which is a result of several females choosing one male with which to associate, is explained by the benefits to the females of proximity to a protective male twice the females' size. Protection is afforded against predators and infanticidal males. Breeding females emigrate to find more powerful males; nubile females emigrate to avoid breeding with relatives.

9.1. Gorilla Female Society

Female gorillas live in groups of mostly unrelated females with a male twice their size (ch. 3.2.2), which is absolutely dominant to them (ch. 4.2). Instead of remaining in the group or area of their birth, most females leave at maturity to join other males; even breeding females leave (ch. 4.2). The emigration explains why gorilla groups consist of unrelated females. But why do the females group with a dominant male twice their size? Why do they not stay in the area or group of their birth, with kin, like many mammals, maybe most? If they are going to leave their birth or breeding group, why do they then not travel alone, or in parties of continually changing size and composition, as do the other great apes (ch. 4.2)?

We summarize our answers to these questions and then expand on issues raised by our analysis. Note that when we say "our" answers, we mean our interpretation of answers that the fields of primatology and socioecology have produced. Throughout our interpretation, a paramount concept is the distinction between what the socioecological environment allows females to do, and what it forces females to do, in the sense of providing obvious benefits of the strategy.

9.1.1. Food and society

Almost all attempts to understand why animals live in groups conclude that grouping necessarily increases competition between individuals for resources, because it necessarily means that animals are often near the same useful resource at the same time (ch. 2.2.3.1). Gorilla females are no exception (ch. 5.1.1). (Note that we do not think that any of the several hypotheses that suggest various foraging benefits to grouping, which offset the competitive costs, apply strongly to primates, and certainly not to gorillas [box 2.3].)

If grouping increases competition, it also allows cooperation in competition with third parties (ch. 2.2.3.2). Gorillas, again, are not an exception (ch. 5.2.1). The benefits of that cooperation could then be a cause of grouping: animals join others to obtain the benefits of the cooperation; they stay with others to continue to reap the benefits of cooperation; or they accept the presence of others in part because of the benefits of cooperation (ch. 2.2.3.2).

Gorillas now are an exception. Although female gorillas definitely compete over resources (ch. 5.1.1), the evidence indicates that they do not do so either frequently or intensely enough to result in any long-term consequences (ch. 5.1.2). For instance, intense aggression does not increase in frequency with increase in group size, and it is quite often the case that gorilla groups show no obvious dominance hierarchies among the females (ch. 5.1.2). Concomitantly, cooperation appears not to be important to gorilla females (ch. 5.2.2). Evidence for that lack of importance is the fact that while subordinate animals certainly benefit from cooperation, the help that they receive does not result in permanent change in competitive ability.

The reason for the lack of importance of either the competition or the cooperation surely lies in the nature of the gorilla's fallback food, the food it eats when preferred foods are in short supply. Unlike the chimpanzee and orangutan, which attempt to continue to eat ripe fruit throughout the year, the gorilla easily switches to a diet high in foliage (ch. 4.1.3; 4.1.8). Foliage is not only abundant, but of low quality. It is thus not only indefensible, but not worth defending (ch. 2.2.2).

Moreover, and here is the first indication of the strong influence of the male on female society, the average female gorilla is supplanted from food nearly ten times as often by the male as she is by the median female in her group (ch. 6.1). With an adult male present, competition between females is irrelevant. Also, adult male gorillas more or less negate any advantage of dominant competitive ability among females by intervening in aggressive contests on behalf of subordinate females and losing coalitions of females (ch. 6.2; 6.3). A major reason for the influence of the male, besides the fact that he is twice the size of the females, is the usual small size of gorilla groups: their average is only three or four females (ch. 4.2.1.1), one-third the number needed *in toto* to match one male. The male with which a female perpetually associates effectively makes both competition and cooperation among the females ineffectual.

Females *can* group with one another and with a male, because competition for resources is not important enough to so adversely affect their ability to survive, mate, or rear offspring that it prevents grouping. Females *can* emigrate, because costs of lost opportunity to cooperate with kin in competition over resources are not great. But the distribution of food does not explain why they do either, and therefore does not, in and of itself, explain the distribution of females.

9.1.2. Protection and society: The male's influence

Females group with a male for protection, we suggest, and group with each other because several females make the same choice of male (ch. 7). But what are they being protected against, predators or infanticidal males? The climate of primatological opinion appears to be infanticidal males. However, we suggest that the benefit to females of using a male for protection against predation could be especially pertinent in gorillas, because their large body size makes them unusually terrestrial and unusually likely to fight rather than flee as a means of defense against predators (ch. 7.1).

Evidence for the antipredation hypothesis for females associating with a male is that gorillas are preyed upon; males defend females (and their offspring) against predators; and females act as if males defend them against predators (ch. 7.1). At the same time, gorillas are a prime example of adaptive infanticide by males, with one population having one of the highest recorded rates of infanticide among primates (ch. 4.2.1.3; ch. 7.2); gorilla males defend against infanticide; and quantitative modeling indicates that females that associate with a male suffer less infanticide than they would if they roamed alone (ch. 7.2).

So, evidence and logic exists to support both the antipredation and the anti-infanticide hypotheses for female-male association. Once a female joins a male, he can, of course, protect against both predators and infanticidal males (ch. 7.3). But which threat do females most strongly avert by association with a male? In other words, which is the more likely evolutionary cause of the gorilla females' association with males in the first place (ch. 7.3)? Which is the greater threat? Is it the infrequent but potentially lethal predation of a female by a predator? Or is it the perhaps more frequent but relatively unimportant loss of an infant (replaceable within a year)?

Across primate species, predation appears to be a powerful influence on, for example, group size (ch. 2.3.3). Infanticide explains no more, and maybe less, variation in group size than does predation (Nunn and van Schaik 2000). By contrast, within species, protection from infanticide appears to be a powerful explanation for female-male association (Nunn and van Schaik 2000; Palombit 2003). Nevertheless, for gorillas, we suggest that at present, the data do not allow a decision one way or the other about which hypothesis is correct. The answer will probably have to be obtained by more quantitative modeling, contrasting predation's infrequent but potentially permanent halt to reproduction with infanticide's more frequent temporary hiatuses (ch. 7.3).

9.1.3. Emigration by females: The male's influence

Animals, especially females, should not leave the group or area in which they were born, for the chances of finding a better place to raise offspring than they were raised in are slight compared to the cost of leaving known areas (ch. 2.2.3.2; 2.4.2.6). Nevertheless, females of many folivorous primate species emigrate, and the gorilla is one of those species (ch. 4.2.1.2).

If gorilla females join a powerful protector (ch. 7), they will also, surely, leave one that they somehow sense is not powerful enough (ch. 8.1). Choice of a better male might be the explanation for why breeding female gorillas leave their current male, especially those females that leave after an infant has died or has been killed. There is also the possibility that females are increasing the genetic variety of their offspring, or searching for males with better sperm or better genetically influenced traits than their current male (ch. 8, Conclusion), though they are not possibilities that are going to be demonstrated in gorillas in our lifetime.

Another explanation is needed for females that leave the group of their birth, given that almost all leave. Most gorilla groups are one-male groups. Nobody has yet seen a resident male ousted from the group. If powerful males retain breeding tenure of a group for longer than the time it takes their daughters to mature, the daughters must emigrate to avoid inbreeding (ch. 8.2).

"The daughters must emigrate to avoid inbreeding," we say. That is true only if the costs or probability of inbreeding are greater than the costs of emigration, or the probability of suffering those costs. Several gorilla females in two populations have stayed to produce their first infant in their natal group (ch. 4.2.1.2). They have done this only when a male other than their father is present (ch. 8.2.1). Given that gorilla males hardly ever immigrate, females that do not emigrate from multi-male groups might avoid mating with a father, but the other males will often be relatives of some sort, including brothers. Incestuous mating occurs (ch. 8.2.2). Does it occur because multi-male groups are rare in most gorilla populations, and therefore female gorillas have not evolved the necessary mechanisms to detect kin? Or is some balancing of costs of inbreeding with costs of emigration occurring? We do not know.

9.2. Gorilla Society: An Unusually Strong Influence of Males?

The females-to-food part of the females-to-food, males-to-females framework is powerful enough to provide a fairly full description of the society of many primate species, with little to no mention of the males (ch. 2.1.2) (Isbell 2004; Isbell and Young 2002; Sterck et al. 1997; van Schaik 1983; Wrangham 1980). This is true even in several group-living species in which females are half the size of males. While the male is an obvious influence on females' behavior in several of these species (Dunbar 1984; Seyfarth 1978a, b; Smuts 1985; van Schaik et al. 2000), yet understanding of much of the society of females in these species does not need understanding of the males' influence (Cheney 1977; Hinde 1983). The male's influence seems almost temporary, even in populations in which females compete for access to the male, and presumably ultimately for his support and protection (Palombit, Cheney, and Seyfarth 2001).

That is not the case with gorillas. Females are in stable groups with other females because they are more or less independently attracted to powerful males (ch. 7.1.2). And female gorillas are in groups of largely unrelated females because males force emigration to avoid inbreeding (ch. 8.2).

Why is the gorilla different? Where it is different, it seems that the explanation might have its origins in its large body size. Its size enables, even necessitates, that it be a folivore. Its size necessitates that it be largely terrestrial. Large body size and folivory appear to correlate with low density and short daily distances traveled (Clutton-Brock and Harvey 1977b). Compared to other primates, the gorilla will therefore live at relatively low density and travel short daily distances. In fact, the gorilla lives at unusually low densities for its body mass ($1/km^2$ instead of $4/km^2$, and travels unusually short daily distances for its group mass (1 km instead of 2 km) (stats. 9.1). This statement holds true whether total species or genera, only diurnal species or genera, or only African diurnal species or genera are included in the comparative analysis. Those two factors of low density and short daily distance traveled prevent females using polyandry and wide ranging as a means to avoid infanticide (ch. 7.2.2).

And we have argued that large body size and terrestriality promote fighting as a means to combat predators. If so, females gain unusual advantages from association with a protective male, which is twice their size (ch. 7.1.1).

In sum, female gorillas, more than the females of many other primate species, benefit from associating with a male. Predation and infanticide favor the association; folivory allows it.

Of course, the influence is not all one-way. By several routes and for several reasons, the nature of gorilla society can be understood only with appreciation also of both direct and indirect influences of females on male strategies—as we discuss in Part 4, on male strategies and the nature of gorilla society.

9.3. Gorilla Society: An Unusual Product from Interaction of Usual Rules

We want to emphasize here a point that we have repeatedly made, namely, that gorillas are not in the sort of society they have because they are in some way genetically programmed to take on the roles that produce their society. Rather, gorilla females (and the males that influence them) act according to rules that all primates (including humans), mammals, vertebrates, even all animals, follow (ch. 2) (e.g., Alcock 2001; Boyd and Silk 1997; Krebs and Davies 1993; Trivers 1985). Gorillas, like all these others, follow rules such as compete over scarce resources, cooperate in that competition for mutual benefit, cooperate more with kin than with non-kin, protect offspring and mates, group with protective individuals, avoid inbreeding, and so on.

The result is suites of behaviors and relationships among individual gorillas, in other words, aspects of the structure of society, that are seen in many animals. These aspects are even seen in primate species in which females remain in the group of their birth, apparently because of benefits from co-operation with kin in competition over resources. Thus, females in some gorilla groups have quite stable dominance hierarchies (ch. 4.2.2.2; ch. 5.1.2); female gorillas form cliques of friendly, cooperative kin (ch. 4.2.2.2; ch. 5.2.1); females doing poorly with one male or in one range emigrate to a better male or range (ch. 8.1); and females emigrate to avoid inbreeding (ch. 8.2).

However, and this is where the study of the structure of societies is so interesting and important, what we see with the gorilla is that the same evolutionary rules that work in all species, the same behavioral ecological rules operating on all individuals' tactics and strategies, produce in a different sort of animal that uses the environment in a different sort of way a different sort of society. Those general rules about competition and cooperation, operating in a large-bodied, terrestrial, folivorous, slowly reproducing species that relies on ground vegetation and defense against predators to survive, produce a society of one-male harem groups of unrelated females that largely merely tolerate one another, which groups are stable until such time as the male dies, a female loses an infant, or a daughter experiences her father as the only male with which to breed.

We are in an academic community, anthropology, many of whose members (not necessarily in our own department) still oppose the attempt to investigate how far evolutionary biological hypotheses can explain human behavior (Dagg 1999). One explicit fear behind the opposition is the deeply mistaken inference that we are somehow trying to prove that human behavior is genetically programmed and unalterable. For instance, Anne Dagg criticizes evolutionary arguments for infanticide in humans (in other words, arguments about its benefits to non-father males' reproductive success [ch. 2.4.3] by saying "Yet little or no attempt has been made to explain why ALL [emphasis Dagg's] human cultures do not have this same base [killing by stepparents] if it is biological rather than cultural."

Our emphasis on the role of the environment in shaping behavior and society obviously negates Dagg's argument. There is no reason whatsoever why all societies should exhibit infanticide by males—if the societies and their environment happen to be such that infanticide is not advantageous to males as a mating strategy. Not all female gorillas leave the group in which they were born (ch. 4.2.1.2; ch. 8.2.2). Just because they do not all emigrate does not begin to mean that our evolutionary explanations for this aspect of gorilla society are wrong. The variation across individuals does not begin to mean that a biological functional explanation in terms of survival, mating, and rearing is wrong. Individuals and their circumstances differ. Not even in animal populations, let alone humans, are individuals genetically programmed with unalterable behavior that appears irrespective of circumstances.

Physical anthropology certainly tries to explain the universals of nonhuman primate species' behavior, for then we might understand some universals that apply to humans (Darwin 1871, 1872; Hinde 1974). But physical anthropology also tries to explain the variation, especially as it relates to use of the environment, for then we get at a different sort and level of universals (Borgerhoff Mulder 1991; Hinde 1974; Smith, Borgerhoff Mulder, and Hill 2001). "Help kin" might be an example of the first level (Dugatkin 1997; Silk 2002a). But some individuals do not help kin. "Help the potentially most useful partner" (Chapais 1995; Seyfarth 1977) might then be an example of the second level, for of course, the group member that is potentially most useful differs among group members, and might not even be kin.

As we said in chapter 1, the gorilla is not unique in the nature of its society (ch. 1.3). An apparently odd assortment of other species have societies similar to the gorilla's. One is the Thomas langur (or leaf monkey). Although the Thomas langur is an Asian monkey, ten times smaller than the African gorilla, an ape, and arboreal instead of terrestrial, nevertheless, the same fundamental behavioral ecological rules (those governing behavior between individuals as they struggle to survive, mate, and rear their offspring) have produced a species with a society remarkably similar to the gorilla's (ch. 1.3) (Steenbeek et al. 2000). What we now have to confirm is precisely which aspects of the two species' biology, which aspects of the environment, which aspects of their use of the environment, are the crucial ones that in interaction with the common behavioral ecological rules produce their similar societies.

And then we need to do the same for the plains zebra and the hamadryas baboon (ch. 1.3). These species' societies are similar to the societies of the gorilla and the Thomas langur, except that in addition, the one-male harems of unrelated females that merely tolerate one another join into large herds, within which the one-male level of society stays very stable. What explains the similarities; what explains the differences (Rubenstein 1986; Rubenstein and Hack 2004; Watts 2000a; Wrangham and Rubenstein 1986)?

Statistical Details

Stats. 9.1. Gorillas live at lower density than expected and travel shorter daily distances than expected. They are always below the median, often in the lowest 25%, sometimes in the lowest 10%, and sometimes the lowest value of all, for residuals from regression of density against body mass or adult group mass, and daily distance traveled against body mass or adult group mass. (Data from primary literature, sources too numerous to give).

For example,

Total species: density by body mass: lowest 10% of residuals, $N = 144$; density by adult group mass: lowest 5%, $N = 88$.
Total genera: density by body mass: lowest 10% of residuals, $N = 60$; density by adult group mass: lowest 5%, $N = 45$;
Diurnal genera: density by body mass: lowest 5% of residuals, $N = 44$; density by adult group mass: lowest value, $N = 37$;
African diurnal genera: density by body mass: less than the median, $N = 14$; density by adult group mass: second lowest value (after *Pan*), $N = 10$;
Total species: daily distance traveled (DDT) by body mass: lowest 25% of residuals, $N = 76$; DDT by adult group mass: lowest 25%, $N = 54$;
Total genera: DDT by body mass: less than the median, $N = 43$; DDT by adult group mass: lowest 25%, $N = 35$;
Diurnal species: DDT by body mass: lowest 30%, $N = 71$; DDT by adult group mass: lowest 25%, $N = 54$;
African diurnal species: DDT by body mass: lowest value, $N = 19$; DDT by adult group mass: lowest value, $N = 13$.

PART 4

MALE STRATEGIES AND GORILLA SOCIETY

SUMMARY

Although gorilla society might seem one permeated by the influence of males on females, in fact the relatively thin distribution of females in the environment affects the distribution of males. It forces the males to stay with females instead of roaming in search of them. Concomitantly, mating competition among males, particularly the use of infanticide, forces females to associate with a protective male (which also protects them against predators), instead of roaming alone. Hence, gorilla society is one of stable, cohesive groups.

While gorillas are a classic single-breeding-male society, with all the associated behavioral and anatomical correlates, nevertheless, some questions remain. A minority of groups, almost all among eastern populations of the species, are multi-male, because some males remain in the group in which they were born. The causes of differences between populations and species in number of males per group is a long-standing puzzle. The different costs and benefits of staying versus leaving in eastern populations are quite well established, but the understanding does not help explain the contrast between the almost 100% one-male society of western populations and the mixture in eastern populations of one-male and multi-male groups. The contrast raises the issue of where answers to these and other questions about the nature of animals' societies should be sought, whether in comparisons within species or between them.

Gorilla males attract females and remain with them. (Mountain gorilla.) © A. H. Harcourt

CHAPTER 10

Male Strategies and Society: Influences of the Environment and of Females

Summary

Male animals have a variety of options for finding a fertilizable female. The male can roam in search of the females, defend areas that females frequent, attract females to himself, or stay with females once they are found. The strategy that the male adopts strongly determines the nature of the species' society. The gorilla's folivorous diet allows a short daily distance traveled, even necessitates it. At the usual low population densities of this large-bodied species, the short daily distance means that gorilla males are (a) unable to use territorial defense as a means of access to females, and (b) committed to permanent association with females if they are to reliably find females when they are in estrus. Once a male is committed to remaining with the females, he is then committed to protect them and their offspring from harm.

10.1. Association as a Means of Access to Females

10.1.1. Means of access to females

In the classic theoretical framework of mammalian mating systems, males associate with females (and vice versa) only to mate, or when infants will not survive with a mother's care alone (ch. 2.4.2.1; ch. 2.5.1). In contrast to avian mothers, mammalian mothers are usually able to raise their offspring without male care, and commonly do so. Thus it is in only a minority of mammalian taxa that males provide substantial paternal care (Kleiman 1977).

In a greater proportion of primate species than in mammals in general, males stay and provide substantial care. Nevertheless, it is a small minority, only about 15% of primate species, almost all of them small New World species (Kleiman 1977). So why in most primate species do males stay with females? The usual association of males with females, including outside breeding seasons, and including species in which males show little to no obvious paternal care, is an unusual facet of primate society by comparison to other mammals (ch. 1.2.4). That association needs to be contrasted with the alternatives of roaming in search of females, defending a territory in which the male travels fairly independently of the females, defending a territory and associating, and finally, letting the females come to the male, maybe with some behavior to attract them (ch. 2.4.2).

So far, we have discussed male-female association of gorillas in terms of the benefits to the female obtained by associating with a powerful protective male (ch. 7). However, if it were not advantageous for the male to associate with the females, the association would not occur: he would walk away from them (ch. 2.4.2). Does the male obtain more access to more females by association, or raise more offspring by association than by use of other tactics (Wrangham 1979)? Clearly he does or he would not associate. So, what are the payoffs of association?

To answer that question, we first ask about the strategies that males could use to get access to females other than permanent association. First, could gorillas obtain access to more females by being territorial than by traveling with females, in other words, by resource defense rather than mate defense (Clutton-Brock 1989b; Davies 1991; Greenwood 1980)? Second, could they obtain access to more females by roaming in search of them than by permanent association? Richard Wrangham (1979) asked both these questions (though the first for females defending food), and answered "No" to both after a qualitative analysis. We answer them quantitatively.

To answer the first question, we use the Mitani-Rodman index of defensibility (Mitani and Rodman 1979). John Mitani and Peter Rodman found that primate species were territorial only if their normal daily distance traveled was at least the diameter of the annual home range. In effect, if the animals could easily travel from one side of their range to the other in a day or less, they could be territorial. Some that could cover the diameter in a day were not territorial, but none that could not do so was territorial. The one-diameter threshold that Mitani and Rodman used was purely empirical. They had no preconceived ideas about what the threshold would be; the primates told them.

To answer the second question, we use the same formula that we used in chapter 7.2 when we investigated the chances of estrous or nursing females meeting males, namely, the Maxwell gas molecule equation for frequency of encounters between moving bodies. Peter Waser (1976) was, as far as we know, the first to treat primates (in his case, primate groups) as randomly moving molecules to ask questions about the nature of primate societies (see ch. 7.2.2.3). And as far as we know, Richard Barnes (1982) was the first to ask about the chances of a randomly moving male finding females compared to a male that stayed with females, as the density of females and their group size varied. His study species was the African elephant. He concluded that at low densities or when groups are very large, it might pay male elephants to stay with females, but otherwise they should roam—as most do.

Robin Dunbar (1988, ch. 13) then used the gas molecule equation to ask Barnes's question about the apes. He found that chimpanzee males would usually find more estrous females by roaming than by long-term association, while the opposite would be the case for gorillas, because of the gorilla's short day-journey length. He subsequently substantiated the suggestion with comparisons across all the apes, finding, for example, that a gorilla would encounter ten times as many estrous females by association as by roaming (Dunbar 2000).

We repeat both Mitani-Rodman and Dunbar with more and other data for the values for the parameters that they used. In so doing, note that we are strongly applying the second half of the females-to-food, males-to-females framework (ch. 2.1.2): we are assuming that the distribution of females determines the males' mating strategy.

10.1.2. Gorillas cannot be territorial as a mate access strategy

Male animals can use territoriality as a strategy for access to mates if females range in areas small enough for a male to defend (Clutton-Brock 1989b; Davies 1991). Using the Mitani-Rodman one-diameter threshold, territorial defense is impossible for gorillas. Home range sizes and daily distance traveled vary across sites, from 4–25 km^2, and 0.5–3 km (ch. 4.1.4; 4.1.5). They vary in rough proportion to each other, with larger annual ranges where the animals travel farther each day. In no site does the longest annual average daily distance reach anywhere near the diameter of a circular home range of the size of the smallest home range at the site. For instance, the smallest home ranges recorded are from the Virunga Volcanoes, at about 4 km^2; the longest daily distances there are 1 km. But the diameter of a circular 4 km^2 range is over 2 km, that is, twice as long as the longest average daily distance recorded in the region. The gorillas in Lossi Reserve, near

Odzala National Park in Rep. Congo, have the combination of longest daily distance (nearly 2 km) and potentially the smallest home range size known for western gorillas, maybe as little as 7 km^2. But a circular 7 km^2 range has a diameter of 3 km.

10.1.3. Gorillas cannot roam, but must associate permanently

What about roaming as a means of getting access to females? We ran the gas molecule model (sec. 10.1.1; ch. 7.2.2.3) to investigate how varying the parameters might affect the result. The model incorporates density of females, area over which the male ranges to find females, and the male's daily distance traveled. It also incorporates female reproductive biology, specifically the annual number of days that a female is normally fertilizable (meeting unfertilizable females is irrelevant to the male in this context).

The results of the model look conclusive, and largely confirm Dunbar's analyses (1988, ch. 13; 2000). At all normal densities, normal home range sizes, and normal daily distances (and with other normal aspects of gorilla biology), gorilla males would find far fewer estrous females, five times fewer, by roaming than by permanent association (fig. 10.1[a], [b]). Even at abnormal densities, home ranges, daily distances traveled, and detection distances, they still would almost always do better to stay with females than to roam (fig. 10.1[a]). Only those males in a population of females at over twice the normal density of less than one per square kilometer, that also travel more than twice the normal daily distance of around 1.5 km, and that can detect each other at twice the modeled 0.5 km, would approach doing as well by roaming as by staying with an average-sized group of females (fig. 10.1[b]).

An obvious tactic in the roaming strategy is to temporarily increase the daily distance traveled, which is what sexually active male African elephants apparently do in some populations (Barnes 1982). Lone male gorillas go out of their way to meet females (Watts 1994c), but even if a male gorilla could speed up to 6 km/day, over four times the gorilla's average, it would still pay him to remain with any females he found, except in an extraordinarily high-density population. In short, male gorillas are extremely unlikely to find a female in estrus by roaming. To get access to fertilizable females, they therefore need to stay with any that they find, or persuade females to join them and stay with them.

The easiest parameter for a gorilla male to change would be detection distance. He has little to no control over female density, and changing daily distance traveled or his normal home range area would be energetically costly. Increasing detection distance by loud calling is certainly not without energetic cost. However, on the assumption that estrous females benefit

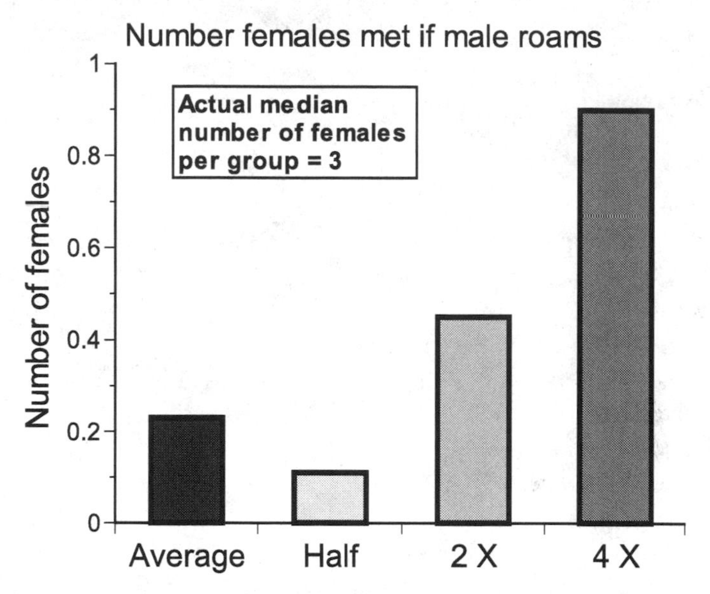

Fig. 10.1A. Male gorillas that roam would, under most realistic scenarios, meet fewer fertilizable females than do males that associate permanently with the median number of females in a group, 3. Data show model's results when density, detection distance, and daily distance traveled are entered as the average value for Africa, or half, twice, or four times the average. *Details at end of chapter.*

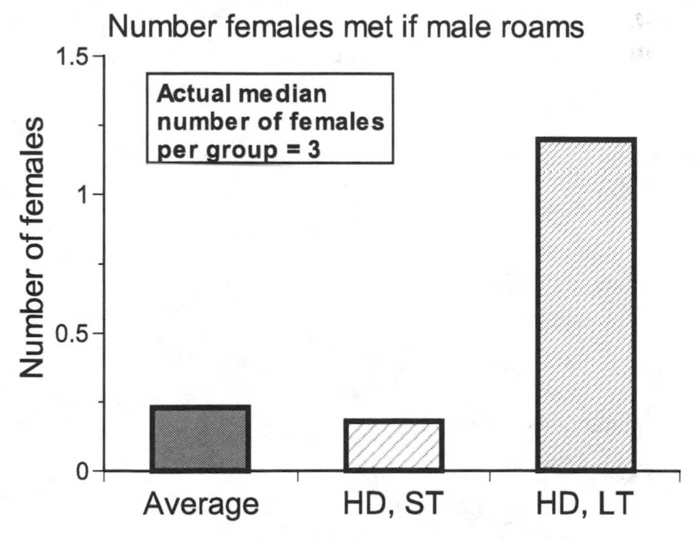

Fig. 10.1B. The same results for male gorillas that roam, in average conditions and under two extreme real scenarios. Average = average values for density and daily distance; HD, ST = high density, short daily travel (Virungas); HD, LT = high density, long daily travel (swamp forest of west-central Africa). *Details at end of chapter.*

from approaching and mating with powerful males, the orangutan's tactic of loud calling would be possible for the gorilla. The problem is that if all other parameters in the model stayed the same, the gorilla male's loud call would need to be reliably heard at distances of over 5 km in the average population for him to meet and mate with the number of females in the average male's harem.

David Watts (1996) has suggested that this whole roaming argument for gorilla males associating with females is circular if females are already clumped with males because the females themselves benefit from being with a male. Females already being clumped with the male makes the argument not circular, but simply unnecessary. However, were it not the case that males also benefit from being with females, then the male could leave the females, however beneficial it might be for the female to be with the male. Half the male's size, the female does not have much opportunity to coerce an unwilling male into staying. The opposite is the case for the male (ch. 11.1.3.1). Thus both partners need to agree to associate, and the payoffs of doing so for both need to be understood if the nature of gorilla society is to be fully understood.

10.2. Predation, Infanticide, and Association with Females

While a main reason for gorilla females to associate with a male is for protection against predators and infanticidal males (ch. 7), the females will not receive that protection unless it is also advantageous for the male to not only associate with them but also protect them. In general, if it is advantageous to remain with a resource, any resource, then up to a point, it must be advantageous to protect that resource from damage. In the present instance, there is little to no point in a male staying with a female, unless she or her offspring survive.

10.3. Comparison with *Pan* and *Pongo*

10.3.1. Finding females: Roam or stay?

Pan and *Pongo* contrast greatly with the gorilla with respect to male-female associations. While successful gorilla males associate permanently with females, in other words, they are in close proximity to the same females for months, even years at a time, *Pan* males do not associate anywhere near that faithfully and nor, especially, do orangutan males: *Pan* parties are continually changing composition, even if community composition is stable, while orangutans cannot even be considered to be in parties (ch. 4.2.3).

Dunbar used the gas molecule equation to test how far the chance of finding a fertilizable female influences male apes' association with females (Dunbar 1988, ch. 13; 2000). The model's results matched reality quite closely. In other words, it looks as though across all the great apes, the proportion of time that males spend with groups of females is proportional simply to the probability of there being a fertilizable female in a party of females. Dunbar's model's results were that orangutan males hardly ever associate with females; common chimpanzee males sometimes do; bonobo chimpanzee males usually do; gorilla males always do.

We reran the model to investigate how variation in values of parameters affected the results (fig. 10.2). The findings were close to Dunbar's, namely, that a bonobo male would not benefit from adopting the roaming strategy, while under some circumstances a chimpanzee male would (fig. 10.2). In reality, male bonobos usually associate with groups of females, while male chimpanzees sometimes do so, and in some sites, for example, Taï Forest, Côte d'Ivoire, often do so (ch. 4.2.3.1). However, it is not always the same group of females with which a male *Pan* associates, because the parties perpetually change composition. In some ways, the chimpanzees use a multiple

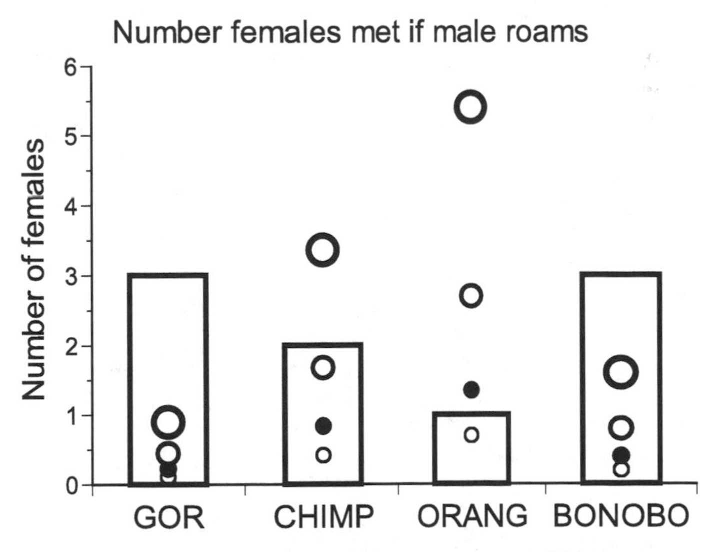

Fig. 10.2. Whether male great apes associate permanently with females or roam in search of them is explained by the probability of finding fertilizable females under either strategy. Gorilla and bonobo males meet more females by association (bars) than by roaming (circles); chimpanzee and orangutan males meet about as many with each strategy. Circles show number of females met with roaming strategy at average values for density of females or daily distance traveled (filled circle), half values (smallest circle), twice average (medium open circle), four times average (largest open circle). Density and daily distance were varied separately. *Details at end of chapter.*

mating tactic (i.e., female-finding tactic). They roam and associate, but associate with continually changing parties of females. For orangutan males, the model indicates that in the average population, the payoffs of association and roaming are very similar (fig. 10.2). That is, the model predicts more frequent long-term association than is seen.

If the model is producing realistic answers, which in this case include an indication that long-term association could be an option for chimpanzee and even occasionally orangutan males, male chimpanzees and orangutans should in reality associate with females over longer periods than they do. One explanation for the disparity between model and reality is that feeding competition makes association disadvantageous for females in these frugivorous species (ch. 6.4). The difference between chimpanzees and orangutans in frequency and duration of association could then arise because female orangutans are more able than female chimpanzees to escape males, traveling as they do faster than the males, and probably on smaller trees and branches (ch. 6.4). Additionally, powerful male orangutans might attract females with their "loud calls," or "long calls" (Delgado and van Schaik 2000). With respect to the model, the value for the parameter of detection distance is greater than we allowed. This attraction is essentially a variation on the roaming strategy, although the idea that the male orangutan calls attract females still needs testing (Delgado and van Schaik 2000).

As in the infanticide model, rough agreement of the roaming model's results with reality across other species than the gorilla give faith in the model, that is, both the hypothesis behind it and the mathematical expression of the hypothesis (see box 10.1). If the model predicted the society of only gorillas, we would be worried about its usefulness. At the same time, it must be remembered that all that the results of any models indicate when they match reality is that the hypothesis could be correct. To show that it is correct, or rather to increase the likelihood that it might be correct (no hypothesis in science can ever be shown forever to be correct), the hypothesis needs to be pitted against other hypotheses that make different predictions (box 10.1).

10.3.2. Predation and association

We have already argued that the small body size of female *Pan* and *Pongo* make fleeing up a tree as a means of escaping a predator more easy for them than for a female gorilla, and therefore protective association with a male less beneficial; and we have already suggested that any such association is especially unhelpful for a *Pan* female as the male is not much larger than she is (ch. 7.1.5). If a male cannot make much difference to a mate's survival, then the scales are presumably tipped more toward the advantage of the

Box 10.1. The use of models

Quantitative models (equations, computer simulations) have two main uses. They test hypotheses, and they allow us to ask which part of a hypothesis is most important, that is, which factor produces the most change in the result. In the case of testing hypotheses, the particular advantage of models over verbal arguments is that they force the hypothesis to be stated in a quantitative way (an equation) and they allow quantitative predictions (Dunbar 1988, ch. 12; 2002). The hypothesis and its prediction, therefore, have to be precise. That precision helps toward clear thinking and allows more rigorous testing of the hypothesis.

The simpler the model is, the better, for it can then be more easily tested. If the model's results do not match observations, normal scientific practice comes into play. The data are checked, and if they are correct, then the hypothesis or the model need refining, perhaps even discarding.

Models can often be most useful when they do not produce answers that match reality. Thus, van Schaik and Dunbar's models showed that the more or less monogamous gibbon males and callitrichid (marmoset and tamarin) males would obtain access to far more estrous females if they roamed instead of, as in reality, remaining with a female (Dunbar 1988, ch. 12; van Schaik and Dunbar 1990). van Schaik and Dunbar used the disparity to argue that a major parameter was missing from the model. The omission both reinforced and extended understanding of primate society.

For the callitrichids, the model omitted the inability of females to raise offspring on their own (Dunbar 1988, ch. 12). A roaming male that mated and left would never produce any live offspring. Here, the model reinforced existing theory about the reasons for male association with females in callitrichids. For the gibbons, the model ignored the danger of infanticide, and the protection against it that a male could give (van Schaik and Dunbar 1990). That was a new hypothesis for gibbon monogamy. So, a model's wrong answer (males should not do what they in reality do) reinforced one hypothesis and helped the genesis of another (even if in the case of male-female association in gibbons the anti-infanticide hypothesis has been challenged [Palombit 1999]).

Similarly, the roaming model would presumably show that zebra, hamadryas baboon, and gelada baboon males would do better to roam, because the males are often closely surrounded by extremely high numbers of females, even hundreds of them (Dunbar 1984; Kummer 1968). Yet they do not roam. Could it be that the very high density of males means that a female cannot mate with all of the males when she is in estrus? If so, at birth of any offspring, she will find herself

(*continued*)

Box 10.1. continued

closely surrounded by large numbers of unmated and hence infanticidal males. No offspring will survive. The roaming strategy of the male is then pointless, for he produces no offspring. At high densities of males, he must remain with females to protect them against infanticidal males.

In these cases, the additional factors, additional ideas have been highlighted by a disparity between the model and reality. However, we have to be careful, as repeatedly suggested in this chapter, to test additional hypotheses even in the case of a fit. We cannot be sure that a hypothesis is correct until we have compared the model to alternatives, however much a model's results fit reality.

roaming strategy as a means to find fertilizable females than to permanent association as the means, given that one benefit of permanent association is diminished. The suggestion needs modeling, though, before we can be sure that a hypothesis that concerns only chances of finding females is the best explanation for male apes' association (or not) with females.

Conclusion

A robust model, and therefore the right answer?

The argument of this chapter is that female gorillas occur at so low a density, and that they and males travel so slowly, that a male is forced to associate permanently with females in order to ensure that he is in the presence of a female when she is in estrus, that is, is potentially fertilizable. It then pays the male to protect the resource that he is tied to, the females and their offspring, especially as he is so much more capable of doing so than are the females and offspring. The fact that the model that we employed produced near-realistic answers for all the other great apes across a range of realistic values for their demography and behavior gives us faith in the model, and therefore in the hypothesis it was designed to test. In sum, it looks as if the chances of finding fertilizable females might determine whether it is beneficial or not for males to associate with females.

Added variations

Many details are not incorporated in the roaming model. If all were, the model would be too complex to be useful (box 10.1). Nevertheless, an important variation is the possibility of mixed strategies, for example, roam until an estrous female is found, and then remain with her and defend access to her for the duration of her estrous cycle, as chimpanzee males sometimes do (Mitani et al. 2002b; Tutin 1980; Tutin and McGinnis 1981; Wallis 1997).

Additionally, several other payoffs of both the roaming strategy and the stay-and-defend strategy are not so far incorporated. Absent from the model of the roaming strategy were the costs of competition imposed on the females (ch. 6.1). Missing from the association strategy were the costs of sperm competition (box 11.1). Missing from both strategies were the payoffs of defense of females against other males and predators (ch. 7; ch. 11.1).

We could go on with adding variations and payoffs to the model. A complete account surely necessitates that we do so in order to test what difference they make. It is possible, though, that payoffs of the additions might cancel one another out. For instance, sexual dimorphism is best explained by intense competition between males for access to females, which includes their defense (ch. 2.4.1.1). Roaming males do not defend, yet males of the roaming elephant and the associating gorilla are twice the size of females, implying similar competitive payoffs to the two strategies. A third strategy of attracting females, perhaps used by orangutan males, is also associated with extreme sexual dimorphism. In other words, if degree of sexual dimorphism reflects intensity of competition between the males, this added payoff is similar in all the mating strategies, leaving mainly the male's chance of finding females as the primary influence on which strategy males adopt.

Are female strategies irrelevant to males?

Dunbar (2000) pointed out that female strategies are nowhere in the model that he used to test the payoffs of roaming compared to association, the results of which appeared to match reality. An apparent absence of an effect of females does not, however, mean that female strategies are irrelevant. In fact, the model subsumes female strategies, for female group size (a fundamental component of female strategies) is in models that Dunbar used, and used as comparison by which to judge the result of the model that we used.

Additionally, where the male strategy is to associate, the males will find it far easier to implement that strategy if it is also advantageous for the females to stay with a male; and males will find it far easier to stay with a group of females if the females for other reasons find it advantageous to travel as a group. With respect to gorillas, while it is advantageous for females to associate with a male, females clearly choose between males (ch. 4.2.1.2;

ch. 8.1), and can exert that choice despite resistance from males that is evident in their herding of females during intergroup encounters (see ch. 11.1.3.1). When females have the option of leaving, it might pay the resident male to provide extra benefits to the females that will keep them with him, such as good quality protection (ch. 7; ch. 8.1), and babysitting of immature animals, which allows the mothers to feed unencumbered (ch. 4.2.2.4; ch. 12.1). If providing these services indeed is a response to females' ability to emigrate (ch. 5.2), female strategies directly affect male strategies. The society is not simply a result of the males mapping onto the females. Rather, the society must be described and understood as a two-way interaction between female and male strategies.

Figure Details

Fig. 10.1(a), (b). The equation used to calculate the number of females that a roaming male will encounter is based on the Maxwell gas molecule equation. The chances of encounter (number of females met) are determined by the density of "molecules" (females), the speed of movement of the molecules (daily distance traveled), the size of the box containing the molecules (area over which the male travels), and the distance at which molecules affect one another (detection distance). Thus:

$F = (2av/H)Yz$ (different coding than for the infanticide model in chapter 7.2, but the units are different),

where F = number of females encountered,
2 = speed and detection distance of *both* "molecules" (male and female) affect encounter rate,
a = detection distance,
v = daily distance traveled,
H = size of home range (larger H = smaller proportion searched/day),
Y = total density of females, and
z = number of fertilizable days per cycle. (It is the number of fertilizable days that matters here, not the duration of the estrous period as when we were considering infanticide, on the assumption that it is offspring actually produced by the male that will affect his behavior over evolutionary time, rather than simply matings. The number of days that a female is fertilizable is a function of both life span of ova [only about one day] and of sperm [about two days in mammals, except bats] [Gomendio et al. 1998]. We allowed three fertilizable days in the model.)

Base values in the model are given in figure 10.2 details.

Fig. 10.1(b). The values for density and daily travel distance entered are HD, ST = 0.75 female/km^2, 0.4 km/day; HD, LT = 1 female/km^2, 2 km/day.

Fig. 10.1(a), (b). Note that we have not statistically tested the difference between number of females met by roaming compared to associating. We have simply assumed that the difference is great enough to indicate that roaming would not work.

Fig. 10.2. Number of females encountered. Data for average population: *B* is in days; detection distance (*a*), daily distance traveled (*d*) in km; male home range area (*H*) in km^2; female density is individuals/km^2; female group size is individuals; female group density is /km^2 (these two parameters were used to explore the effects of group size and density, but neither made a difference, because a larger group, which reduces the probability of finding any female [it effectively reduces their findable density] multiplies exactly proportionately the chances that once a group is encountered, a female in it will be in estrus [e.g., three times the chance of a female being in estrus in a group of three than a group of one]); number of fertilizable days per conception per female (*z*) is the same for all the species (3 fertilizable days per cycle [compared to Dunbar's 5 days [Dunbar 2000, 2002]; 3 cycles per birth; = 9 fertilizable days per birth interval per female, = for gorillas 9 *d*/4 yr = 9/1,460 = 0.0062]). Differences in values of parameters from infanticide model of chapter 7.2 are because, for ease of comparison, we here used less precise values. Data are from relevant references in chapters 3 and 4.

Species	Birth interval (*B*)	Detec. dist. (*a*)	Daily travel (*d*)	Male range (*H*)	Fem. density (/km^2)	Fem/ group	Group density	No. fert. days (*z*)
Gorilla	1,460	0.5	1.5	15	0.25	3	0.125	9
Chimp	1,825	0.5	3.5	15	0.5	2	0.25	9
Orang	2,190	0.5	1	5	0.5	1	0.5	9
Bonobo	1,460	0.5	2	22	0.5	3	0.25	9

Adolescent (blackback) and fully adult (silverback) males from the same group.
(Mountain gorilla.) © A. H. Harcourt

CHAPTER 11

Male Mating Strategies and Gorilla Society

Summary

Male mating strategies have a large impact on gorilla society. They help to explain the stability of gorilla groups (long male breeding tenure and female group tenure), as well as group composition, in particular the number of males. Male mating competition influences both male and female dispersal and female residence decisions.

As indicated by the extreme sexual dimorphism of gorillas and the single-male, multi-female composition of groups, male gorillas compete intensively for exclusive, long-term access to females. At its most serious, this rivalry includes damaging, sometimes fatal, aggression between males, and infanticide. Competition between males, therefore, serves not just to attract and retain females, but to protect infants as well.

While male-male competition is an important determinant of social structure, so too is female choice. The freedom of gorilla females to move between silverbacks means that the females strongly influence the size of breeding groups, as well as the nature of males' mating strategies. The bases on which females choose mates are not clear, but they are likely to include male qualities such as strength or experience, which enhance his ability and willingness to protect a female and her offspring, and possibly more subtle behaviors such as a male's tolerance or affiliation toward her infants.

Although contest between non-group males is the primary stage for mating competition, it is not the only one. Males residing in the same group also compete to mate. The intensity of that competition, and the benefits that dominants might gain from tolerating young subordinates, influence whether young males emigrate or remain in the group with other males. Here they must wait in line for top rank and the breeding advantages this position affords. Once again, female choice plays a role in the relative benefits to males of dispersal or philopatry, specifically, females' preferences for single or multi-male groups.

The difference between gorilla populations in proportion of multi-male groups relates to the myriad ecological and social variables that influence group size, male and female dispersal, and females' choice of males and groups.

11.1. Competition to Be Sole Breeder

11.1.1. Gorilla society as a one-male mating system

The data from most gorilla populations indicate that they have a single-male mating system. Across Africa, the majority of groups contain only one fully mature male, which remains the resident silverback for years at a time (ch. 4.2.1). Paternity studies from one western site, Mondika in CAR, and from mountain gorillas in the Virungas, show that females breed with only resident males and do not sneak matings with outsiders (Bradley et al. 2004, 2005). Thus, the only populations in which mating with multiple males might occur to any significant degree are those of mountain gorillas in which groups with more than one male are common (ch. 4.2). The extent to which a multi-male group structure correlates with multi-male breeding is considered later in this chapter.

Anatomical evidence also indicates that gorillas evolved to be single-male breeders. Firstly, extreme sexual dimorphism is characteristic of societies with intense competition between males for sole access to fertile females (stats. 11.1) (Harvey and Harcourt 1984), and the gorilla is one of the more sexually dimorphic of primates (ch. 3.2.2).

Secondly, the sexual anatomy and physiology of both sexes points to a system in which multi-male mating and, therefore, sperm competition, has played a small evolutionary role. Across many taxa the males of species in which females mate with a single male have smaller testes and inseminate fewer sperm than do the males of species in which females mate with several males during any one fertile period. Gorillas have some of the smallest testes compared to body mass of any primate (box 11.1). In addition, female primates in one-male mating systems have relatively short estrous periods and inconspicuous or no sexual swellings, as is the case for female gorillas (ch. 2.4.4.1; ch. 3.2.3).

Rather than competing for fertilizations at the time of estrus, as do, for example, chimpanzees, male gorillas compete primarily by attracting and

Box 11.1. Form and function: Mating system, reproductive anatomy, and sperm competition

Seventy years ago, Adolph Schultz traveled around the world collecting primates—by shooting them. Back then, the seemingly endless forests were full of monkeys and apes, and the conservation ethic was not part of Western fieldworkers' consciousness. Schultz, an evolutionary anatomist, dissected scores of carcasses and noticed that the testicular weight of primates relative to their body weight differed enormously across species (Schultz 1938). He had no idea why. It was not until thirty years later, in the 1970s, that enough observers had collected enough behavioral data to show the connection between Schultz's results and the primates' mating system.

Roger Short, a widely read reproductive biologist interested in evolutionary questions, put the two fields of functional anatomy and socioecology together. He argued that in species in which several males mated with the same female during the same fertile period, the male likely to succeed at fertilization would be the one that bought the most tickets in the breeding lottery, in other words, inseminated the most sperm. The way to achieve this is to possess the most sperm-producing tissue, that is, the largest testes (Short 1979). If only a single male is mating, he need inseminate only enough sperm to ensure that at least one reaches and fertilizes the egg. Short demonstrated the argument with a comparison across the great ape family, the gorilla, common chimpanzee, orangutan, and humans.

Subsequent work quickly confirmed the correlation: multi-male primates have relatively large testes; single-male primates have relatively small ones (Harcourt et al. 1981c). And confirmation has continued unabated (box 11.1, fig. 1) (Birkhead and Møller 1998; Dixson 1998, ch. 8; Gomendio et al. 1998). Multi-male species not only have larger testes but indeed produce more sperm, a greater proportion of healthier sperm, faster sperm, more energetic sperm, all carried by more semen from larger seminal vesicles. All these correlations apply not only to primates but to a wide range of taxa, from mammals in general to birds to butterflies. One of the more interesting recent findings is that the multi-male primates also have a larger midpiece to their sperm. This component is a sperm cell's powerhouse, packed with mitochondria (Anderson and Dixson 2002).

While Geoff Parker might have been the originator of sperm competition theory (Parker 1970), it was Roger Short's seminal contribution that set the field on its way, with its empirical demonstration that in apes and humans, testicular size correlated with mating system. If not for the previous decade of intense field study by primate socioecologists, Adolph Schultz's puzzle would have had to have waited longer for a solution.

(*continued*)

Box 11.1. continued

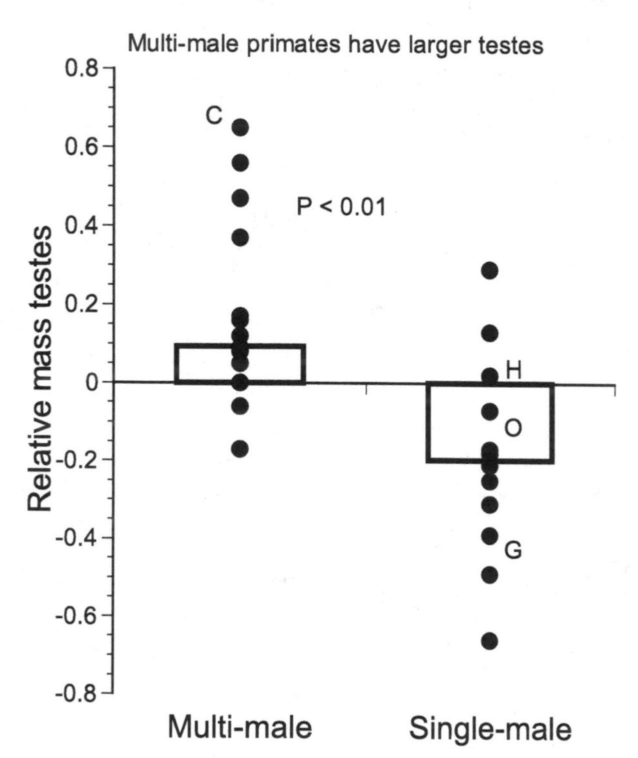

Box 11.1, Fig. 1. Primate taxa in which several males mate with an estrous female have larger testes for their body mass (relative mass testes) than do primate taxa in which, typically, only one male mates. Circles show values for genera; bars are medians. C, H, O, G, = chimpanzee, human, orangutan, gorilla. *Details at end of chapter.* Source: Data from Harcourt et al. (1995).

retaining females with which they associate permanently, and by repelling outside silverbacks.

11.1.2. The reproductive payoffs of infanticide

Killing the infants of rivals is one extreme way in which gorilla males compete for breeding opportunities (ch. 4.2.1.3). While the theory of infanticide as a sexually selected strategy is now widely accepted (box 2.4), data from primates demonstrating the reproductive payoffs of the behavior are relatively few (van Schaik 2000b). Research on mountain gorillas has provided some of the strongest support yet for the theory.

In his analysis of 13 incidents of infanticide, Watts (1989a) showed that, by and large, infanticidal males benefited reproductively from the tactic. All victims were unrelated to their killers, and most were young infants (less than 18 months), which means their removal would have had a relatively large impact on females' reproduction (ch. 3.2.3). In most cases for which there were data, mothers remained and bred with the male that killed their infants (6/8), and gave birth sooner than they would have if their infants had survived to weaning. Of 11 males in habituated groups observed as silverbacks for at least five years, at least 5 have killed or attacked unrelated infants (Robbins 1995; Watts 1989a). In the Virungas, it seems likely that all males would be infanticidal given the right circumstances. Observations of eastern lowland and western gorillas in which males spare unrelated infants are puzzling in light of the data from mountain gorillas (ch. 4.2.1.3). We consider this issue later on in this chapter (sec. 11.3.2).

Most documented cases of infanticide involved vulnerable mothers, that is, those that became separated from their mate, usually because he had died (ch. 4.2.1.3) (Watts 1989a; Yamagiwa and Kahekwa 2004). Clearly, the main safeguard against infanticide is the protective presence of an infant's potential father or other close relative. Conflict between non-residing silverbacks, therefore, needs to be understood not just as competition over females but also as defense of related infants.

11.1.3. Male competitive tactics: Mate acquisition versus mate retention and offspring protection

11.1.3.1. Contests during interunit encounters

The main stage for mating competition between silverbacks is interunit encounters. Variation in the nature of encounters (ch. 4.2.1.4) reflects in part different male tactics, either mate acquisition, or mate retention/offspring protection (Harcourt 1978a; Sicotte 1993; Watts 1994c).

Solitary silverbacks actively seek out groups and force confrontations. Their daily travel is erratic, with relatively sedentary periods interspersed with long forays far outside their normal range. They sometimes track groups silently, following their trail for days on end, behavior not shown by breeding silverbacks. These mate-seeking behaviors of lone males have been described in mountain and western gorillas from several sites (Bermejo 2004; Caro 1976; Cipolletta 2004; Tutin 1996; Watts 1994c; Yamagiwa 1986).

Once silverbacks have acquired mates and sired offspring, they decrease their active pursuit of other groups and switch to either avoidance or defense, behavior more suited to retaining females and protecting offspring.

Some studies have shown that the nature of encounters varies with the types of units involved. Confrontations between groups and lone males or small units with no infants are either longer (Sicotte 1993), or more antagonistic, than those between two established groups (i.e., those with infants) (Bermejo 2004; Harcourt 1978a). For example, in western gorillas at Lossi, Rep. Congo, the main study group normally responded to lone males by fleeing or displaying, whereas it was frequently tolerant and even indifferent in its meetings with other groups (fig. 11.1) (Bermejo 2004).

The adventures of lone silverback Tiger, a mountain gorilla in the Virunga Volcanoes, illustrate the different tactics of solitary and group-living males and the threat that unmated males pose to breeding silverbacks (Watts 1994c; Yamagiwa 1986). Over a five-year period, Tiger steadily expanded his range, meeting groups at a frequency closely tied to his monthly travel distance. During one particular month, he pursued the breeding group Nk (Nunki), continuously for a week, chasing them day and night. In this one week, both units covered over three times the area that either normally covered in a month. In the course of this routing, several of Nk's females and their offspring became separated from the group, some for several weeks,

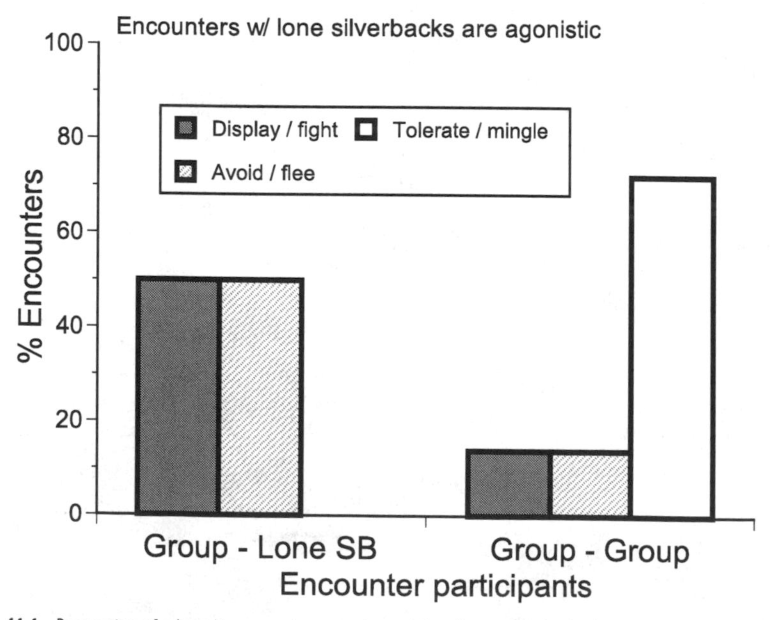

Fig. 11.1. Proportion of a breeding group's encounters with solitary silverbacks (left set of bars) and other groups (right set of bars) during which the group male avoided or fled, aggressively displayed, or peacefully tolerated (including mingling with) the other unit. *Details at end of chapter.* Source: Data from western gorillas at Lossi, Rep. Congo (Bermejo 2004).

before rejoining it. Tiger almost certainly killed the infant of one of these "lost" females, who then transferred to another group outside the study area. By the end of the month, the silverback Nk had lost two females (one of them to Tiger) and one infant to infanticide. All animals appeared exhausted (Watts 1994c).

After acquiring a female, Tiger dramatically reduced his monthly travel distances, retreating to use only about half of the range he covered when solitary. By contrast, Nk shifted his range in the opposite direction, abandoning Tiger's neighborhood and entering a completely new (to the group) area.

Group males also vary their behavior during encounters according to the likelihood that their females will transfer. In mountain gorillas, silverbacks are more likely to fight the higher the number of potential migrants, that is, females without dependent offspring (stats. 11.2) (Sicotte 1993). Furthermore, a male is more likely to "herd" a potential migrant than a lactating mother (Sicotte 1993). Herding during interunit encounters is a form of mate-guarding in which males display, charge, and sometimes bite a female when she moves toward the rival silverback. However, herding is rare, and occurs in only 10% or so of encounters between males (Harcourt 1978a; Sicotte 1993), possibly because in the thick vegetation, the male cannot easily see females or keep track of their movements.

This variation in male tactics emphasizes that males compete both to acquire and retain females.

11.1.3.2. Why are there no male takeovers? The influence of females on the stability of male-female associations

The long tenure of dominant males has a major influence on gorilla society. It limits breeding opportunities for subordinate males (see sec. 11.1.4.1) and means that when daughters mature, their putative fathers may be their only mates, an important impetus for females to emigrate from their natal group (ch. 8.2). One reason male tenure is so long is that extragroup males do not take over breeding groups and oust the resident male. Why not? They certainly do so in other group-living primates, for example, langurs (Rudran 1973; Sugiyama 1984) and geladas (Dunbar 1984).

We suggest that the answer lies partly in female choice. In a female-resident species characterized by one-male groups, as in some populations of Hanuman langurs, for example, an outside male that fights his way into a breeding group, defeating and ousting the resident dominant male (and sometimes juvenile males as well), is effectively guaranteed access to all the females in the group. He may have to kill a few infants or wait for juvenile females to mature, but the package he fought for remains intact (Koenig and Borries 2001; Sommer and Rajpurohit 1989). This is not the case with gorillas. Females have the option to transfer, and most do so more than once in their lives (ch. 4.2.2). Group membership, therefore, depends heavily on females' choice of male or group, which means that a silverback that has invested a lot of energy ousting the resident male might find himself with a hard-won resource that then disappears of its own accord.

If a male loses all of his females to another silverback during an encounter, then the distinction between transfer and aggressive takeover is a subtle one. Such incidents, however, are rare since it is usually only non-lactating, non-pregnant females that transfer, thus making it unlikely that all the females in a group would be ready to move at the same time (ch. 4.2.1.2) (Sicotte 1993; Watts 1990b). When all females transfer voluntarily to the same silverback, they usually do so incrementally, over a series of encounters. An example was recorded in Grauer's gorillas in Kahuzi-Biega when, in three encounters over a period of six months, the weaker (and wounded) silverback gradually lost all twelve of his adult females to the same male (Yamagiwa and Kahekwa 2004).

Observations like this are rare, but they indicate that even when one male loses all his mates to another silverback, he does so mainly because of female transfer decisions. In Thomas langurs, a species similar to gorillas, with predominantly one-male groups and female transfer, aggressive takeovers do sometimes occur but are far less common than in female-resident Hanuman langurs . More usually, Thomas langur males acquire females gradually, as in gorillas (Steenbeek et al. 2000).

The rarity of either male immigration or takeovers leaves maturing males with two options for becoming successful breeders. The reproductive consequences of these alternatives, and the reasons why males might choose one or the other, are considered next.

11.1.4. Mating competition in multi-male groups

A main challenge to the description of gorilla society as a one-male mating system has come from the studies of mountain gorillas. Not only are multi-male groups relatively common (about 30%–45%, Ch 4.2), but such groups are an enduring aspect of the society and not just a transitional phase. For example, two study groups in the Virungas have had multiple males continuously for fifteen years and counting (Robbins 2003).

Early observations on relatively small multi-male groups concluded that these were effectively single-male mating systems because dominant silverbacks were able to prevent younger subordinates from mating (Harcourt et al. 1981b). However, studies of larger groups during the late 1980s and early 1990s (Robbins 1999, 2003) showed that mating by more than one male was more common than previously suggested. To what extent do dominant males control access to receptive females? How do males in the same group compete for mating opportunities?

11.1.4.1. Breeding success and mating competition: The importance of being dominant

Studies of two multi-male groups in the Virunga Volcanoes showed that most adult sexually cycling females—13 out of 18 in one study period and 11 out of 14 during the other—mated with more than one male (Robbins 1999; Watts 2000a). Nevertheless, dominant males copulated more often than did subordinates and were significantly more likely to mate with adult fertile females (fig. 11.2A). In contrast, lower-ranking males mated more frequently with subadult females that were receptive (sexually cycling) but not fertile (Harcourt 1979c; Robbins 1999; Schaller 1963; Watts 1990b). Genetic data reflect these behavioral observations (fig. 11.2B). In a paternity analysis of 48 offspring born into four groups (including the two in fig. 11.2A), Bradley and colleagues (2005) showed that the dominant silverbacks sired 83%–100% of offspring that survived past weaning, while in three of the groups, the second-ranking male sired about 15% of the infants. Because two younger males usurped top rank from older dominants, breeding success was significantly related to rank, but not age (stats. 11.3). The total reproductive skew in one group (Group Pb in fig. 11.2B) may have been due to incomplete data, since there was a relatively high proportion of infants (one-third) in this group whose paternity was not resolved.

Recent genetic work on the mountain gorillas of Bwindi, Uganda, revealed similar findings: dominant silverbacks in multi-male groups enjoy a huge breeding advantage but lose a small proportion of fertilizations to lower-ranking males (Nsubuga 2005, cited in Robbins 2006).

The biggest constraint on subordinate males' mating activities is aggression from older silverbacks. While dominant males sometimes tolerate copulations between subordinates and subadult females, they vigorously try to prevent younger rivals from mating with adult fertile females (Harcourt 1979c; Robbins 1999; Schaller 1963). The only times dominant males tolerate such matings are when the females are their putative daughters. This indifference is presumably due to a lack of sexual attraction between fathers and daughters. Very few completed matings in such pairs have ever been observed (Harcourt et al. 1981b; Watts 1990b).

Males harass each other during copulation, usually with mild aggression such as cough grunts or charges. In a study of two multi-male groups,

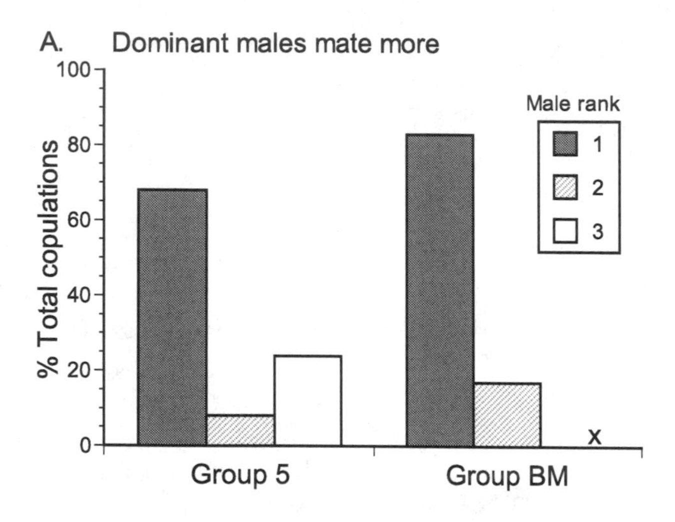

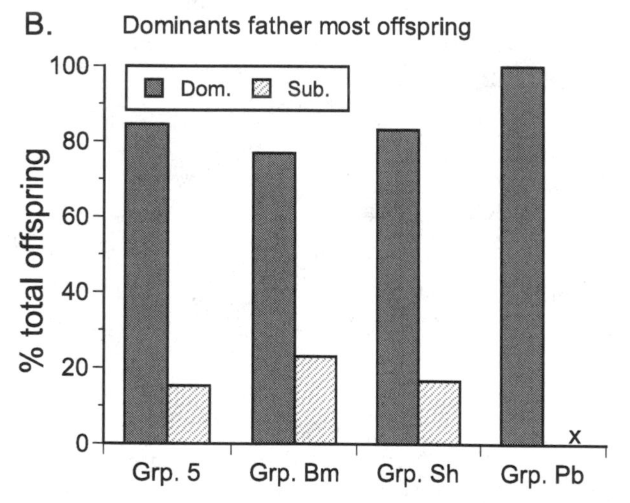

Fig. 11.2. Dominant males do most of the breeding. (A) Proportion of copulations with *adult fertile females* (not lactating or pregnant) performed by silverbacks of different ranks in two groups. (B) Proportion of offspring fathered by dominant and lower-ranked males. (Mountain gorilla.) *Details at end of chapter.* Sources: Data from (A) Robbins (1999, 2003); (B) Bradley et al. (2005).

Robbins (1999) recorded harassment during 22%–30% of copulations. Dominant males were usually successful at interrupting mating attempts by lower-ranking males, but they themselves did not go unchallenged. Subordinate males also harassed their superiors, but with less success at stopping the proceedings (fig. 11.3).

Overt mating competition is not confined to harassment during copulation. In some multi-male groups, general aggression between males (usually from dominant to subordinate) is more frequent on days when at least one female is in estrus than on days when none is. The number of males in a group vying for matings, along with the competitive differences between them, probably influence the level of aggression between them over estrous females (Robbins 1996, 2003; Sicotte 1993).

Researchers have speculated that the ability of subordinate males to steal fertilizations should vary with group size and competitive differences between the males (Robbins 2003). When the balance of power is large and groups are small, the alpha male should have a relatively easy time keeping his younger rival away from mates (Harcourt 1979c), while in larger groups with more evenly matched males, dominants have a harder time monitoring sexual shenanigans (Robbins 1999). Surprisingly, the genetic data showed that the subordinate's breeding success did not vary with number of cycling females or the age difference (i.e., power difference) between the first- and second-ranking males (Bradley et al. 2005).

In sum, subordinate males in multi-male groups have limited breeding opportunities until they become dominant. Their chances of doing so affect

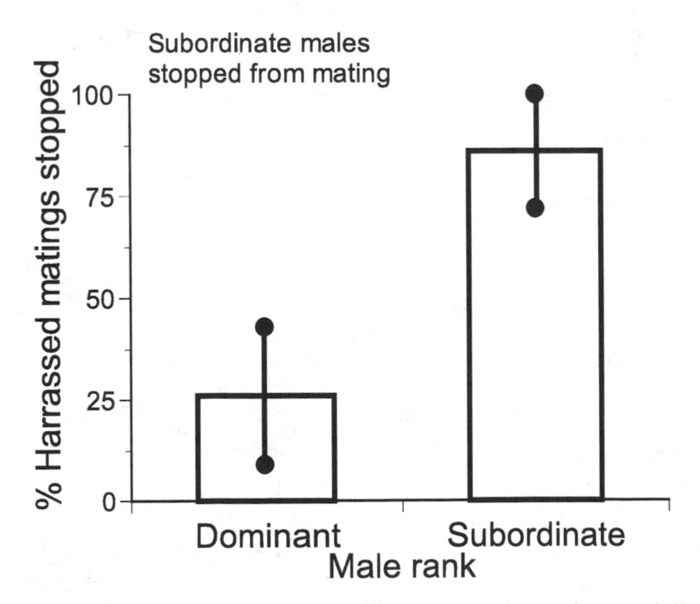

Fig. 11.3. Proportion of mating harassments to which dominant and subordinate males responded by ceasing to copulate. Bars are medians; lines are range of values for two dominant and two subordinate males in two mountain gorilla groups. *Details at end of chapter.* Source: Data from Robbins (1999).

the probability of multi-male groups persisting in a gorilla population. We examine those chances in section 11.2, when we consider the reproductive payoffs to males of staying in or emigrating from the group in which they mature.

11.1.4.2. Mating competition and male-female interactions: Coercion and mate-guarding

Co-resident males compete for breeding opportunities not only by directly challenging each other but also by trying to influence females' behavior (ch. 2.4).

Male gorillas sometimes use coercion or mate-guarding to persuade females to mate, behaviors that are most obvious in multi-male groups. In one study, for example, both dominant and subordinate males were significantly more aggressive toward estrous than non-estrous females. Much of this aggression, excluding that directly related to copulation, consisted of displays and was not linked to any specific context. Rather, it seemed to be attempts to assert males' dominance over females (Robbins 2003). By contrast, in groups with only one silverback, males did not increase aggression toward estrous females (Harcourt 1979a).

Silverbacks in multi-male groups intensify their efforts to stay near receptive females (i.e., mate-guarding) by persistently following them and sometimes "neighing." This is a specific vocalization given by males when females are moving away from them. Its consequence is more often than not a maintenance of proximity, either because the female ceases her departure, or the male follows her. Males neigh over four times as often when females are potentially fertile than when they are lactating or pregnant (Sicotte 1994). The fact that this vocalization has been recorded from only multi-male groups indicates its association with mating competition (Fossey 1972; Harcourt et al. 1993; Sicotte 1994).

11.1.5. Female choice and male mating competition

Female choice is considered to be a major influence in gorilla society, and an important component of mating competition between non-resident males, as we have already discussed (ch. 8, and see earlier discussion). It also plays a role in mating competition within groups.

The dominant male's mating success results in part from female preferences. In most groups, adult females initiate over 63% of their copulations, and solicit most often from dominant males (Harcourt et al. 1981b; Watts 1990b, 1991b). The degree to which female choice influences the mating success of subordinate males is not clear. While subordinates initiate over 80% of their copulations, most of these are with subadult (i.e., infertile) females. However, adult females also copulate with subordinate males (fig. 11.2A, B), and in one study, initiated 45%–50% of these matings (Sicotte 2001).

In a variety of primate species, female promiscuity (mating with more than one male during any one cycle) has been functionally interpreted as a means of creating paternity confusion, and thereby ensuring protection against infanticide (ch. 2.4.4.1) (Hrdy 1979; van Schaik 2000b). In gorillas, mating with multiple males in a group certainly has the potential to serve this function (Robbins 1999; Watts 2000a). When the dominant male of a multi-male group dies, group disintegration and consequent infanticide is often averted because the subordinate male ascends to top rank, retains the females, and the group carries on (ch. 4.2.2). The reason this works is because successors treat group infants as their own. In many cases, the young males will have mated with the infants' mothers and will be at least potential sires (figs. 11.2 and 11.3; ch. 4.2.2.1).

11.1.6. Long-term reproductive strategies

11.1.6.1. The wooing of females by subordinate males

While subordinate males are usually unsuccessful in direct competition for fertile matings, they may use other longer term tactics to improve their chances of future breeding. Some younger males actively foster friendly associations with females, including those that are lactating. In a reverse of the typical gorilla pattern in which females take the initiative in friendly interactions with silverbacks (ch. 4.2.2.1), subordinate males seek the proximity of particular females, groom them, and interact affiliatively with their infants (Stewart 2001; Watts 2003; Watts and Pusey 1993). These friendly overtures may represent attempts to win over females as future mates, as Smuts (1985, ch. 8, 9) has argued for male baboons.

Evidence that long-term friendly interactions may influence female mate choice comes from data on the fission of one of the Virunga study groups, Group 5. At the time of the split, the dominant silverback had recently died, leaving three mature males, ten adult females, and two subadult females. The group divided evenly into two. The oldest, newly dominant male and the youngest third-ranking male ended up with six of the females and their immature offspring, while the other six females joined the second-ranking silverback. The rates of affiliation and close proximity during the three years before the fission (when the new group leaders had been subordinate males) predicted reasonably well which silverback the females would join (Robbins 2001; Watts 2003).

11.1.6.2. Control of female aggression

Some long-term aspects of males' social relationships with females are usefully interpreted as reproductive strategies because they may influence the probability that a female will stay in the group rather than transfer to another male.

Specifically, males control aggression between females by intervening in their disputes (ch. 6.2; 6.3). In all study groups in the Virungas, silverbacks, especially dominant males, were the primary interveners in females' conflicts (ch. 4.2.2.1) (Harcourt 1979a; Harcourt and Stewart 1987, 1989; Watts 1997).

Most significantly, the majority of males' interventions are neutral, favoring neither of the contestants, but simply stopping the conflict. When males do give obvious support, it is to aid the victims or targets of the aggression (ch. 6.2; 6.3). Because males are so much bigger and stronger than females, they make effective policemen. Watts (1997) found that silverbacks successfully ended, with no argument, 80%–85% of the conflicts in which they intervened.

Policing behavior serves to even out competitive differences between females (ch. 6.2; 6.3). Whether a silverback supports a victim or simply breaks up a fight, his interventions prevent stronger females from consistently outcompeting weaker ones and protects subordinate females from escalated aggression. Male protection may be especially important for the integration of recent female immigrants (Harcourt 1979b; Watts 1991c, 1994a).

Controlling female aggression might help males retain mates that would otherwise leave due to competition from other female residents. Ultimately, male interventions perpetuate emigration and transfer as a general female strategy (ch. 8; ch. 9). By reducing the costs of being a poor competitor, males minimize the value to females of living with kin (ch. 5) as well as the potential costs of moving into a new group (ch. 2.2.3.2; ch. 2.4.2.6; ch. 8) (Watts 1997).

11.2. Maturing Males: Stay or Emigrate?

When gorilla males begin to silver, they must decide whether to remain in the group in which they mature or disperse to find females elsewhere, a decision that affects the rest of their reproductive careers. Males that stay have to queue for top position in the male hierarchy before beginning their breeding tenure. Because neither group takeover nor immigration into a breeding group are viable options for mature males, those that emigrate have to acquire mates from other groups (ch. 4.2.1.2).

In western and eastern lowland gorillas, the vast majority of males disperse before breeding, hence the low proportion of multi-male groups in their populations. In mountain gorillas, philopatry is far more common (ch. 4.2.1). What are the reproductive consequences for mountain gorillas of different life history pathways? Why do some males leave and some males stay? The question is relevant to understanding both the general impact of male mating strategies on female grouping (ch. 2, ch. 8), as well as the variation in social structure across populations (see sec. 11.3).

Section 11.2.1 first considers the main influences on males' lifetime breeding success and then contrasts the reproductive fortunes of males that stay in the group in which they mature with those of males that emigrate. The findings reviewed in this section summarize work by Watts and Robbins. Their analyses are based on decades of demographic data from intermittent censuses of the Virunga population as well as detailed long-term data on known individuals in three to five habituated groups (Robbins 1995; Robbins and Robbins 2005; Robbins et al. 2001; Watts 2000a).

We refer to males that mature in a breeding group and remain there into adulthood as "stayers." Sexually mature males that disperse from breeding groups before reaching full adulthood are "dispersers" (called "followers" and "emigrants" or "bachelors," respectively, in previous publications [Robbins and Robbins 2005; Watts 2000a]). In single-male groups, "breeding tenure" refers to the time that a mature male resides with adult females. In a multi-male group, breeding tenure refers to a silverback's reign as the dominant male (see previous discussion).

11.2.1. Breeding tenure and number of mates

Male gorillas vary widely in their breeding success. Some individuals produce no surviving offspring (infants that live past three years) while the most successful male so far has produced at least thirteen (Bradley et al. 2005; Robbins 1995). The most important determinants in this variation are the duration of a male's breeding tenure and the number of adult females in his group during that time. More females for a longer time translate into a greater number of offspring surviving past weaning (fig. 11.4) (Robbins 1995).

11.2.2. Why do dominant males tolerate younger rivals?

The whole question of dispersal decisions comes about because male gorillas appear to emigrate voluntarily. Despite somewhat tense relations, serious

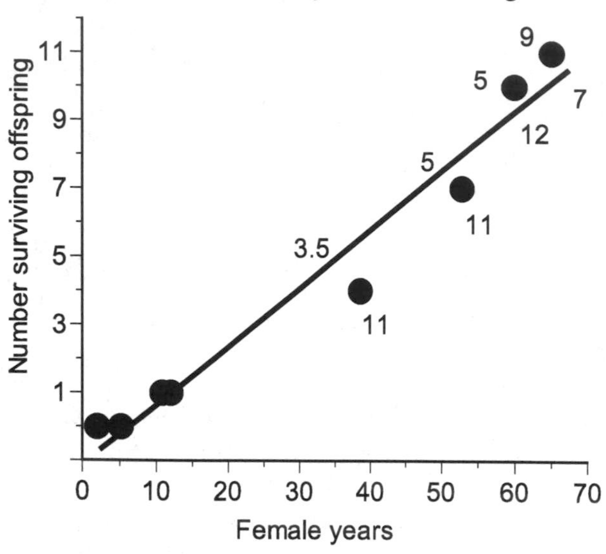

Fig. 11.4. Number of surviving offspring sired by males as a function of female group size and duration of males' tenure. Female years = mean number adult females in group per year of tenure × duration of tenure. Numbers above the line show mean female group size for the four most successful males, and numbers below the line show these males' tenure length (yrs). (Mountain gorilla.) *Details at end of chapter.* Sources: Data from Robbins (1995) and Bradley et al. (2005).

aggression between co-resident males is rare (ch. 4.2.2.3), and there is no evidence that dominant silverbacks forcibly evict subordinates, even in populations with almost universal male dispersal (Parnell 2002; Yamagiwa and Kahekwa 2001). Given that dominants face mating competition from lower-ranking males (fig. 11.2A, B; fig. 11.3), as well as the risk of being usurped, why don't they force them out before they become serious opponents, for example, when they are still subadults or blackbacks? What do silverbacks gain from having young males stay in the group?

The primary long-term benefit of having a successor is that of offspring survival. If the dominant male dies, his offspring are protected from infanticide by his successor, who can retain females and prevent group disintegration (ch. 4.2.1.2). Of 11 certain cases of infanticide in the Virungas, 9 (80%) occurred after the death of a dominant male, and in all these cases, the silverback was the group's only fully mature male (fig. 11.5). In 3 instances, a blackback or young silverback (about 12 years old) was present, but it appears that males this young have a hard time retaining females (Robbins 1995; Stokes et al. 2003; Watts 1989a).

In addition, because of the protective benefit of multiple males, transferring females in the Virungas preferentially join multi-male rather than single-male groups (ch. 8.1, fig. 8.2). Thus, co-resident males benefit in the currency of mate attraction as well as offspring survival. This is a striking example of the interactions of male and female strategies. Male mating tactics (infanticide) influence female residence choices, which reflect in part their rearing strategies, that is, choosing a group in which their offspring are protected (ch. 7; ch. 8.1), which in turn influence male dispersal decisions.

In the shorter term, but related to infant defense and mate retention/acquisition, dominant silverbacks benefit from subordinates' participation in interunit encounters. Co-resident males jointly defend the group against outside threat (Fossey 1983; Harcourt 1978a; Sicotte 1993, 2001). While the males are not necessarily acting together in a coalitionary sense, nevertheless, aggression from multiple silverbacks must be more effective than aggression from just one. Furthermore, as males age past their prime, the aggressive prowess of a fit younger silverback may become increasingly important for group defense (fig. 11.6).

Finally, because co-resident males are often relatives (with fathers or full brothers less common than more distant kin, for example, uncles or half brothers [Bradley et al. 2005]) dominant silverbacks may gain inclusive fitness by tolerating maturing males, and therefore helping them to become established breeders (Watts 2000a).

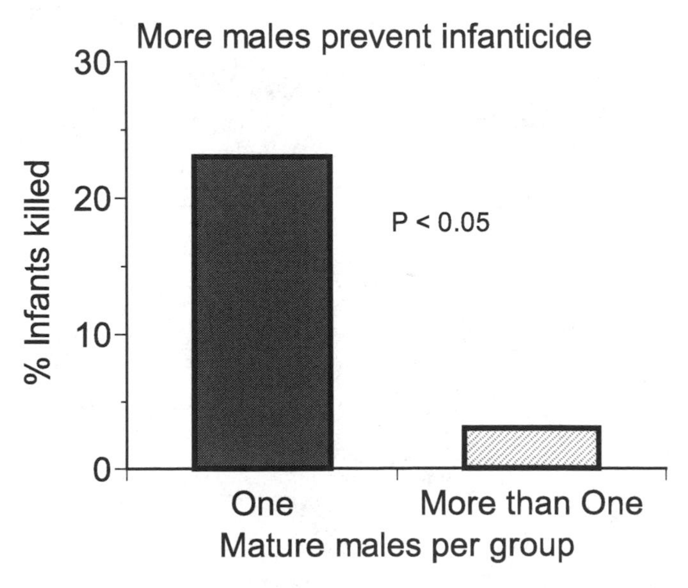

Fig. 11.5. Proportion of infants in single- and multi-male groups that died from infanticide. (Mountain gorilla.) *Details at end of chapter.* Source: Data from Robbins (1995).

Fig. 11.6. Joint defense of the group against potential danger (human), with younger male taking most active role (closer to human and in more aggressive stance) (Mountain gorilla.) © A. H. Harcourt

A recent model assessed the fitness consequences of dispersal decisions to both dominant and subordinate males across a wide range of conditions (Robbins and Robbins 2005). The model showed that in general, whether a subordinate male leaves or stays has little impact on the dominant silverback's lifetime reproductive success. The fitness costs of mating competition from a younger rival, including the risks of being usurped, were slightly offset by the increase in offspring survival that comes from having a successor. While this slight fitness gain vanished with the presence of more than one subordinate male, the model did not take into account the possible inclusive fitness benefits of tolerating a younger relative.

11.2.3. Staying versus dispersing

The earliest difference in reproductive opportunities for dispersers and stayers comes during their pretenure years. Most dispersers emigrate at around 13.5 years (Robbins 1995; Watts 2000a) and are likely to be alone until they acquire females (ch. 4.2.1.2). Stayers, on the other hand, have a slight chance of stealing fertilizations as subordinates, even when they are still blackbacks (Bradley et al. 2005; Nsubuga 2005, cited in Robbins 2006).

After emigration, dispersers then face a greater chance than stayers of total reproductive failure. While data on lone silverbacks are few because they are hard to keep track of, it seems that dispersers have difficulty gaining females and retaining them for long enough to breed. For example, of six males that left their groups and became solitary, only two managed to produce surviving offspring. In contrast, seven out of nine stayers became dominant breeders (Bradley et al. 2005; Robbins 1995).

Dispersers gain females and stayers become dominant at around the same age, about 15–16 years (range = 14–19) (Bradley et al. 2005; Robbins 1995). However, because dispersers usually acquire females a few at a time, they typically start off their breeding tenure with fewer mates than do stayers, and never catch up (Watts 2000a). It is estimated that stayers have, on average, 4.7 females in their group during their tenure, while dispersers average only 1.8 (Robbins and Robbins 2005).

Finally, most males that opt for staying will end up residing with at least one other mature male, either the silverback they succeed, or younger males waiting in line. They therefore benefit from living in a multi-male group in ways discussed in section 11.2.2.

In dispersers' favor, they are likely to be the sole breeder because they have started their group from scratch. By doing so, they avoid the costs of reproductive competition inherent in multi-male groups (see sec. 11.2.2) (Robbins and Robbins 2005; Watts 2000a). Finally, because dispersers are less likely than stayers of being deposed, the duration of their breeding tenure is slightly longer. Watts (2000a) estimated ten years versus nine years for dispersers and stayers, respectively.

11.2.4. Reproductive payoffs of different male strategies

How do the variables discussed previously interact with each other and with reproductive and demographic factors? Is it possible to calculate which decision, dispersing or staying and queuing, will produce the most surviving offspring during a male's lifetime? Watts (2000a) and Robbins and Robbins (2005) developed models to tackle this question, and despite differences between the models, they both come to the same conclusion: staying is dramatically more successful than dispersing. On average, stayers produce 3.2 surviving offspring whereas dispersers produce only 1.6. Furthermore, 45% of dispersers fail completely at breeding, compared to only 22% of stayers (Robbins and Robbins 2005).

The two most important factors accounting for this difference are a stayer's greater chances of holding breeding tenure and his access to more females during that time. The benefits that successful dispersers gain from uncontested mating access are compromised by the greater risk of losing offspring to infanticide (table 11.1).

Table 11.1. Benefits of different dispersal decisions. ++ = most important correlate of differences in reproductive success. (Mountain gorilla.)

Important variables	Stayers	Dispersers
Age begin breeding tenure	~ =	~ =
Duration tenure		+
Chance of breeding before age 15	+	
Chance of holding breeding tenure	++	
Average # FF during tenure	++	
% adulthood in multi-male group	+	
Fewer infants lost to infanticide	+	
Inclusive fitness benefits	+	
Less breeding shared		+
Lifetime reproductive success	++	

Sources: Robbins and Robbins (2005) and Watts (2000a).

Table 11.2. Males that become dominant through succession within a group achieve higher reproductive success than those that become solitary and form their own group. Mean # females = mean number of females per year of male's breeding tenure. Under Surviving offspring for Ziz, value in parentheses is the number of offspring if breeding had been divided equally among mature males in group. (Mountain gorilla.)

Male	Strategy	Dom. tenure (yrs)	Start # Females	Mean # Females/year	Surviving offspring
Tiger	Disperse	2	1	1.0	0
Nunki	Disperse	11	2	4.8	7
Ziz	Stay	7	6	9.3	11 (8.5)

Source: From Robbins (1995).

The differences in reproductive payoffs of the two strategies are illustrated by the fortunes of one stayer and two solitary males, considered here to be the equivalent of voluntary dispersers (table 11.2).

Tiger became a solitary silverback in 1981 at age 13, after years in a bachelor group following the disintegration of his natal group. He was alone for almost four years, persistently forcing encounters with breeding groups until he finally acquired a female in 1985. For the next two years he resided with one to two females. He sired one infant that died, his females deserted, and Tiger died in 1987, a few months after a serious fight with a group-living silverback (Robbins 1995; Watts 1994c).

Nunki was a more successful disperser. He appeared in the study area as a solitary silverback in 1974 and began acquiring females from breeding groups. Although we don't know his previous history (i.e., whether or not he had emigrated voluntarily from a breeding group), becoming a solitary silverback is the most common fate for dispersers. During his eleven-year tenure (four years longer than *Ziz's*) ending with his death, he resided with an average of 4.8 females per year and sired 7 infants that survived to weaning. The number would be 9, except that his last 2 infants were killed by other males when he died.

When the young male, *Ziz,* matured in his natal group (Group 5) and became the dominant silverback in 1984, there were two younger males (one a full brother) in the group who became silverbacks over the next five years and who competed with Ziz for access to estrous females (see figs. 11. 3 and 11.4). During the seven years of Ziz's nine-year tenure for which there are data on surviving offspring, there was an average of 9.3 breeding females in the group. With them, Ziz sired 11 surviving offspring (Bradley et al. 2005). Even if breeding had been split evenly with the other silverbacks, Ziz would have produced 8.5 surviving offspring (table 11.2). Thus, he still would have done better than Nunki because he started out with more females, retained more throughout his tenure, and lost no infants to infanticide (Robbins 1995).

Recent genetic data on two other stayers in two different groups showed that, by 2001, each male had sired at least 11 surviving offspring and were still the dominant silverbacks in 2005 (Bradley et al. 2005).

11.2.5. Why leave?

The data from the Virunga mountain gorillas indicate strong advantages to young males of staying and queuing for dominant position rather than emigrating. Why then, do males disperse when it appears to be by far the inferior strategy?

It should first be pointed out that these conclusions are based on very few data from dispersers, since solitary males usually disappear from study sites. Their predicted breeding prospects may improve with more information. In addition, many males end up as solitaries through no choice of their own. For example, if the dominant silverback dies and his females desert, the subordinate males are left no option but to travel alone, or join a group of bachelors (ch. 4.2.1.2, and see *Tiger,* previous section). It's estimated that only about half of the males in a population ever even face a dispersal decision (Robbins and Robbins 2005).

Furthermore, the models' predictions are based on averages and, as Watts (2000a) pointed out, individual animals rarely face average conditions. Breeding prospects in a group may be so limited that it isn't worth a male's while to wait in line for dominant position. Such might be the case if, for example, a male is low down in a long queue, the dominant male is in his prime, or there are few females in the group (Robbins and Robbins 2005). With data from the Virungas, Watts showed that stayers lived in groups with fewer other males (i.e., a shorter queue) (fig. 11.7A) and with more females per male than did dispersers (fig. 11.7B). Stayers also tended to be the second-ranking male, that is, the next in line (9 of 13 stayers), and were more likely than dispersers to reside with older alpha males (fig. 11.8). In other words, they had less time to wait before becoming dominant (Watts 2000a).

11.3. Variation Across Gorilla Populations

Understanding the causes and consequences of male strategies in mountain gorillas can be applied to other populations and can help explain the variation in gorilla social structure across Africa. Specifically, male strategies are

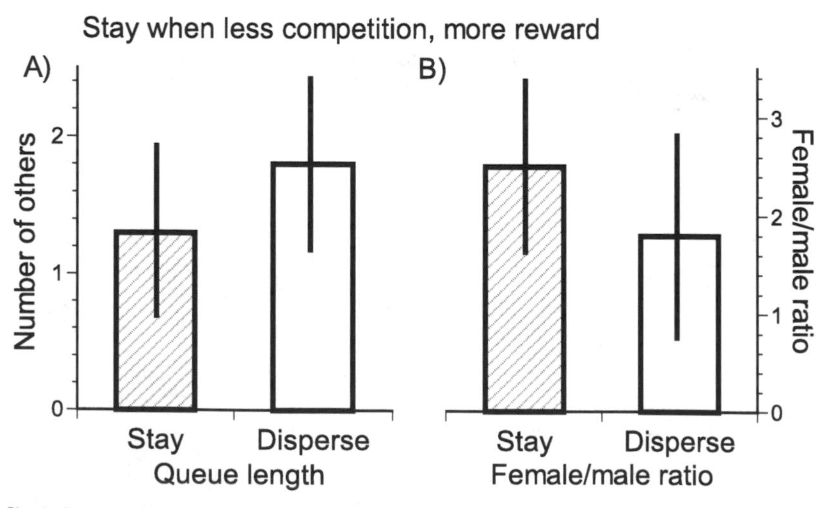

Fig. 11.7. Males stay when they face (A) less competition (fewer other males), or (B) more reward (more female per male). Bars are means; lines are one standard deviation for 11 stayers and 12 dispersers. (Mountain gorilla.) *Details at end of chapter.* Source: Data from Watts (2000a).

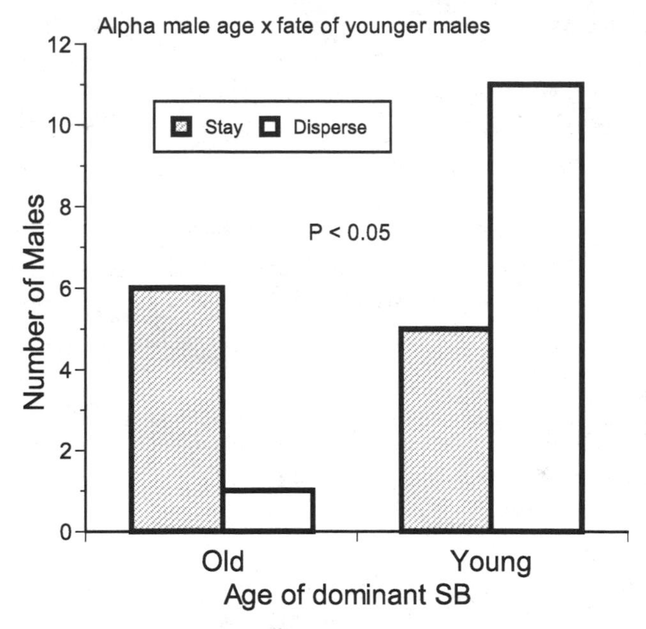

Fig. 11.8. The number of stayers (Stay) and dispersers (Disperse) that, on reaching 12–13 years of age, resided with an old (silverback for ≥ 10 yr) or young (silverback for < 10 yr) dominant male. (Mountain gorilla.) *Details at end of chapter.* Source: Data from Watts (2000a).

relevant to population/subspecific differences in the proportion of groups that contain multiple males and may also relate to apparently different rates of friendly versus hostile encounters between social units and in the predictability of infanticide.

11.3.1. Differences in rates of male dispersal

The most well-established societal difference between gorilla subspecies is the relatively high proportion of multi-male groups in mountain gorillas. For example, of 124 groups sampled at seven western and eastern lowland sites, only 3% had more than one silverback, compared to 30%–45% of mountain gorilla groups in Bwindi and the Virungas (table 4.2, ch. 4.2.1.1). It is possible that demographic factors contribute to this difference. Because of the gorilla's long life history (ch. 3.2), it takes many years for a multi-male group to develop. From the time a solitary male gains a female and begins breeding, it will be at least fifteen years before there will be another mature silverback (Robbins and Robbins 2004). Variation in birth rates, mortality rates, or infant sex ratios could lead to differences in the likelihood of multi-male groups developing. However, there is no strong evidence so far that this is the case (ch. 3.2) (Robbins et al. 2004). Thus, since the processes of group formation and decline are similar across populations (ch. 4.2.1), the contrast in proportion of multi-male groups is most likely due to variation in male dispersal at maturity.

The data from mountain gorillas indicate that males benefit reproductively from residing in groups with more than one silverback. It seems extraordinary, therefore, that male dispersal in other subspecies is almost universal. What explains this difference? The most commonly suggested reason relates to habitat.

11.3.1.1. Ecological constraints on group size, male mating competition, and male emigration

Maturing males will be more likely to remain in groups with favorable breeding opportunities (fig. 11.7A, B; fig. 11.8). Since the numbers of adult and immature females represent current and future breeding possibilities, large groups, with their large number of females, should increase the likelihood of male philopatry (ch. 4.2). In mountain gorillas, single-male groups have fewer weaned individuals than do those with more than one male. This holds true for a small sample of research groups in the Virungas (Robbins 1995) and for a broader comparison of weaned group size based on census data from Bwindi and the Virungas (fig. 11.9).

These findings support the suggestion that multi-male groups will flourish primarily when ecological conditions allow the development and persistence of large groups. The differences in diet and foraging effort across gorilla populations indicate that mountain gorillas face fewer ecological constraints than do other populations (ch. 4.1). Furthermore, in the Virunga gorillas, group size, and therefore level of feeding competition, makes little difference to females' transfer decisions (Watts 1990a, 2000a). In contrast are the claims from one western site, Mbeli in Rep. Congo, that females preferentially transfer from larger to smaller groups, suggesting that they are trying to minimize levels of feeding competition (Stokes et al. 2003). However, since larger groups have more females available to transfer, this finding is precisely what one would expect were females moving randomly with respect to group size. It remains to be seen if western gorilla females show a preference for group size in their transfer decisions.

Having said that, there is another factor unrelated to habitat that is implicated in the persistence of large multi-male groups in mountain gorillas. That factor is human disturbance, especially the extreme violence to which the Virunga region was subjected during the 1990s. Warfare led to habitat

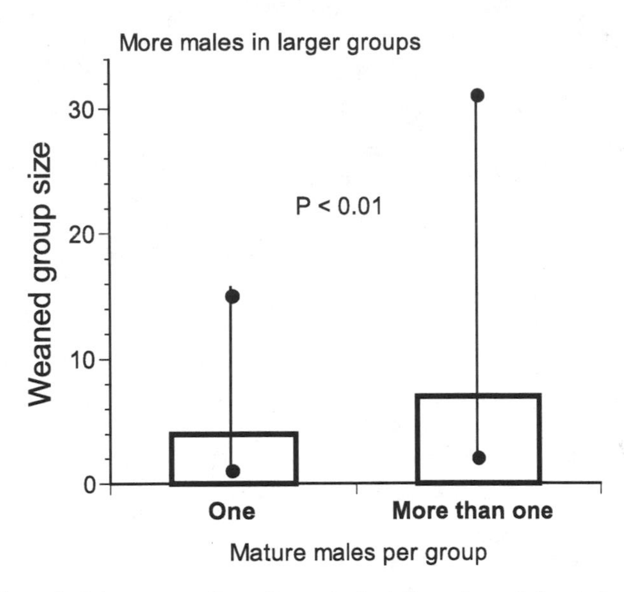

Fig. 11.9. More males in larger groups. Weaned group size (excluding mature males) against number of males per group (one vs. more than one). Bars are median group size; lines are total range of values. (Mountain gorilla.) *Details at end of chapter.* Sources: Data from Kalpers et al. (2003) and McNeilage et al. (2001).

degradation, military activity inside the park, and the direct killing of gorillas (Kalpers et al. 2003; Plumptre and Williamson 2001). While gorillas throughout the Virungas suffered, some were worse off than others. The seventeen habituated research and "tourist" groups in Rwanda and DRC received better protection than did other groups (Gray et al. 2005). Of these, three research and one "tourist" group with adjacent ranges were monitored almost continuously. This differential protection affected mortality rates as well as emigration, creating safe havens that animals did not want to leave. As a result, the four groups receiving the highest protection had a median group size of 29 (r = 22–47), and a median number of 3 silverbacks per group (range = 2–6) in contrast to the other relatively well protected habituated groups with a median size of 11 and one mature male (range = 1–2) (Kalpers et al. 2003).

This possible human influence is important to keep in mind when considering male dispersal decisions. The models predicting large advantages to males in staying versus dispersing (see earlier discussion) depend heavily on data collected from gorilla groups that might be enjoying "artificial" prosperity due to differences in protection levels. However, while human disturbance might help explain the supersized groups in some regions of the

Virungas, it does not account for the consistent occurrence of multi-male groups in this population since the 1960s when censuses began, nor the equally common occurrence of multi-male groups at Bwindi.

11.3.1.2. Interaction of male and female strategies and the perpetuation of group structure

Rates of male dispersal may influence females' residence and transfer decisions in ways that perpetuate a multi-male or single-male group structure. When male philopatry is common, natal females will be less likely to disperse because they have access to a mate other than their putative father (ch. 8). When females do transfer, they preferentially join multi-male groups (ch. 8.1, fig. 8.2) where their infants are protected from infanticide (fig. 11.5). Finally, a high proportion of multi-male groups in a population will decrease the likelihood of group disintegration and multiple female transfers (ch. 4.2.1.2). These effects on female dispersal reduce the reproductive chances for solitary males, thus increasing the relative benefits of philopatry and the persistence of multi-male groups (Parnell 2002; Robbins 1995; Robbins and Robbins 2004; Stokes et al. 2003). If the process is followed through its logical progression, it results in groups that grow larger and larger until they fission. This is precisely what has been observed in the Virunga population, where multi-male groups have increased to quite extraordinary sizes and where fission has been documented six times (Kalpers et al. 2003).

On the other hand, when males habitually disperse and multi-male groups are rare, female transfer is relatively frequent. Natal females are likely to emigrate to avoid inbreeding, and groups that lose their silverback are likely to disintegrate (ch. 4.2.1.2, but see exceptions in Kahuzi-Biega, DRC, in following discussion). At Mbeli, Rep. Congo, for example, where all breeding groups are single-male, groups disintegrate at roughly three times the rate as do those in the Virungas: five disintegrations at Mbeli in 63 group-years of observation versus five in the Virungas in 200 group-years (Robbins et al. 2004). Attracting females from a disintegrating group may actually be a lone silverback's best chance for breeding. Such events provide an opportunity for him to gain multiple females in an "en masse" transfer, rather than single females incrementally (Parnell 2002; Stokes et al. 2003).

Relatively high rates of female transfer may thus improve the breeding prospects for solitary males, thereby increasing the relative benefits of male dispersal. Is there any evidence that solitary males in western or Grauer's gorillas have an easier time forming groups than do mountain gorilla males? The only data on regularly observed lone western males come from Mbeli, where 2 out of 10 solitaries managed to acquire females long enough to breed successfully, in a study of 6.5 years. In the Virunga population, the success rate of solitary males was also 2 out of 10 over 33 years of observation (Robbins et al. 2004). Because of the very different observation conditions at the two sites, it's hard to know how comparable these data are. All we can say currently is that there are no obvious differences between populations in the success with which solitary males gain mates. At both Mbeli and Kahuzi-Biega (Yamagiwa and Kahekwa 2001), solitary males sometimes spend years alone, as do mountain gorillas.

11.3.1.3. Lower risk of infanticide

Some researchers have suggested that multi-male groups do not persist in eastern lowland and western gorillas because infanticide isn't as great a risk as in mountain gorillas (Yamagiwa and Kahekwa 2001). In this case, neither males nor females gain the reproductive benefits conferred by better offspring survival in multi-male groups, which would reduce the advantages of male philopatry to both subordinate and dominant silverbacks (Robbins and Robbins 2005). If this is the case, then the question is, what would lower risk of infanticide in other populations?

11.3.2. Differences in predictability of infanticide and nature of interunit encounters

Although infanticide has been either observed or strongly suspected in Grauer's and western gorillas, in some populations it appears to be less universal than in mountain gorillas (ch. 4.2.1.3). Some western sites also report relatively high rates of non-aggressive interunit encounters (ch. 4.2.1.4). While these findings are still tentative, pending more data, they are worth considering in light of the differences in male dispersal and proportion of multi-male groups.

11.3.2.1. Demographic influences on mating competition

One reason for an absence of infanticide could be low numbers of unmated males and, therefore, less intense male-male mating competition. For example, the unusual "widow" groups in Kahuzi-Biega may have persisted, unmolested, because they rarely came across solitary males that were actively pursuing females (Yamagiwa et al. 1993; Yamagiwa and Kahekwa 2001). Nevertheless, these widow groups did occasionally encounter lone silverbacks, with no harm to their infants. Regardless of how infrequent the opportunities for infanticide are, a mature male should benefit from killing an unrelated infant when the chance arises.

11.3.2.2. Familiarity and relatedness between males and females of different groups

A possible explanation for variation in the occurrence of infanticide is variation in relatedness/familiarity of extragroup males (Yamagiwa and Kahekwa 2001). Long-term observations of mountain and eastern lowland gorillas show that males emigrate gradually from their groups, sometimes paying visits after long absences of several months. The same process occurs during fission and the gradual separation of subgroups. During their intermittent return visits, emigrating males are often accepted by the resident silverbacks (ch. 4.2.1.2). It can take a long time for dominant males to view former residents as strangers and aggressively repel them. In a recent observation, a subordinate male from the Virunga study population returned to his group after six months' absence and was treated as if he had never left (DFGFI 2005). Had this male not been observed since birth and his emigration documented, researchers might have recorded his return as an unusual case of immigration by an unfamiliar outside male into a breeding group.

If emigrating males establish ranges in the neighborhood of their parent group, then they may not be unfamiliar with the lactating mothers that might transfer to them from those groups. In fact, they might even have mated with those females in previous months, making the infants potential offspring or at least relatives. Yamagiwa and Kahekwa (2001) describe a scenario like this for eastern lowland gorillas. Males that emigrate either as solitaries or with a portion of the group females (fission) remain in the general range of the parent group.

At the western gorilla site of Mondika, CAR, a genetic study found that the silverbacks in neighboring groups were more closely related to each other than were the females in those groups, leading to speculation of "networks" of dispersed, but related males (Bradley et al. 2004). Similarly, among mountain gorillas in Bwindi, neighboring males are often closely related (Nsubuga 2005, Ph.D. diss. cited in Robbins 2006).

If infanticide really is less frequent in western and Grauer's gorillas than in mountain gorillas, then there will be less benefit to females in residing in a group with more than one male, and females may base their choice of group on minimizing feeding competition rather than maximizing protection, a process that will inhibit the development of multi-male groups (Parnell 2002; Robbins and Robbins 2005; Stokes et al. 2003; Yamagiwa and Kahekwa 2001).

Differences in relatedness or familiarity between silverbacks may also explain apparent variation in the nature of interunit encounters. While all studies report displays and contact aggression between extragroup males, tolerant encounters, including mingling and sometimes even nesting in proximity, are more frequent in western than mountain gorillas (fig. 11.10) (Bermejo 2004). These findings are not just from swamp studies where interunit encounters are so common that aggressive responses to strange males might simply be too costly.

If tolerance during intergroup encounters results from familiarity between the males, then the nature of encounters should vary with the individual identities of the groups, a finding that some observers have reported (Bermejo 2004; Levréro 2001).

This discussion has been largely speculative. We need more information to confirm and understand differences in infanticide rates and the nature of interunit encounters across populations. More data on feeding competition in western gorillas will elucidate possible variation in ecological constraints on group size, male dispersal, and female residence decisions. We also need far more long-term data on life histories of males in Grauer's and western gorillas, as well as on relationships between males in the same group. For example, while there is no evidence that young males are evicted by dominants, they may in fact be less tolerated by their superiors than are subordinate mountain gorillas, and may have less of a choice. Obviously too, data on relationships, both social and genetic, between the silverbacks of different groups, as well as those between males and females in different groups, are going to be highly illuminating for future understanding.

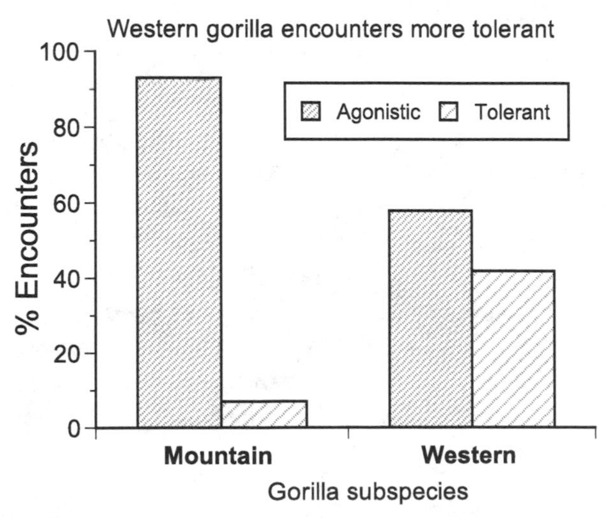

Fig. 11.10. Mountain gorillas are more often agonistic during interunit encounters than are western gorillas. *Details, including data sources, at end of chapter.*

11.4. Comparison with *Pan* and *Pongo*

11.4.1. Common chimpanzees

The male mating strategies of two primate species could hardly be more different than those of gorillas and chimpanzees. Because ecological constraints prevent permanent cohesive groups in chimpanzees (ch. 4.1.8), the harem-style mating system of gorillas could never work for them. Females are too dispersed to be monopolizable (ch. 2.2; ch. 10.1). Instead, a group of chimpanzee males defends a territory within which females range, excluding outside males and sharing community females reproductively among themselves. Unlike gorillas, males do not face dispersal decisions because all males remain in the community of their birth (ch. 4.2.3).

The genus *Pan* is the poster primate for a multi-male, promiscuous mating system. The majority of matings occur in mixed-sex gatherings where one or more estrous females copulate many times (several hundred per conception) with multiple males, and occasionally with all the males in the community (Mitani et al. 2002b; Muller and Wrangham 2001). In some multi-male gatherings, mating is completely promiscuous with no male-male aggression or coercion of females, whereas in others (possessive mating), males try to monopolize access to an estrous female by aggressively preventing others from copulating with her (Watts 1998c). Sometimes, a male and female pair remove themselves from the rest of the community and mate exclusively with each other during consortships (Tutin 1980), but at no site is this the primary tactic. The data from across populations indicate that 75%–94% of conceptions occur in multi-male settings (Boesch and Boesch-Achermann 2000; Constable et al. 2001; Hasegawa and Hiraiwa-Hasegawa 1990; Wallis 1997). Sperm competition, therefore, has played a far greater role in chimpanzee male strategies than in gorillas', a fact reflected in the chimpanzees' much larger testes (box 11.1). The relatively long estrous period (10–12 days, compared to gorillas' 2–3) and obvious sexual swellings of female chimpanzees also indicate a predominantly multi-male mating system (Furuichi and Hashimoto 2002; Wallis 1997; Wrangham 2002).

This is not to say that aggressive contests are not involved in mating competition. In fact, possessive mating may be the most important tactic by which male chimpanzees gain a reproductive edge over other males in their community and is used most frequently when females are in the most fertile phase of their cycle (Watts 1998c). The most dominant males are best able to monopolize access to females (by aggressively guarding them and repelling other males) and at several sites, rank correlates with measures of mating success with maximally fertile females (Muller and Wrangham 2001) or with actual paternity (Constable et al. 2001). This may help to explain chimpanzee males' obsession with status-related contests within communities (ch. 4.2.3) (Muller and Wrangham 2004). Nevertheless, the reproductive advantage of dominance is nowhere near as extreme as in multi-male gorilla groups. For chimpanzees, the two main mating strategies are to achieve high rank, which carries costs in the form of wear and tear (Muller and Wrangham 2004), or to remain subordinate and compete entirely through sperm competition. In contrast, for gorillas, life as a subordinate is not a competitive strategy.

The majority of long-term chimpanzee studies have recorded infanticide by males both within and between communities (Arcadi and Wrangham 1999; Stumpf 2007). Unlike infanticide by gorilla males, infant-killing by chimpanzees is not readily explained by sexual selection (i.e., as a male mating strategy) because it occurs in a variety of contexts. In some within-group cases, however, the victims were infants whose mothers may have mated with extragroup males (Arcadi and Wrangham 1999).

The most serious aggression between chimpanzee males occurs in the context of boundary patrols and attacks on individuals from neighboring communities (ch. 4.2.3). Most evidence suggests that this xenophobic violence serves to defend a territory and its resources by decreasing the fighting capabilities of the other side and does not specifically increase males' access to more females (Watts et al. 2006; Williams et al. 2004; Wilson et al. 2004). It is not, therefore, as clear an example of sexual selection as is interunit aggression between gorilla males. However, territorial aggression by chimpanzees is a form of communal, long-term mate-guarding, since it protects females from harassment by neighbors (Williams et al. 2004) and ensures community males near exclusive sexual access to those females. Genetic analyses of eastern (Gombe) and western (Taï) chimpanzees indicate that fertilizations by males outside the community are rare (Constable et al. 2001; Vigilant et al. 2001).

A remarkable aspect of intergroup aggression in chimpanzees is the extent of coalitionary action among community males (Watts et al. 2006). While co-resident gorillas may cooperate during interunit encounters, their behavior isn't nearly as coordinated or as concertedly violent as that of chimpanzees (ch. 4.2.3). Cooperative aggression is also a component of short-term mating competition in chimpanzees because of its importance in dominance contests and because males sometimes use coalitions as part of a possessive mating tactic (Goodall 1986; Watts 1998c).

While the mating strategies of male chimpanzees have been shaped largely by ecological influences on female distribution (ch. 2.1.2; ch. 4.1.8; ch. 4.2.3) (Wrangham 1979), male-male competition has in turn shaped the reproductive strategies of females. Because it is too costly for females to travel permanently with males, thereby ensuring paternity and benefiting from their mates' protection (Wrangham 2000), they mate promiscuously with a set of males, creating paternity confusion and countering the risk of infanticide (ch. 2.4.4.1; ch. 4.2.3) (van Schaik et al. 2000). This has further promoted the evolution of sperm competition as the primary male mating strategy. Although females might exert mating preferences during their most fertile days of estrus (e.g., at Taï, Stumpf 2005), in general, female mate choice has had a relatively small influence on chimpanzee society. In contrast, gorilla social groups are based on females' choice of males as permanent mates, and male strategies evolved in part to influence this choice.

11.4.2. Bonobos

Male bonobos are generally less aggressive than chimpanzees in many contexts (Hohmann and Fruth 2002), including use of coercion and aggression in their mating tactics. For example, male bonobos do not compete aggressively over mating access to an estrous female the way chimpanzees do in possessive mating (Furuichi 1997; Hohmann and Fruth 2002; Muller and Wrangham 2001). Nevertheless, at two sites, the dominant males were more successful reproductively than others. They were the sons of dominant females whose support may have helped the males achieve high rank (Bradley and Vigilant 2002; Gerloff et al. 1999; Hohmann et al. 1999).

In bonobos, the power differential between males and females is more balanced than in chimpanzees (ch. 4.2.3), and females probably have a greater impact on male strategies (Parish 1994). For example, female choice may be mediating the relative mating success of dominant males and might even have shaped bonobo males' relatively unaggressive mating tactics (Furuichi 1997; Hohmann and Fruth 2002).

The relative power of female bonobos is tied in part to their cooperative alliances with other females, which are possible because sociality is less costly than it is for chimpanzees (ch. 4.1.8; 4.2.3) (Wrangham 1986).

11.4.3. *Pongo*

Researchers have characterized the orangutan's mating system one of "roving male promiscuity" because wandering males depend on opportunistic mating with females that they happen upon. Dominant resident males are unable to exclude these competitors from a region, because of foraging costs, just as they are unable to continually monitor the dispersed females in their range (ch. 4.1.8; 4.2.3) (Rodman and Mitani 1987; van Schaik and van Hooff 1996).

Most evidence suggests that orangutan males compete reproductively primarily through aggressive contest competition, rather than sperm competition, and in this way resemble gorillas. This evidence includes their high degree of sexual dimorphism, relatively small testes, and females' inconspicuous and relatively short periods of receptivity (Delgado and van Schaik 2000). In addition, fully developed or flanged males (box 3.2) are highly intolerant of each other, and either fight seriously or avoid one another. They are more tolerant of unflanged males, except in the presence of an estrous female, when they aggressively repel the smaller subordinates and attempt to maintain exclusive access to the female (Galdikas 1981; Mitani 1985b; Utami and van Hooff 2004). Finally, while females often mate with more than one male during any one reproductive cycle (see following discussion), mating with multiple males on days of peak fertility (i.e., ovulation) is probably rare due to females' strong preference for dominant flanged males as consort partners (Knott and Kahlenberg 2007; Utami and van Hooff 2004).

Because heavy, slow-moving flanged males are unable to track receptive females, the females are the ones that seek out dominant males, possibly locating them from their long calls (ch. 4.2.3), and take the lead in initiating consortships during estrus, which lasts four to six days. Thus, part of the reproductive advantage males gain from being dominant comes from females' preference for them as mating partners over unflanged or subordinate flanged males (Fox 2002; Knott and Kahlenberg 2007; Utami and van Hooff 2004).

Females' mating preferences, however, cannot always be expressed. In a classic example of male sexual coercion, unflanged males and sometimes subordinate flanged males force copulations with unreceptive and often strongly resisting females. These individuals, being smaller and faster than flanged males, are better able to keep up with females that have a hard time evading them (Fox 2002; Galdikas 1981; Mitani 1985b; Utami and van Hooff 2004).

Despite these forced matings, researchers have long assumed that dominant males sire the majority of offspring in an area because of their ability to guard fertile mates, and because of females' mating preferences (Knott and Kahlenberg 2007). However, at one site in Sumatra unflanged males sired 6 of the 10 infants whose paternity was assessed (Utami et al. 2002). One of these males had been unflanged (and breeding) for over twenty-three years. These findings suggest that remaining an undeveloped male (box 3.2) might be an alternative strategy with potentially high payoffs, perhaps especially during periods of rank instability between flanged males (Utami and van Hooff 2004). It remains to be seen if unflanged males at other sites are equally successful breeders.

Orangutan society has all the ingredients for high risk of infanticide. Although female orangutans gain protection from sexual harassment while they are in consort with preferred dominant males, they do not associate permanently with a protector, but nor are they completely promiscuous (Fox 2002; Utami and van Hooff 2004). The fact that infanticide has never been observed in orangutans remains a puzzle (Knott and Kahlenberg 2007; van Schaik and van Hooff 1996).

This is an example of how grouping in apes results from females choosing to associate with males because of the services they provide. If feeding costs did not prohibit long-term associations in orangutans, one can imagine that consortships with dominant flanged males would turn into permanent associations; the alternative strategy of itinerant unflanged male would cease to pay off, and orangutan society would become more like that of gorillas.

Conclusion

Gorillas' habitat and diet allow, but do not select for, group-living. Because of these weak ecological constraints, male reproductive strategies have a relatively large role in shaping gorilla society. The adult membership of breeding groups is based primarily on aggressive competition between males for permanent access to females, and on females' choice of male as long-term mates. The male mating tactic of infanticide influences females' residence decisions, which in turn may affect male dispersal and therefore group structure. The most obvious difference in gorilla society across populations is the proportion of groups that have multiple males. While this contrast might be fundamentally related to ecological differences that influence maximum group size, it comes about through variation in male dispersal decisions, which are based on mating opportunities.

The male mating strategies of the great apes vary enormously, for example in the relative importance of sperm competition, the nature of alternative strategies, and the impact of female choice. At a basic level, these differences reflect variation in feeding competition, the costs of grouping, and the resulting distribution of females (ch. 2; ch. 4.1.8; ch. 4.3). Across the great apes, male mating strategies have had variable influences on shaping society, depending on the relative impact of ecological constraints. What is similar among the apes is that there are few ecological reasons (e.g., enhanced competition over food) for females to live in groups (Wrangham 1979). Instead, females are gregarious in order to benefit from services provided by males, one of which is protection from coercion by other males, including mating harassment or infanticide (ch. 2; ch. 7; ch. 8) (Smuts and Smuts 1993; Wrangham 2002; Wrangham 1986).

Figure Details

Box 11.1, Fig. 1. The *p* value we show is after phylogenetic correction (box 2.2) on species' values ($N = 9$ contrasts, all showing multi-male taxa with larger relative testes mass, $p < 0.01$, Sign test). That humans and orangutans have testes quite a bit larger than the gorilla implies that some multi-male mating during the short period while the female is fertile might be occurring in these two species. Note that it is mating only in the few fertile days around ovulation (perhaps no more than 3) that is evolutionarily relevant, not non-fertile mating.

Fig. 11.1. Encounters with lone males are more agonistic. Group–Lone silverback, $N = 8$; Group–Group, $N = 14$.

Fig. 11.2A. Dominant males copulate most. Total copulations: Group 5 = 37; Group Bm = 48. Number cycling adult females: Group 5, Male 1 = 4; Male 2 = 2; Male 3 = 3. Group Bm, Male 1 = 4; Male 2 =3.

Fig. 11.2B. Number offspring with identified father: Group 5 = 13; Group Bm = 13; Group Sh = 6; Group Pb = 7. Total of 9 additional offspring with identified father, but unknown dominant relationships among males at time of conception.

Fig. 11.3. Effectiveness of harassment by males of different rank. Number harassed copulations for two dominant males, $N= 11–14$, and two subordinate males, $N =3–18$ in two groups. Only males with at least three harassed copulations were included.

Fig. 11.4. Male success proportional to female-years. Data come from eight males in both single-male and multi-male mountain gorilla groups. Calculation assumes that dominant male is siring 85% of offspring when actual paternity data not available (fig.11.2B).

Fig. 11.5. Infanticide is less common in multi-male groups. $N = 26$ infants born in single-male groups; 30 in multi-male groups; data from six study groups. Fisher exact probability test, $p < 0.05$. Statistical result from Robbins (1995).

Fig. 11.7. Factors influencing male dispersal in mountain gorillas. Queue length = number other males (12+ yr) in group; Female/male ratio = adult females per sexually mature male (9+ yr). Queue length: $t = 1.85$, $df = 21$, $p < 0.1$ (*t*-test); sex ratio: $t = 2.16$, $df = 21$, $p < 0.05$.

Fig. 11.8. Natal males more likely to stay when dominant male is old. $N =$ 11 Stayers, 12 Dispersers. Fisher exact probability test, $p < 0.05$.

Fig. 11.9. Multi-male groups are larger. $N = 43$ groups in two mountain gorilla populations, 21 single-male, 22 multi-male. Weaned group size, *t*-test: $t = 2.86$, $df = 41$, $p < 0.01$. Data from Virunga and Bwindi populations.

Fig. 11.10. Western gorilla interunit encounters are more tolerant. "Agonistic" includes fleeing, displaying, or contact aggression. "Tolerant" includes proximity with no displays, mingling of members from different units, or both. Values for western gorillas = median of three populations. $N = 58$ encounters, mountain gorillas; 131 encounters, western gorillas. Data from: mountain gorillas, (Sicotte 1993); western gorillas, Lossi Reserve, Rep. Congo (Bermejo, 2004); Lokoué, Rep. Congo (Levréro 2001); Mondika, CAR (Doran-Sheehy et al. 2004).

Statistical Details

Stats. 11.1. Sexual dimorphism in single-male versus multi-male genera, $t = 2.14$, $N = 12, 18$, $p < 0.05$. (Harcourt and Stewart, unpublished data).

Stats. 11.2. Greater defense with potential emigrant females: (a) fight during encounter, mean of 5.7 females without offspring ($N = 10$); (b) not fight, mean 2.9 females ($N = 48$). Data from Sicotte (1993). Statistical details are slightly complex, but across four intensities of encounter, number of potential emigrant females differ significantly (Kruskal-Wallis test, $H = 16.2$, $df = 3$, $p < 0.01$ (Sicotte 1993).

Stats. 11.3. Dominance rank, not age, was related to reproductive success of males in multi-male groups ($r_s = -0.594$, $p = 0.015$, $n = 16$). Relationship held when controlling for mean age of males ($r_s = -0.5417$, $p = 0.037$), but age was not related to reproductive success when controlling for rank ($r_s = 0.2355$, $p = 0.398$).

A young male that eventually inherited leadership of the group from his presumed father.
(Mountain gorilla.) © A. H. Harcourt

CHAPTER 12

Male Strategies and the Nature of Society: Conflict, Compromise, and Cooperation Between the Sexes

Summary

Although gorilla society might seem one permeated by the influence of males on females, in fact the relatively thin distribution of females in the environment affects the distribution of males. It forces the males to stay with females instead of roaming in search of them. Concomitantly, mating competition among males, particularly the use of infanticide, forces females to associate with a protective male (who also protects them against predators), instead of roaming alone. Hence gorilla society is one of stable, cohesive groups. While gorillas are a classic single-breeding-male society, with all the associated behavioral and anatomical correlates, nevertheless some questions remain. A minority of groups (almost all among eastern populations of the species) are multi-male, because some males remain in the group in which they were born. The causes of differences between populations and species in number of males per group is a long-standing puzzle. The different costs and benefits of staying versus leaving in eastern populations are quite well established, but the understanding does not help explain the contrast between the almost 100% one-male society of western populations and the mixture in eastern populations of one-male and multi-male groups. The contrast raises the issue of where answers to these and other questions about the nature of animals societies should be sought, whether in comparisons within species or between them.

12.1. Gorilla Society: The Influence of Females on Males

The classic socioecological framework of females distributing themselves according to the distribution of food and males according to the distribution of females implies a pervasive influence of females on males (ch. 2.1.2). If females cannot raise offspring on their own, the male stays and helps, and a more or less monogamous society is the result. If females are thinly distributed across the land, so are males. If females live in groups, then males can associate with those groups and defend them against other groups. And so on.

Gorillas fit this framework, or put another way, the framework helps us understand gorilla society. The low density of females, which is determined by the distribution of their food in relation to their body mass (ch. 2.2.1; ch. 4.1.1), and the females' (and males') slow speed of movement through the environment determine that a gorilla male's best mating strategy is to associate permanently with females, rather than to roam in search of them (ch. 10.1).

Additionally, the relatively low quality but wide distribution of gorillas' main fallback food, foliage (ch. 4.1.2; 4.1.3), means that females have the option to leave the male, because females do not benefit strongly from each other's presence in cooperation in competition over their food (ch. 2.2.3.2; ch. 5.2.2). That option to leave could well be the factor that allows long breeding tenure by males (ch. 11.1.5). The fact that dominant breeding males are apparently not ousted is a notable aspect of gorilla society (ch. 4.2.1.1.; ch. 11.1.5). If females can leave, it would not pay a challenger to fight furiously enough to take over a group of females and eject the resident male, when the consequence might be a mass exodus of the females. After all, it was the females that chose to join the original male (ch. 8), so why should they passively accept a new male?

The female gorillas' option of emigration, allowed by their diet, might also affect the behavior of the resident breeding male. Given that females have the option to leave, do males need to work harder than otherwise to retain females (ch. 10, Conclusion)? The breeding male gorillas' tolerance of offspring is noticeable, to the extent of males sometimes apparently being the main parent, rather than females (ch. 4.2.2.4). A possible effect of that care might be evident in the difference between gorillas and chimpanzees in the proportion of juveniles that survive loss of their mother. All nine gorilla juveniles orphaned between three and five years of age survived the orphaning (Watts and Pusey 1993), whereas three of seven chimpanzee juveniles died, including two orphaned at four or five years old, despite being adopted by older female kin (Goodall 1986, ch. 5). Whether in fact the male's babysitting affects females' decisions to stay or leave is unknown, as is whether the babysitting reduces females' periods of lactational anestrus by lengthening bouts between nursing (ch. 3.2.3.3), and hence speeds return to estrus. However, across all primates, species in which females other than the mother carry infants have short birth intervals by comparison to species in which only the mother cares (Mitani and Watts 1997), and the gorilla has shorter birth intervals than does the chimpanzee or orangutan (ch. 3.2.3).

Other facets of females' influence on males exist too, beyond just the females' distribution and movements, or potential movement, in relation to food. The gorilla females' slow primate reproductive schedule means that infanticide is an advantageous mating strategy for males (ch. 2.4.3; ch. 11.1.2). Infanticide is especially advantageous given the fact that duration of breeding tenure is such a strong influence on males' reproductive success, and that the interbirth interval of around four years is over a quarter of even a successful male's breeding tenure (ch. 3.2.3.3; ch. 4.2.1.2; ch. 11.2.1).

12.2. Gorilla Society: The Influence of Males on Females

The use of infanticide by males then affects females' association with males (ch. 7.2). The male's large body size by comparison to the female's, which surely evolved in response to advantages conferred by large size in competition between males over females (ch. 2.4.1.1), allows the male to protect females against infanticidal males and also predators (ch. 7.1). Females therefore benefit from association with a powerful male (ch. 7). Females of other species have evolved other or additional counterstrategies to the males' competitive behavior, such as multi-male mating, which gorilla females might use within groups (ch. 11.1.5), and also evolution of long receptive periods and sexual swellings (ch. 2.4.4.1; ch. 7.2). In other words, while female distribution and strategies affect male strategies, the male strategies in turn affect female strategies.

12.3. Gorilla Society: Conflict, Compromise, and Cooperation

Gorilla society adds some links to the females-to-food, males-to-females framework. While food certainly influences the distribution of females, and the distribution of females influences the distribution of males, at the same time and inseparably, males influence the distribution (and behavior) of females. Just as an anecdote to illustrate the interacting influences of the sexes, we found it difficult to decide on the best order of presentation in this book of the sexes' strategies and their influence on one another, and even now do not entirely agree that we have found the best order.

Additionally, food affects the distribution and therefore the strategies of the males, as well as of the females. The distribution of the gorillas' maintenance food, foliage, means a low density of not only slowly moving females but also slowly moving males. If males could move fast enough, they might be able to roam in search of females as a mating strategy. However, their size and diet constrains them to slow movement, and therefore to permanent association with females as their mating strategy, instead of roaming (ch. 10.1.3).

So, food affects the distribution and strategies of both sexes of gorilla, and the distribution and strategies of each sex affects the other. These interactions produce the conflict, compromise, and cooperation that are the hallmark of so many societies.

Conflict is epitomized by the fact that males are the females' main competitors for food (ch. 6.1), and most obviously by the fact that males kill the offspring of females with which they have not mated (ch. 4.2.1.3; ch. 11.1.2).

Nevertheless, both sexes benefit from association with the other. Despite being outcompeted by males over food (ch. 6.1), female gorillas benefit from the male's protection (ch. 7), and therefore compromise by associating with a male (ch. 7; ch. 8). Despite costs incurred in protection of females and their offspring against other males and predators (ch. 7), male gorillas benefit by associating permanently with females (ch. 10), protecting the females and their offspring (ch. 7; ch. 10.2), and even in a mild sense caring for females' offspring in the temporary or permanent absence of the mother (ch. 11.1.4).

And so beneficial to both sexes is the association and the protection that cooperation, both sexes working together, is evident in the tactics both sexes use to remain as a group, for instance actively negotiating with vocalizations coordinated movement of the group (Harcourt and Stewart 2001; Stewart and Harcourt 1994).

12.4. Males and Society: A Familiar Case Study?

The gorilla is a standard mammal as far as male strategies and some aspects of their influence on society are concerned. Males compete intensively to be the sole breeder for as long as possible. As a consequence of the competition between males for sole access to females, and not just sole mating access, but long-term sole access, the gorilla is effectively a single-male society (ch. 4.2.1.1; ch. 11.1.1). Gorilla males display all the classic behavioral and anatomical correlates of the intense competition between males for sole breeding access to females (ch. 2.4; ch. 11.1). They are more frequently and intensely aggressive than are females, for instance; they use infanticide as a mating strategy; they are far larger than females; they coerce females into continued association; they are differently colored; and they have extremely small testes for their body mass. These behavioral and anatomical traits are at one and the same time manifestly a result of the intense contest competition, and evidence for it.

But, of course, uniformity is not complete. Individuals differ: some males are calm, others apparently easily irritated. Groups differ: most are single-male, but some are multi-male. Different animals, groups, and population behave differently, a point made over and over again in this book—in part to counter the common supposition that unless all individuals have the behavior, it could not have evolved (see ch. 9.3). The few multi-male gorilla groups that exist do so because some gorilla males stay to inherit the breeding group from a relative, rather than following the usual path of emigrating from the natal group and establishing their own group (ch. 4.2.1.2; ch. 11.2). While such a staying strategy is very unusual in gorilla society as a whole, it is common in mountain gorilla society, and the explanations for it are standard ones that apply to societies of many vertebrates in several contexts (ch. 2.2; ch. 11.2). For instance, the young, staying males are queuing in an age-graded mating system as a means to inherit easy access to an established group of females. The older males tolerate them because of the benefits gained in cooperative defense of the group, and to avoid the costs of aggressively ousting them.

So, gorilla males present a familiar story, provide familiar understanding, in part because their study contributed to that understanding (e.g., Fossey 1984). As one of the aims of this book was to be a case study in socioecology, the normalcy of male gorillas' mating strategies and their effects on the nature of society is a good reason to use them as a case study.

That normalcy does not mean that nothing is left to understand about gorilla males and the structure of gorilla society. Why are more eastern lowland gorilla groups multi-male than are western lowland groups, and why are nearly half of both populations of mountain gorilla groups multi-male (ch. 4.2.1.1; ch. 11.3; ch. 13.3.3)? We will bet large sums of money that genetic differences have nothing to do with it. The difference, we are sure, must be explained by ecological contrasts between mountain and other gorilla populations' environments. However, exactly what the socioecological contrasts are and why they exist, we do not yet know. The route to the knowledge will be via understanding of the nature of the payoffs to the strategies, and of the connections between the strategies and the resulting nature of the society within the mountain gorilla populations compared to the western ones.

12.5. Males and Society: An Unfamiliar Case Study?

We have just argued that the strategies of male gorillas and the influence of the males' strategies on the nature of the gorilla's society is unexceptional. And yet one of the stated aims of this book was to use an unusual species and society to investigate socioecology in general—exceptions being useful as means to test rules (ch. 1.3).

Where gorillas add to the paradigm is in the perhaps unusually strong influence of male strategies on female strategies. For instance, socioecological schemes that explain contrasts between species in the nature of their society based on the females-to-food, males-to-females framework either largely ignore males because they are not relevant to the distribution of females (Isbell and Young 2002; Sterck et al. 1997; van Schaik 1989; Wrangham 1980), or have females benefiting from association with males through males' active help in raising of offspring (e.g., by feeding or carrying it), or defending a territory (Clutton-Brock 1989a; Davies 1991). Otherwise, association is to the males' benefit. The male gorilla's protection as a cause of females associating with a male is an addition to the scheme (ch. 7; ch. 9).

The females-to-food, males-to-females framework is, as we have said, to quite a large extent a comparison of the strength of influence of ecology (females to food) and sexual selection (males to females) on the nature of vertebrate societies (ch. 2.1.2). Nunn and van Schaik (2000) suggested that ecology (food and predation) better explain differences between species in the nature of their societies, while social factors (including infanticide) better explain differences within species. If so, primatologists will need to think carefully about whether their predictions are better tested with comparisons across species, or comparisons within species.

PART 5

GORILLA SOCIETY: THE FUTURE

An infant studies, a little nervously, its reflection in the camera lens. (Mountain gorilla.)
© A. H. Harcourt

CHAPTER 13

Gorilla and Primate Socioecology: The Future

Summary

The gorilla is a folivore that lives in one-male, multi-female social groups, between which females transfer, and from which no breeding male is ousted by an immigrant outsider. This society with its many additional complexities is the outcome of compromise between male and female tactics and strategies of survival, mating, and rearing. The environment affects the optimum tactics and strategies of both sexes. These tactics and strategies interact with one another in ways dependent on the payoffs to each sex, the outcome of which interactions is the society that we see.

What do we need to add to our current knowledge and understanding of gorilla society in particular, primate socioecology in general? We suggest nine areas of future work.

1. While we have placed a very strong emphasis throughout this book on the influence of the environment on society, nevertheless, strong physiological contrasts exist between species (and therefore genetic and phylogenetic contrasts). These physiological differences correlate with contrasts in the nature of the species' societies. Primatologists need to know a lot more about the physiological and neurological influences on behavioral propensities and how they ultimately correlate with contrast between species in the nature of their society.
2. We see very little difference in the society of western and eastern gorillas, despite the far more frugivorous diet of the western populations. But is that because we know too little about western gorilla

social behavior in the fruiting season? In general, despite several well-known and well-explained contrasts between predominantly frugivorous and folivorous primate species, many questions remain about why they react differently to the environment. Perhaps as part of that ignorance, we poorly understand the reasons for apparent contrasts between species in readiness to use cooperation as a means of competition. Would our understanding be improved by more standardization of methodology and by concentration on diet in the poor season?

3. Why does a greater proportion of eastern than of western gorilla groups contain more than one male? What is the correlation between, on the one hand, consanguinity of males of different groups and, on the other hand, the frequencies and intensities of aggression between them? And why do we not see male gorillas immigrating into groups and ousting the resident male? These are also general topics for primatology for which answers are needed. Connected to them are the related questions of:
4. the fate of dispersers, and
5. the nature of community structure, if any, in primate populations.
6. Protection against predation cannot yet be separated from protection against infanticide as a main benefit to female gorillas of associating with a male, because of the lack of testable predictions that separate the two hypotheses. More generally, while infanticide and the protection that males can or cannot provide against it can explain some male-female association within groups, how much does it explain contrasts between species in male-female association? Why do primate species apparently differ so greatly in frequency of infanticide? More generally still, what are the relative roles of environmental influences and sexual selection in determining the nature of society?
7. How precise are the socioecological descriptions and concepts that primate socioecology uses to explain the nature of primate society? Are they even precise enough for testing? If not, they are not scientific. A major advantage of quantitative modeling as a means of exploring hypotheses is that descriptions and hypotheses have to be extremely precise, which helps greatly in the ability to form testable predictions from the hypotheses. Primate socioecology now has a lot of data and hypotheses. It is time that it moved more strongly than hitherto to modeling and field experimentation as means of investigation.

8. Primate socioecologists must pay far more attention that they do at present to the fields of human behavioral and social ecology (and vice versa).
9. And finally, much as we know already about gorilla and primate socioecology, we still need more information from the field.

13.1. Gorilla Society Yesterday

In the beginning of his career, in independent life, the gorilla selects a wife with whom he appears to sustain the conjugal relations thereafter, and preserves a certain degree of marital fidelity. From time to time he adopts a new wife, but does not discard the old one; in this manner he gathers around him a numerous family, consisting of his wives and their children. Each mother nurses and cares for her own young, but all of them grow up together as the children of one family. There is no doubt that the mother sometimes corrects and sometimes chastises her young, which suggests a vague idea of propriety. The father exercises the function of patriarch in the sense of a ruler, and the natives call him ikomba njinai, which means gorilla king. To him the others all show a certain amount of deference. Whether this is due to fear or to respect, however, is not certain, but there is at least the first principle of dignity.

The gorilla family, consisting of this one adult male and a number of females and their young, are within themselves a nation. There do not appear to be any social relations between families, but within the same household there is apparent harmony.

The gorilla is nomadic, and rarely ever spends two nights in the same place. Each family roams about in the bush from place to place in search of food, and wherever they may be when night comes on they select a place to sleep and retire. The largest family of gorillas that I have ever heard of was estimated to contain twenty members. But the usual number is not more than ten to twelve. . . . When the young gorilla approaches the adult state, he leaves the family group, finds himself a mate, and sets out in the world for himself. I observed that, as a rule, when one gorilla was seen alone in the forest it was usually a young male, but nearly grown; it is probable that he was then in search of a wife. At other times two only are seen together, and in this event they are usually a pair of male and female, and generally young. Again, it sometimes occurs that three adults are seen with two or three children; often one of the children two or three years old, and the others a year younger, which would indicate that the male had had one of his wives much longer than the other. In large families young ones of all ages, from one year old to five or six years old, are seen; but the fact is plain that the older children are much fewer in number. I have once seen a large

> female with her babe, quite alone; whether she lived alone or was only absent for the moment I cannot tell.
>
> . . . In the matter of government, the gorilla appears to be somewhat more advanced than most animals. He leads the others on the march, and selects their feeding grounds and places to sleep; he breaks camp, and the others all obey him in these respects. Other animals that travel in groups do the same thing; but in addition to this, the natives aver that the gorillas from time to time hold palavers or rude form of court or council in the jungle. On these occasions, it is said the king presides; that he sits alone in the centre, while the others stand or sit in a rough semicircle about him, and talking in an excited manner. Sometimes the whole of them are talking at once, but what it means or alludes to no native undertakes to say, except that it has the nature of a quarrel. To what extent the king gorilla exercises the judicial function is a matter of grave doubt, but there appears to be some real ground for the story.
>
> As to the succession of the kingship there is no certainty, but the facts point to the belief that on the death of the king, if there be an adult male he assumes the royal prerogative, otherwise the family disbands, and they are absorbed by or attached to other families. Whether this new leader is elected in the manner that other animals appoint a leader, or assumes it by reason of his age, cannot be said; but there is no doubt that in many instances families remain intact for a time after the death of their leader.
>
> Richard Garner (1896, p. 214–217)

Over a century after Garner described gorilla society, we do not have much to add to his description. Most of the key points are there. So, what have gorillaologists been doing for the last century?

Garner's description was based on accounts from local people. He himself saw hardly any wild gorillas. (His field methods are famous—he shut himself in a cage in the forest for six months and observed what passed [Garner 1896, ch. II, III].) As Robert and Ada Yerkes (1929, p. 431) wrote, "Garner . . . indulged freely and rather uncritically in the unchecked report of hearsay." But the accounts from the local people happen in this case to be extraordinarily accurate. In other instances, they can be extraordinarily inaccurate. Talk to local people anywhere in the world about the habits of the local fauna (and flora), and you will hear some fantastical stories. Even if nothing else, the last fifty years of gorilla research have confirmed Garner's description.

More importantly, they have confirmed it with quantitative data, and they have confirmed it across many other sites in Africa than the small area of western Gabon that Garner knew. The comparison across sites could have demonstrated contrasts in the nature of gorilla society as the environment changed. However, our reading of the evidence is that the crucial environmental influences are largely similar across sites, and therefore the gorilla populations have largely similar societies. Hence, the lack of importance we attach in this book to distinguishing species or subspecies of gorillas.

We can also see in gorilla society a lot of similarity at various levels to the societies of other primates, even to the apparently very different societies of female-resident Old World cercopithecines. Those similarities and the inevitable differences, combined with the sometimes amazing detail that primatologists collect on the environment and the animals' use of it—such as counts of mouthfuls of food—has enabled the final major advance over the last fifty years in understanding of gorilla society. Fifty years ago, neither Garner nor anyone else could have told us why gorillas had the sort of society that they did, or why it differed from that of chimpanzees, or differed from the society of any other primate species. Now we have a very good idea. Fifty years of socioecological field study have produced a well-substantiated body of theory. More needs doing, of course—that is what most of the rest of this chapter is about. But we have a lot to go on.

13.2. Gorilla Society Today

The gorilla is a large-bodied folivore that lives in one-male, multi-female social groups, averaging about three females. Females, especially nubile females, transfer between males. Breeding males compete furiously for access to females, including by use of infanticide. Once successful in that access, they appear to hardly ever be expelled by an immigrant outsider. This brief description of gorilla society is comprehensible as the outcome of interaction and compromise between the survival, mating, and rearing strategies of the two sexes.

Female gorillas compete for food and cooperate in the competition, tactics that have been hypothesized to favor group-living and residence with kin. However, the competition is not severe enough, nor the cooperation useful enough to keep females in the group in the face of the costs of continued residence after maturity with their father or other close male kin. Therefore, females emigrate from their birth group in this species in which breeding males are seemingly hardly ever ousted from their group by immigrant males.

But having emigrated, why do females always immediately join a male? Why not avoid competition with a dominant partner twice their size and travel independently of males, as do orangutan females? All evidence, including from quantitative modeling, points to protection from predators and from infanticidal males as the main benefits to gorilla females of associating with a male. Whether protection from predators or infanticide is paramount will depend on the relative costs of loss of the female's own life to a predator (uncommon but, of course, extremely high cost) compared to loss of an infant to infanticide (potentially common, but low cost). We do not yet know the answer.

And what is the benefit to a male of associating with females? Quantitative modeling indicates that a male that associates permanently with females is more likely than one that roams in search of females to be in the presence of a female when she is in estrus, especially if he is powerful enough to attract many females. An additional potential benefit to the male is the ability to protect his mating and reproductive investments (the females and offspring) against both predators and against infanticidal males.

13.3. Gorilla and Primate Society Tomorrow

The gorilla is the largest primate living (ch. 1.3). Yet only this century, in November 2002, was a previously undiscovered population of gorillas found by scientists (Morgan, Wild, and Ekobo 2003; Oates et al. 2003). And it was less than twenty years previously that the estimated number of gorillas in west-central Africa was more than doubled to about 40,000 as the result of the first countrywide census there by Caroline Tutin and Michel Fernandez (1984). And only ten years ago, the estimated number of western gorillas was again more than doubled (Harcourt 1996), with the discovery and report of high densities over huge areas of flooded forest in, especially, the Republic of Congo (Fay 1989; Fay and Agnagna 1992; Fay et al. 1989; Mitani 1992). We might add that along with so much unknown about extant gorillas, paleontology still appears to know almost nothing about the origins of the African great apes (Fleagle 1999, ch. 15).

We wrote in our introductory chapter (ch. 1.3) that one advantage of concentration on the socioecology of one primate species, instead of all, was that the concentration might make gaps in socioecological understanding more obvious. We now discuss some of those potential gaps. Before we do, though, let us emphasize a point raised in chapter 1. Primatology has made a massive contribution to tropical biology, because we know so much about such an extraordinarily little-known part of tropical biology, the biology of tropical forest mammals. Yes, there are gaps in the understanding of primate society, but tropical biology, and therefore all biology, is very much the richer because of the work of field primatology.

13.3.1. Phylogeny, environment, and society?

We have placed strong emphasis in this book on the influence of the environment on the structure of gorilla society, or perhaps more precisely, on the influence of the interaction of the nature of the environment with the nature of the animals on the structure of gorilla society. Gorilla society is as it is, not because gorillas are gorillas, not because the species is neurophysiologically programmed to have the sort of society that it does, but because being large animals, and hence being able to live on a relatively abundant, widespread, but not very rich food source, foliage, the animals can and do live in groups . . . and so on—we have given our description of gorilla socioecology in the previous section (sec. 13.2).

One of the main reasons why we insist so strongly on the influence of the environment rather than of phylogeny on the nature of gorilla society is because we see aspects of it that are very similar to what one sees in the society of other primates, despite the fact that the species' society is unusual in some ways. In most New World primates, both sexes leave the group of their birth, as do both gorilla sexes (ch. 4.2.1.2). In Old World cercopithecines, female kin are friendlier to one another than they are to non-kin, as are female gorillas (ch. 5.2). In many primates, males compete for dominance, with the main evolutionary or functional reward being majority access to mating with females, and hence majority reproduction, as with gorilla males (ch. 11.1). In Old World cercopithecines, as in the gorilla, periods of uncertain contrast in competitive ability are the periods associated with the greatest mating and hence reproductive success of erstwhile subordinate males (ch. 11.1.4). In sum, again and again as we attempt to understand gorilla society, we see the operation of general socioecological principles. We see the gorilla, not really as a unique species (though of course all species are unique), but as a primate whose society can be explained by general vertebrate socioecology.

This argument about similarities to other species and the operation of general socioecological principles is different from the usual one used to emphasize the influence of the environment, which is that if a species' society changes with change in environment, the environment must be having a large influence (Doran and McNeilage 2001). We agree completely with this argument too, and discuss later several contrasts between gorilla populations.

At the same time, we would never argue that there is no phylogenetic signal in any primate species' society, by which we mean a genetically influenced physiological or behavioral signal. Contrasts in anatomy correlate with contrasts in behavior and society (ch. 2.2.1). Form fits function, and the form (such as degree of sexual dimorphism) did not develop in the lifetime of the individual as a result of its life experiences. For instance, titi and squirrel monkeys react in fundamentally different ways to strangers of each sex. The differences can be tied, and must be tied, to fundamental differences between the species in both the nature of their society and their physiology (Mendoza 1991). And, of course, if there were no phylogenetic signal, it would not be necessary to take account of phylogeny in comparative studies, as we strongly argue is in fact necessary (box 2.2).

Some behavioral and societal differences between taxa cannot yet be explained, and indeed some traits appear to be quite resistant to modification (Kappeler and van Schaik 2002). For instance, Old World cercopithecine females seem intrinsically readier to form kin groups through residence and sometimes also coalitions and alliances than do New World monkeys, without it being obvious that the environment of cercopithecines is more conducive to acquisition of the benefits of cooperation than is the environment of other taxa (Di Fiore and Rendall 1994; Nunn and van Schaik 2000; Sterck et al. 1997; Strier 1999). There appear to be differences among groups of macaque species in how aggressive or tolerant they are in the same circumstances (de Waal 1989a; Matsumura 1999; Ménard 2004; Thierry 2004, 2007), although how absolute those differences are remains to be determined. And lemurs seem fundamentally different in many ways from other primates, for instance showing effectively no variation of degree of sexual dimorphism with nature of mating system (Kappeler 1993, 1999; Kappeler and Heymann 1996).

One of the best examples now known of the influence of phylogeny via physiology (or vice versa) on social systems comes from Sue Carter and co-workers' (1993; 2004) contrasts of prairie and meadow voles. In simplistic brief, prairie voles are largely monogamous with lots of paternal care; meadow voles are standard polygynous mammals. The contrast correlates nicely with a variety of physiological differences between the males of the two species, whose behavior can be altered by appropriate experimental manipulation of the physiology of the males. For example, treatment with the neuropeptide arginine vasopressin can increase an individual's readiness to be in social bodily contact with other animals, while blocking its action decreases the contact.

Similarly, in a variety of New World species in which the fathers actively care for infants, the fathers have relatively high levels of the hormone prolactin in their blood, which seems to cause the obvious paternal behavior, rather than be caused by it (Schradin and Anzenberger 2002). Nevertheless, neither for these species, nor for the titi and squirrel monkeys, is there yet available as much experimental study as for rodents to conclusively demonstrate cause and effect, as opposed to correlation (Schradin and Anzenberger 2002).

The other area of "internal" contrast between species that is of interest is neurological. Clever chimpanzees, with their large brain for body size, use tools to get at termites in termite mounds (Goodall 1968). Stupid gorillas, with their smaller relative brain size, merely smash the nest open (Tutin 1983). Of course, that's a simplistic comparison, but how much of the difference between species in brain size or complexity determines differences in complexity of use of either the physical or social environment? Do gorillas smash termite nests, rather than fish for termites with specially fashioned twigs, not because they are too stupid to use the tools, but because breaking the nests open is more efficient, and they are strong enough to smash? The fact that primates have larger brains for their body size than do other mammalian taxa, and that within not only primates, but other taxa too, larger brained taxa might be cleverer by various measures (e.g., tool use, social learning, innovation) than small-brained taxa (Lefebvre, Whittle, and Lascaris 1997; Reader and Laland 2005), makes it possible that there is some reality to the supposition that primates are indeed more socially intelligent than other taxa (Seyfarth and Cheney 2002). But how much reality?

Exploiting the full benefits of cooperation—and the societal consequences that flow from the exploitation—could well benefit from advanced information processing capabilities (intelligence?), even depend on them (Barrett and Henzi 2005; Bergman et al. 2003; Byrne and Whiten 1988; Cheney and Seyfarth 1990; Dunbar 2003; Seyfarth and Cheney 2003). While most examples of the complex use of cooperation in competition still come from Old World species, yet the long-lived, large-brained, socially clever, tool-using capuchin monkey (de Waal 2000; Judge and Carey 2000; Moura and Lee 2004) belies any suggestion that Old World primates might in some ways be generally more socially intelligent than are New World monkeys. It remains to be seen whether lemurs in particular, strepsirrhines in general, are more socially simple than monkeys and apes because of insufficient information processing capacity (Jolly 1966).

We need to be careful about inferring social complexity and intelligence in "higher" primates because we expect it (Harcourt 1992; Marler 1996). Fiddler crabs, a species not notable for the size of its brain, cooperate with neighbors in at least one of the same apparently clever ways as do monkeys, namely, helping smaller partners in fights against opponents smaller than the helper (Backwell and Jennions 2004). If fiddler crabs can do it, why not lemurs? Not only do we need to tighten our concepts of exactly what we mean by social intelligence (Seyfarth and Cheney 2002), we have surely nowhere near sufficiently explored intrinsic differences between species in social ability as a correlate of differences between them in the nature of their society.

13.3.2. Diet, competition, cooperation, and grouping?

13.3.2.1. Frugivore-folivore differences

A main socioecological correlate of a species' society is its diet. As a simple example, we know that western gorilla ranging and feeding behavior changes between the fruiting and non-fruiting seasons in ways that match some overall differences between western and eastern (including mountain) gorilla populations and their environment (ch. 4.1). A more frugivorous diet seasonally or regionally correlates with longer distances traveled each day (ch. 4.1.4). However, we still do not know enough about daily social behavior of western gorillas, especially as it relates to a rich, frugivorous diet (Doran and McNeilage 2001). Most published research on western gorilla social behavior has come from observations at the swamp clearings (ch. 3.1.3), where the gorillas do little other than feed on the abundant herbaceous vegetation. Therefore, some obvious predictions have yet to be tested about, among several others, more cooperation in competition over the richer, more clumped frugivorous diet in the west, and hence a society of stronger alliances among kin (Doran and McNeilage 2001; Harcourt 2001).

Some of the differences between western and eastern gorillas, and some of our ignorance about reasons for the differences, reflect for this one species a lack of understanding that Charlie Janson (2000) highlights for frugivore-folivore differences in general. Why, he asks, do frugivores increase foraging effort with group size (effort largely judged as daily distance traveled) and yet escape a drop in fecundity, whereas the opposite appears to be the case with folivores (with, of course, some exceptions)? Is it really because the frugivore diet allows an increase in intake that prevents a drop in reproductive output, whereas gut fill (and maybe lower metabolism) prevents the folivores traveling farther, and thus they suffer reproductively? And why might frugivore and folivore biomass change differently with change in annual rainfall (Janson 2000; Janson and Chapman 1999)?

13.3.2.2. Cooperation

We have a very poor grasp of why in female-resident species some appear to use coalitions and alliances while others do not (Janson 2000). Why do Colombian woolly monkey females apparently not cooperate with kin, even though they live in cohesive groups within which there is intense competition (Strier 1999)? Part of the problem is surely that we do not yet have adequate data on the precise nature of competition in so many of the species (Koenig 2002). Janson adds the question of why we see so little evidence of alliances among male primates (Janson 2000), a question that Clutton-Brock (1989b) asked of male mammals in general. The answer is presumably that sole access to a rich resource for a likely short period in a climate of intense competition provides more benefit than does shared access. But how would we test the idea?

13.3.2.3. Some methodology

We will close this section with three methodological points relevant to the questions that we have raised.

First, we suggest that quantitative testing of reasons for contrasts in group size between species or environments should always use adult group size. We have three reasons for advocating this. Not only must it be mainly adults (or near-adults) that are making the decisions about grouping that socioecologists are trying to understand, but the numbers of immature animals (and hence total group size) can vary enormously from year to year, even season to season, given the generally far higher mortality rate of immature animals than of adults (e.g., Dittus 1977; Harcourt et al. 1981a). Also, for those studies that combine nocturnal and diurnal species (largely a strepsirrhine-catarrhine contrast), an additional confound exists, in that many nocturnal strepsirrhines park their infants (van Schaik and Kappeler 1997), and are therefore counted as having a relatively smaller group size than do diurnal species that carry their consequently visible infants.

The second area calling for standardization is in feeding. If food is so important as the basis for socioecological arguments (food determines the distribution of females, which determines the distribution of males), primatology needs a renewed push for standardization of definitions and measures. Are we in a better situation now than thirty years ago, when Alison Jolly commented on different recording methodologies as a difficulty in comparing studies of feeding (Jolly 1972, ch. 3)? We suspect not. Primatology could surely get a better handle on reasons for differences in societies if measures and indices were the same.

One measure we would strongly suggest is to categorize species by their fallback, poor-period diet rather than by their annual diet (ch. 4, Introduction). In the fruiting season, the most folivorous of primates can be largely frugivorous. The gorilla is a prime example (ch. 4.1.3). However, what separates the gorilla from the common chimpanzee is that in the poor season, the gorilla becomes a folivore while the chimpanzee remains a frugivore (ch. 4.1.8). At the very least, all ecological studies should separate diet according to season, so that socioecology has a chance to investigate the influence of variation in quality of the environment on society in more precise detail than if only annual diet is presented.

Third, the size and quality of patches of food has for decades been fundamental to investigations of socioecology (ch. 2.2). Animals that use small, rich patches have different societies than those that use large, poor patches. Until recently, it is the socioecologist who has decided what a patch is, measuring, for example, the size of crowns of fruiting trees or diameter at breast height of the trees (Janson and van Schaik 1988). Is it time that we more often devise means to get the animals to tell us how they perceive the environment, in this case perceive the patchiness of their environment? Isbell, Pruetz, and Young (1998) did so by measuring time spent at a feeding site (a measure of size of patch), and distance the animals traveled between sites (dispersion of patches). Such measures can even indicate that patchiness differs between the age-sex classes (Fossey and Harcourt 1977), or between individuals of different competitive ability (Whitten 1983), a possibility barely considered in primate socioecology up to now. And we probably still need a lot more information on the nature of use of patches and movement between patches (Chapman and Chapman 2000).

13.3.3. Intermale competition?

13.3.3.1. Number of males per group

Despite all the work on gorillas, despite all the years of questioning variation within and between mammalian species in number of males per social group (Clutton-Brock 1989b; Kappeler 2000), we have neither found in the literature, nor ourselves produced, a good explanation of why mountain gorilla populations should have so many more multi-male groups than do western gorillas (ch. 4.2.1.1; ch. 11.2). The answer should be that mountain gorillas live in groups with more adult females, and that with a larger resource to defend, any one male finds it harder to keep others out (ch. 2.4.2.5). However, no correlation exists within or between gorilla populations between number of females in a group and number of males (ch. 4.2.1.1).

Alternatively, is it the case that a benefit of multi-male groups among mountain gorillas, protection from infanticide (ch. 7.2; ch. 11.1), is not important in the west, because infanticide is less frequent in the west (ch. 4.2.1.3)? The problem we have with that explanation is that as one-male groups of polygynous species appear to be the recipe for infanticide in a population (ch. 2.4.3), we do not see why infanticide should be less frequent in the west. However, the association between one-male groups and infanticide occurs mainly in species in which the breeding male is ousted by the incoming male, a phenomenon that is surprisingly rare in gorillas (ch. 4.2.1.2).

Hanuman langurs provide a similar conundrum. Obvious and apparently long-standing differences exist between populations in the proportion of groups that are multi-male (Koenig and Borries 2001). Some of the differences can be explained by contrasts between the populations in the number of females in groups, with more females being associated with more males (Koenig and Borries 2001; Sterck and Van Hooff 2000). But that correlation simply shifts the question to why the number of females should differ between populations—to which nobody seems to have a good answer (Koenig and Borries 2001).

Demographic accident is a possibility that should not be ignored by any of us. Maybe small, one-male groups are simply younger than large multi-male groups (Harcourt 1978a; Sterck and Van Hooff 2000; Yamagiwa 1987b).

13.3.3.2. Intergroup aggression and kinship

Students of western gorillas have commented on the lack of aggression between the leading males of groups (Bermejo 2004; Bradley et al. 2004; Doran-Sheehy et al. 2004). We have seen this lack of aggression among western gorillas ourselves. We watched as a mother carrying a young infant approached close to a non-father male. We were convinced that we were about to see an attempted infanticide and a major fight between the two males. Yet both males ignored the female. This incident was in one of the open swamp clearings. Our explanation for the equanimity of the males in the clearings is that, on the one hand, all animals are concentrating on the excellent feeding opportunity, nothing else; and on the other hand, because visibility is so good, males do not have to react until the last minute of escalation of a situation. However, it appears that within the forest too, groups of western gorillas are remarkably pacific (Bermejo 2004; Doran-Sheehy et al. 2004).

Bradley and her co-workers (2004) suggest another explanation for the lack of aggression. It is that neighboring males are related to one another. The suggestion is similar to Dorothy Cheney and Robert Seyfarth's (1983) explanation for why dispersing male vervets usually went to the same group: they received less aggression from the familiar males, possibly quite close kin, already there. Similarly, Juichi Yamagiwa and John Kahekwa (2001, 2004) suggested consanguinity and familiarity between neighboring males as the reason for lack of infanticide in the eastern lowland gorilla population of Kahuzi-Biega in DRC, which suggestion brings us back to consideration of causes of the infrequency of infanticide in western populations (sec. 13.3.3.1). Why, though, should neighboring western breeding males be any more likely to be related than neighboring eastern ones? Is the answer that all western males leave the group of their birth (almost all groups are one-male), and therefore that there is a larger population of neighboring kin?

We are not sure we accept that solution. In the first place, overt aggression is common between what are estimated to be more or less closely related males in multi-male groups of Virunga gorillas (Harcourt and Stewart 1981; Robbins 1996; Veit 1983; Yamagiwa 1987b). Secondly, while some western gorilla males in the clearings might be related to one another, we find it hard to believe that in all the combinations of the many groups that visit the clearings in a year (over twenty [Robbins et al. 2004]), kinship is close enough to be an important reason for lack of aggression. Also, the argument begs the question of where the loop of causation starts. Western males leave the group of their birth; therefore neighbors are kin; therefore they are pacific and do not commit infanticide; therefore males do not need to stay to cooperate in defense of infant kin; therefore males leave the group of their birth; therefore neighbors are kin. . . .

13.3.3.3. Duration of male tenure

In both western and eastern gorilla populations, it seems that males retain tenure for life, or at least remain for life in the group in which they became the dominant breeder (ch. 4.2.1.2). The long tenure is surprising, given the manifest intense competition between males for females (Harcourt 2001). The corollary of long tenure is absence of takeovers. Gorilla males do not obtain females and offspring by ousting the resident male, killing his offspring, and breeding with the then estrous females. They do so, instead, by attracting females. Our explanation for the absence of fights intense enough to lead to takeovers is that the effort could be wasted, indeed dangerous, if in this female-emigrant species, the females abandoned the male as soon as he ousted the resident male (ch. 11.1.3.2). Are we correct?

13.3.4. The fate of dispersers?

We know extraordinarily little about the fate of dispersers of any species (Koenig, Van Vuren, and Hooge 1996; Macdonald and Johnson 2001). Emigrants are often never seen again; immigrants are often complete strangers in the records. How then can we test the idea that the reason why several Virunga gorilla females delayed emigration until after the birth of their first offspring is that first-time gorilla mothers are more successful in their natal group than in a new group (ch. 8.2)? How can we test the idea that the female chimpanzees that return to breed in their natal range might be ones that experienced the greatest drop in quality of range when they dispersed (ch. 8.3)?

The open swamp clearings in western Africa have provided a wonderful opportunity to increase the sample size of dispersing gorillas over that previously available from studies of eastern and mountain gorillas (Sicotte 1993; Yamagiwa and Kahekwa 2001), because of the excellent visibility at the clearings and the large numbers of gorilla groups that visit them (ch. 4.1.2, box 4.1). However, gorillas that move beyond the region occupied by the groups that use the clearings are lost. The chances of such movements occurring might be calculable if primates do the same as North American mammals, and disperse an average of about seven home range diameters (Bowman, Jaeger, and Fahrig 2002), and if the range of groups outside the clearings were known. The problem at present is that information on the groups once they leave the clearings is so minimal. And that information could dry up if Ebola has the envisaged devastating future impact on gorilla populations (Walsh et al. 2003; Walsh, Biek, and Real 2005). Other studies that have managed to follow dispersers include those of red howler monkeys (Pope 2000) and vervet moneys (Cheney and Seyfarth 1983). But such studies are very few.

Use of DNA surveys to identify the distribution of relatives and nonrelatives through the population has already been used with some success and will be a powerful method in the future (Bradley et al. 2004; Di Fiore 2003; Hoelzer, Morales, and Melnick 2004; Janson 2000), but of course answers only a subset of the questions of interest.

13.3.5. Within-species community structure?

Primatology has very little knowledge at present of community structure in any population. However, one example is Bradley and co-workers' (2004) study that indicated that neighboring gorilla males are more closely related to one another than more distant ones, and raised the question of whether that relatedness was a cause of the apparent pacific relations of neighboring breeding males. Another is the Cheney and Seyfarth (1983) study that suggested that vervet males might find it easier to enter groups into which familiar group members, maybe even kin, had previously immigrated. These studies raise the possibility that primate society needs to be described not just with respect to what happens within groups depending on the relationships and strategies of individuals but also with respect to what happens between groups as it is affected by those relationships. After all, we would not consider that we had described the society of a human population without considering between-village relationships as well as within-village relationships. With dispersal of individuals being such a potentially crucial component of community structure, DNA methodology will surely be a powerful investigative tool (sec. 13.3.4).

The study of forest mammals is extraordinarily difficult. Indeed, by comparison to what primatology knows about its mammal of interest, effectively nothing is known about any other order of forest mammals. Gorillas are no exception to the difficulty of study, even if they are easier to study than many other primate species. If anyone really wants to get huge amounts of data on a one-male, multi-female harem society in which males defend females, and females move between males, and especially on the structure of the community, the plains zebra could be the species to go for (Rubenstein 1986; Watts 2000a).

13.3.6. Sexual selection, environment, and society: Predation versus infanticide?

The question of whether the main benefit that female gorillas obtain from joining a male is protection against predators or protection against infanticidal males (ch. 7; ch. 9) is an example of a general distinction that Janson (2000) has identified as important for the next decade or so of primatology, namely, the relative roles of sexual selection and environment in determining the nature of society. For instance, if infanticide (sexual selection) is so influential, why do infanticide rates vary so much across species, even populations (Harcourt 2001; Janson 2000)?

We are surely getting closer to the answer (Janson and van Schaik 2000b; Nunn and van Schaik 2000), but if we cannot explain differences in infanticide rates between western and mountain gorillas, we have not crossed the finishing line yet.

Similarly, do primatologists know enough yet about the interplay of food, predation, and infanticide to model and hence predict the range of sizes or compositions of groups of any species in any one environment? We suspect not—for two reasons. We do not have enough data on especially predation (Cheney and Wrangham 1987; Stanford 2002), and we have not modeled quantitative predictions that would separate the various hypotheses. An excellent theoretical base on which to build is Nunn and van Schaik's (2000) quantitative, comparative analysis of the relative roles of predation, competition between females, and infanticide as determinants of various aspects of primate societies. In the building, primatology needs to think hard about whether the tests should be based on comparisons within species or between them (ch. 12.5) (Nunn and van Schaik 2000).

13.3.7. Schemas, quantification, modeling, and experiments?

13.3.7.1. Categorization, quantification, and definition

The gorilla can be categorized as a folivore (rather than frugivore) that lives in a one-male (not multi-male), harem society (not dispersed), in which females emigrate between groups (instead of remaining in the group of their birth), and no breeding male is ousted by an immigrant outsider (by comparison to many one-male species in which breeding males are ejected by newcomers). The development of socioecology has from the start involved a mixture of categorization and quantification, seemingly alternately, as a means of advance. In primatology, we first had the classic Crook and Gartlan categorization (1966), followed by Eisenberg and colleagues' (1972), followed in turn by the classic Clutton-Brock and Harvey (1977a, b) quantitative analyses.

Categorization identifies obvious contrasts between societies: female-resident versus female-emigrant (Wrangham 1980), within-group versus between-group competition (van Schaik 1989). Quantification clarifies perhaps subtler contrasts and, maybe more importantly, allows us to interpret a world that is more realistically graded than it is categorical. Yes, some categories exist—male versus female, forest versus savanna. But frugivore versus folivore is the scientist's categorization, not the world's. Few species are 100% either frugivore or folivore, and probably no primate. Quantified analysis can incorporate graded differences and can provide insights not apparent from the biologists' imposition of their own categories (Clutton-Brock and Harvey 1977a; Janson and Goldsmith 1995).

In primate socioecology, the quality, size, and distribution of patches of food have been major components of the interaction between environment and society. But, of course, patches vary continuously in quality, size, and distribution, that is, distance apart. Is it really the case that primate societies can be divided into those in which within-group competition is paramount and those in which between-group competition is paramount? Probably not (Koenig 2002). Rather, if for example, between-group competition (BGC) depends on size of patch relative to size of group (patches large enough to feed a group are defended), the likelihood of BGC for any one group will surely vary dependent on the nature of the food patch at the time. Thus Peter Fashing (2001) scored BGC as ranging from common, but not universal in about half a dozen of the roughly twenty-five species for which he found adequate data to occasional, but not absent, in another half dozen or so.

But when, exactly, is BGC occurring and when is it not? A problem in analysis is that many of the concepts used in primate socioecology are currently vague enough that one cannot objectively test several of socioecology's hypotheses (Isbell and Young 2002; Koenig 2002). That is a damning critique of the field, for a hypothesis that cannot be tested is not a scientific hypothesis. A great advantage of quantification and (next section) modeling is that hypotheses and measurements have to be precise, which greatly helps in making them testable.

We might add that within that precision, we need to incorporate phylogenetic correction (box 2.2). A simple and valuable student project would be, we suggest, to redo the classic 1977 Clutton-Brock and Harvey analyses, this time with full phylogenetic correction, and of course with all the updated data of the last thirty years of socioecological research—such as our own correction of our previous estimates of the time that gorillas spend feeding (Harcourt and Stewart 1984).

13.3.7.2. Modeling

Before a model can be produced to test a hypothesis, the hypothesis must be precise, and some accurate and more or less precise data must be available. (Note that by "model," we mean a quantitative expression of the hypothesis [box. 2.1]). Modeling therefore comes late in the development of a scientific field. Modeling has certainly come to socioecology (e.g., Dunbar 1996, 2002; Harcourt and Greenberg 2001; Janson and Goldsmith 1995; Robbins and Robbins 2004). However, the field needs more of it (Harcourt 2001; Janson 2000). Primate socioecology is replete with ideas, hypotheses, and theories (ch. 2). It is time to model them, and we can do so usefully because we have data to put into the models. Janson (2000) suggested that primatology needs to apply more of the sort of game theory that human behavioral ecology uses. We suggest that in general, the application of modeling should perhaps be the next major area in the development of primate socioecology. We surely should not leave this step to human behavioral ecology (Winterhalder and Smith 2000), especially as primatology makes so little use of human behavioral ecology (Harcourt 1998b; Janson 2000).

The point of modeling is to simplify (box 10.1), and the models that we used to investigate payoffs of female gorillas associating with males (ch. 7.2) and male gorillas associating with females (ch. 10.1) were very simple indeed. We mentioned in those two chapters several additional payoffs that would need to be incorporated in a full analysis, or in additional models. For instance, what risk or rate of predation (rare but lethal) would be needed to make protection from predation the reason for females to associate with a male more likely than protection from infanticide (more common, but less costly) (ch. 7.3)? What levels of protection would be needed to outweigh the competitive cost to females of associating with a male? How do the benefits and costs of protection of mates and offspring affect the payoffs to males of remaining with females compared to roaming in search of them? What amount of change in any of the parameters would produce a changed society?

These sorts of questions can be answered only by modeling. The larger question is, if primatology cannot answer these sorts of questions, have its hypotheses been fully tested, or do they remain hypotheses of adequate fit rather than best fit? Do they remain the equivalent of correlations rather than causes? If they do, then primatology is wasting one of its great strengths, namely, the amount of detailed, high-quality, field data that it has—and not just field data on any species, but field data on one of the least studied groups of organisms, tropical forest mammals (ch. 1.2.1).

13.3.7.3. Experimentation

A male English robin will treat a tuft of red feathers on a stick as a live male, and attack it (Lack 1939). Primates are somewhat more difficult animals to experiment on, especially in the wild. Nevertheless, experiments have been conducted. Cheney and Seyfarth's long-running series of experiments on wild vervets and baboons to investigate mechanism, function, and evolution in primate communication and social knowledge is justly famous (1990; 2003). Playback of calls (of the primates themselves, or of predators, or other species) is perhaps the most common form of field experimentation (e.g., Wilson et al. 2001; Zuberbühler, Noë, and Seyfarth 1997). For instance, Ronald Noë and Redouan Bshary have shown how red colobus monkeys join groups of diana monkeys in response to playbacks of chimpanzee pant hoots, but not in response to leopard coughs or control loud noises. By contrast, Garber (2000) and Janson (1998b) have investigated primates' foraging abilities by setting out arrays of baited trays. But field experiments are as famous for their rarity as their elegance and results. We suspect that no other primatologist has performed even a tenth of the number of field experiments that the Cheney-Seyfarth team has. Primatology needs more (Harcourt 2001; Janson 2000).

A criticism of field experiments is that many unknown confounding factors are probably operating. Yes, but we suspect that a problem with laboratory experiments is the assumption that confounding factors have been removed. A greater problem with primates is simply the practical difficulty of experimentation, even in captivity. Primates take a lot of management. Vertebrate (and therefore primate) socioecology needs a common, easily manageable, fast-living, group-living species. We suggest that guinea pigs will be the answer to many questions in mammalian socioecology.

By experimentation, we mean active manipulation of the environment. However, modeling is another form of experimentation. In our model of the consequences for infanticide rates of lone ranging and promiscuous mating by female gorillas, we in effect created an experimental gorilla female in an experimental gorilla population (Harcourt and Greenberg 2001), something that of course could not have been done in reality.

13.3.8. Communication between primate socioecology and human socioecology?

While we have concentrated on the society of the gorilla in this book, we see the book as fundamentally about primate socioecology. However, one primate is conspicuously absent from our discussion—humans. Part of the reason is that there is not an obvious field of human socioecology. The absence of the field arises, we suspect, because understanding of human society is seen as depending on so much more than the simple understanding that animal socioecology can bring (Ingold 1989; Layton 1989). Also, the field simply has not coalesced from its various subfields, such as human behavioral ecology; indeed it might have splintered into subfields (Rodseth et al. 1991).

By contrast, the field of human behavioral ecology appears to be thriving (Winterhalder and Smith 2000). The two fields of primate socioecology and human behavioral ecology are obviously closely allied (Rodman 1999). Just down the corridor from us work two of the top human behavioral ecologists in the United States, Monique Borgerhoff Mulder and Bruce Winterhalder. Yet communication between our offices, and between the disciplines in general, is minimal. A body of workers—primatologists—who have thought harder than perhaps any other taxonologists about relationships between social behavior, individual variation, and the structure of society ignores and is ignored by perhaps the most relevant like-minded group, the ecological anthropologists.

We are certainly guilty in this book of the ignorance: chapter 14 on conservation is the only place in which we substantively mention humans and their behavior. And primatology in general is guilty. The new *Primates in Perspective* (Campbell et al. 2007a), modeled on *Primate Societies* (Smuts et al. 1987), does not list any facet of human socioecology or behavioral ecology in its index and has no chapter on the topic. Janson (2000) wrote about primate behavioral and social ecology in one of two millennial articles in *Evolutionary Anthropology*. He referred to two articles on humans, of 122 in his reference list (in which we could identify the taxon concerned). Adjacent to Janson's article, Winterhalder and Eric Smith (2000) wrote about human behavioral ecology. They referred to one article on primates and five on birds, out of 297. And neither referred to each other's article, implying that the *Evolutionary Anthropology* editorial board too was not interested in interaction between the disciplines.

Some direct comparisons have certainly been made (Foley and Lee 1989; Rodseth and Wrangham 2004). However, the disciplines are far from automatically using each other's field for ideas and data to incorporate into their own field. Whatever the causes of the chasm, both disciplines would surely benefit from more bridges to the other (Harcourt 1998b; Janson 2000; Rodman 1999).

13.3.9. More fieldwork?

Is there any discipline anywhere whose practitioners have said that they have sufficient data? While we have argued that primatology has an immense amount of data, indeed has strongly influenced the development of two fields (vertebrate socioecology and sperm competition) because of the amount of data it has, while we have argued that we have enough data for a strong push on modeling as a route for the future of the discipline, we have to agree with everyone who has argued that we need more data, especially straightforward natural history data from species that might soon go extinct, and therefore be lost forever to us as sources of information and ideas (Campbell et al. 2007b; Greene 2005). Let us give two examples.

13.3.9.1. Three areas of ignorance

Lemurs are a wonderful test of socioecological principles, because they have evolved independently of the taxa from which almost all the principles have been produced. And yet we are distressingly ignorant about the taxon, famous as lemurs are, famous as Madagascar's biodiversity and levels of endemism are. An intense literature search failed to find published data on density (one of the most fundamental items of information about a population) on half of lemur species (Coppeto and Harcourt 2005). In a table of socioecology of lemurs listing sixteen traits of nine group-living species (presumably the best known of the group-living species, given that there are more than nine group-living species), over 60% of the 144 cells were blank, although the table was produced by one of the more experienced of lemur socioecologists, Peter Kappeler (1999).

The second example concerns predation on primates. Primate populations can suffer quite heavily from predation (ch. 2.3.1). Nevertheless, good estimates of predation rate, let alone risk, are few and far between (Cheney and Wrangham 1987; Stanford 2002). The ignorance arises in part because we study the primates not the predators (Janson 2000), and in part because observers of primates probably frighten predators away (Isbell and Young 1993). A consequence of this ignorance is that when Nunn and van Schaik (2000) investigated the relative roles of predation, feeding competition, and sexual selection on primate grouping, they had to separate primates into high, medium, and low categories of risk almost by guesswork. They argued that small-bodied, terrestrial species in open habitat in mainland Africa and

South America are most at risk. But where was the comparative analysis that demonstrates these assumptions? Nunn and van Schaik gave no reference. We know of none. We are absolutely not criticizing the attempt. The point is that as fundamental an assumption about primate socioecology as levels of predation risk had to be guessed, because we do not have the data.

Finally, socioecologists conduct their studies where they can find lots of study animals (Chapman and Peres 2001). That means they study them in sites where the animals are doing well. But natural selection surely happens fastest and most intensely when things are going particularly badly (Grant 1999). If one really wants to understand the evolutionary biology of primates, to really improve understanding of socioecology by investigation of primate societies' responses to environmental change, the place to do it is in marginal habitats—marginal either because at the edge of the species' geographic range, or because humans are threatening it. Some studies are being done in marginal habitats, such as disturbed or fragmented habitats, but almost entirely within the discipline of conservation biology, not socioecology.

13.3.9.2. Gorilla socioecology

What more does socioecology need to know about wild gorillas? We have already indicated in this chapter several areas of investigation that need more field data. We will simply list them here, citing at the end of each question the section of this chapter to which the question mainly refers.

1. What is the behavior, the social relationships, the nature of society of western gorillas away from the swamp clearings in which so much of the observation of western gorilla social behavior has been made (sec. 13.2)?
2. How do western gorillas' competitive and cooperative tactics differ between the fruiting and non-fruiting seasons (sec. 13.2)?
3. How do gorillas perceive and use the patchiness of their environment (sec. 13.2)?
4. What is the fate of dispersing animals—how far do they go, where do they go, how successful are they (sec. 13.3; 13.4; 13.5)?
5. Is there any connection between past history of males (and maybe females) of different groups and their current behavior—are males that have known one another in immaturity less likely to be aggressive to one another (sec. 13.3; 13.4; 13.5)?
6. What proportion of females and infants are killed by predators (including humans) and by non-father male gorillas in the different gorilla populations, and what degree of protection is afforded to females and their offspring by association with a male (sec. 13.6)?

7. All the data need to be collected, measured, and reported in a way that makes quantitative comparison between studies, populations, and species feasible and valid (sec. 13.7).

There will still be gorillas in African forests in a hundred years' time (ch. 14.1.2). Whether at the end of this century our summary here of socioecological understanding of gorillas will be as valid as Garner's description of the society is one hundred years after he published it, our great-grandstudents will see.

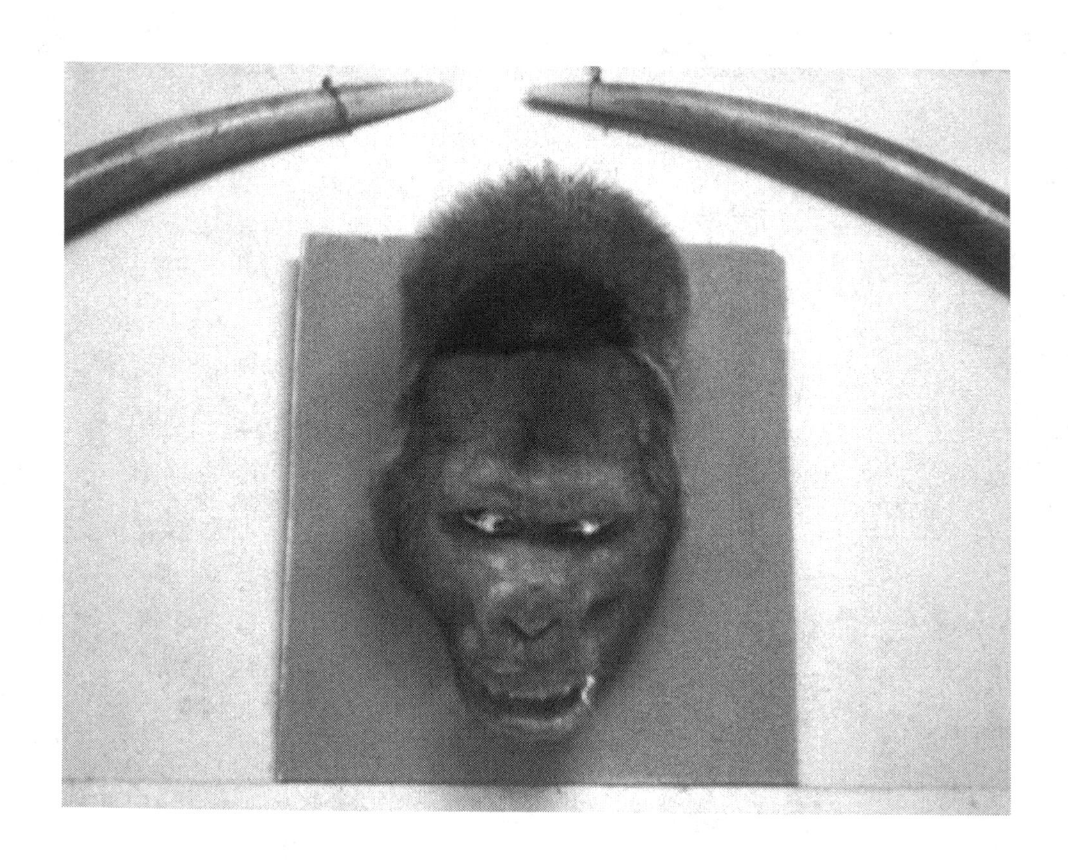

A trophy on the wall of a restaurant in Gabon in 1980. (Western gorilla.) © A. H. Harcourt

CHAPTER 14

Socioecology and Gorilla Conservation

Summary

This chapter has two parts, the first on conservation of gorillas, and the second on the relevance of socioecology to conservation.

1. The why, what, how much, where, and how of conservation. *Why conserve?* Gorillas are threatened because of worldwide increasing human population and increasing demand for forest products, with the former rising faster in Africa than elsewhere. *What to conserve?* Order of threat is (1) Cross River gorillas (tiny, fragmented population amidst dense human population), (2) eastern lowland, or Grauer's, gorillas (only a few thousand in the midst of the chaos of war), (3) mountain gorillas (only about 700, but so well protected that numbers are increasing), and (4) western gorillas (hunting, forest clearance, but still high numbers over large area, mostly sparsely populated by humans). *How much?* A variety of analyses indicate that, very roughly, 5,000 large-bodied animals might allow persistence for several millennia: 5,000 gorillas need 5,000 km^2 of good habitat. For safety's sake, more than one population of 5,000 gorillas must be conserved. *Where?* Seven national parks in Africa are 5,000 km^2 or more and might contain over 5,000 gorillas. Six are in four West African nations; one is in eastern gorilla range; additionally, the two very successful mountain gorilla parks should surely also be considered. *How to conserve?* Africa is extremely poor, and its people in dire

need. Funding for conservation must come largely from outside the continent. Given what first-world governments spend on their armed forces and on subsidizing environmental damage, plenty of money is available. Conservation must concentrate on protecting the five identified national parks, and the two mountain gorilla parks, along with active encouragement of tourism in some sites to provide funds and visibility.

2. Africa's protected areas are on average far larger than Western countries' reserves. Currently, protection of reserves in Africa depends more on economics and politics than science. However, by the time Africa's parks are as small as Western Europe's (6 km^2), and as densely surrounded by humans, science will be vital. Conservation is about the behavior of populations under changing circumstances. Socioecology is relevant to understanding the process by which populations react to change. Socioecology is therefore relevant to conservation. We illustrate the relevance by showing how important socioecological knowledge of the Virunga mountain gorillas was to implementation of successful tourism programs. Additionally, socioecologists are the people on the ground who often fight hardest for conservation, since conservation tends to be done where the field biologists work.

Africa's gorilla numbers are going to crash in this century. However, the dedication of some of Africa's conservationists and the remarkable story of the success of mountain gorilla conservation in eastern Africa are grounds for hope.

Introduction

To the extent that this is a book about socioecology, about how we might explain why a species has the sort of society that it does, a chapter on conservation is out of place. To the extent that it is about a species commonly perceived to be in danger of extinction, it would surely be deficient of us if we said nothing about conservation, especially as we will have nothing to study if we do not conserve (Janson 2000). So, we will close the book with a two-part chapter, first a section on conservation of the gorilla, and second, a discussion of the relevance of socioecology as a discipline to conservation, especially conservation of the gorilla.

14.1. Conservation's Five Questions: Why? What? How Much? Where? How?

Any successful conservation effort needs to answer five questions.

Why conserve? What are the threats that cause conservation to be necessary?

What should we conserve? What ecosystems, communities, species, and populations are most in need of conservation? That is, which are most sensitive to a given environmental change; which are most suffering from the change? Specifically, does the gorilla need special conservation effort? If so, which subspecies or population?

How much do we need to conserve? What area of ecosystem, what number of populations, what size of each population needs conserving?

Where should we implement the conservation effort? To some extent, this question is a summary of the answers to the three previous questions. However, it crucially adds the concept of triage. We must ask, more than we do now, whether we make last-ditch efforts to save even doomed species and populations, or whether we concentrate effort where success is most likely.

Finally, *How should we conserve?* Knowing the threats, knowing what should be conserved, knowing the number and size of populations that need conserving, knowing where effort should be concentrated, how do we implement this knowledge? How do we actually conserve? Protected areas? Community development? Captive breeding? Multiple-use areas? And so on and on.

14.1.1. Why conserve?

Humans try to convert the world to an environment suitable for themselves—as do many species. We are more successful than most. The problem, of course, is that an environment suitable for one species is often not suitable for others. That is undoubtedly the case for humans (Hannah et al. 1994; Millennium Ecosystem Assessment 2005). The Bjorn Lomborgs (2001) of the world are wrong in their contention that we are not rapidly and irremediably damaging the world: species and their habitat are fast disappearing (Ehrlich and Ehrlich 2004; Goudie 1993; Harcourt 1996; Kates and Parris 2003; Pimm et al. 1995; Wilson 2002). And the more humans, the faster the conversion of other species' habitat to human habitat, and the faster the loss of species (Barnes 1990; Cardillo et al. 2004; Harcourt and Parks 2003; Kerr and Currie 1995; Laurance et al. 2002; Liu et al. 2003; McKinney 2001; McNeely et al. 1995). A major problem for African wildlife, including gorillas, is that human populations are increasing more rapidly in sub-Saharan Africa and in host countries of gorillas than in the rest of the world (fig. 14.1).

Primates, including the gorilla, live mostly in tropical forest. Pristine tropical forest is disappearing fast, including in Africa (Achard et al. 2002;

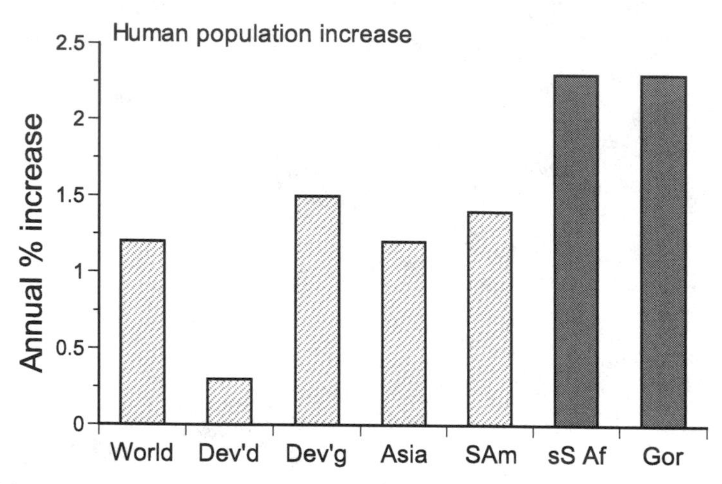

Fig. 14.1. Human population is increasing globally (World), in developed countries (Dev'd), developing countries (Dev'g), Asia, South America (SAm), sub-Saharan Africa (sS Af), and gorilla countries (Gor). Source: Data from World Resources Institute (2004).

Barnes 1990; Grainger 1993; McNeely et al. 1995; Sayer, Harcourt, and Collins 1992; Skole and Tucker 1993; World Resources Institute 2005). This disappearance, this destruction of primate habitat, is a main threat to persistence of the world's primates, and surely a threat for the gorilla too (Butynski 2001; Caldecott and Miles 2005, ch. 2, 6–8, 16; Chapman and Peres 2001; Harcourt 1996; Mace and Balmford 2000).

The ravages of the viral infection Ebola on western gorilla populations has received a lot of attention recently (Caldecott and Miles 2005, ch. 7, 13, 16; Huijbregts et al. 2003; Leroy et al. 2004; Walsh et al. 2003, 2005). Disease in general is increasingly recognized as a threat to wildlife (Chapman, Gillespie, and Goldberg 2005; Daszak, Cunningham, and Hyatt 2000; Deem, Karesh, and Weisman 2001; Harvell et al. 2002; Hudson et al. 2002; Leendertz et al. 2004). However, it seems to us that as long as large expanses of forest remain, the gorilla population will recover from its epidemics.

But aye, there's the rub. Africa's increasing human population is almost certainly going to accelerate the destruction of Africa's forest (fig. 14.2) (Barnes 1990; Harcourt 1996). Add the world's increasing consumption of forest products (Ehrlich and Ehrlich 2004; Myers and Kent 2003), and it is difficult to see how any forest is going to remain outside protected areas.

The other huge threat to African forest wildlife is the bushmeat trade (Anadu, Elamah, and Oates 1988; Asibey 1974; Bowen-Jones and Pendry 1999; Butynski 2001; Caldecott and Miles 2005, ch. 7, 13, 16; Chapman and Peres 2001; Cowlishaw and Dunbar 2000, ch. 9; Fa, Peres, and Meeuwig 2002; Muchaal and Ngandjui 1999; Oates 1996a; Peterson 2003; Refisch and

Koné 2005a, b; Robinson, Redford, and Bennett 1999; Walsh et al. 2003; Wilkie, Sidle, and Boundzanga 1992). The trade is exacerbated by commercial logging, which sends roads deep into the core of forests, so providing easy access to what would otherwise be a sanctuary from human encroachment and exploitation. Satellite photo after satellite photo demonstrates that hard on the heels of the logging roads comes small-scale settlement (Achard et al. 2002; Grainger 1993; Skole and Tucker 1993), while on-the-ground observation demonstrates that along with the logging comes subsistence and commercial hunting (Caldecott and Miles 2005, ch. 7, 13, 16).

In Gabon in the 1980s, we saw gorilla meat openly offered on restaurant menus. John Fa and colleagues (2002) have estimated that 60% of mammal species in the Congo Basin are possibly being hunted unsustainably. The bushmeat trade in Nigeria in the 1980s was estimated at $200 million (Anadu et al. 1988). Primates were a minor part of the overall trade, as in other parts of Africa (Fa, Ryan, and Bell 2005; Fa et al. 2006; Wilkie and Carpenter 2000), and were one of the least favored items of game in Nigeria (Anadu et al. 1988). Nevertheless, we estimated that in Nigeria in the 1980s, before later improvements in protection, gorillas were being killed at one-and-a-half times the rate that they were being born (Harcourt et al.

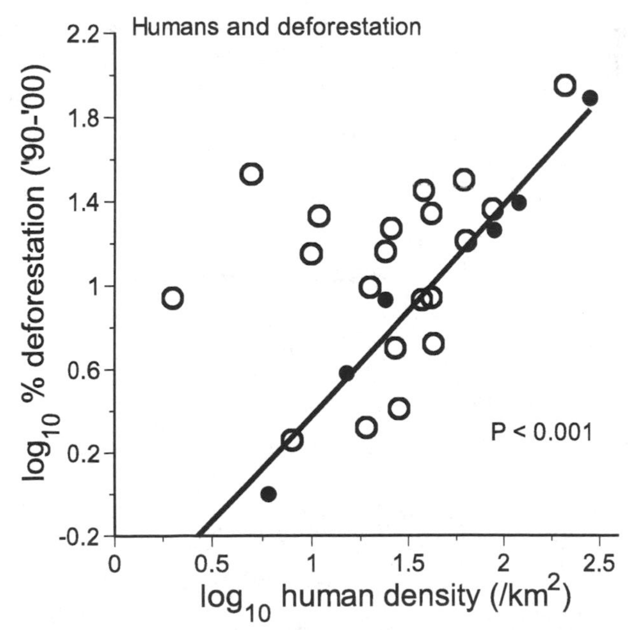

Fig. 14.2. Human density correlates with (causes?) destruction of tropical forest. Data are for sub-Saharan African countries. Closed circles and regression line are gorilla countries. *Details at end of chapter.* Source: Data from World Resources Institute (2004).

1989b). Both there and in south-west Cameroon such unsustainable intensities of hunting of gorillas continue (Fa et al. 2005, 2006).

Other studies too have indicated hunting of primates in Africa, including the chimpanzee and gorilla, at a rate that is greater than recruitment (Bowen-Jones and Pendry 1999; Kano and Asato 1994; Refisch and Koné 2005a; Wilkie and Carpenter 2000). Different species are susceptible to different threats. Among primates, the gorilla is a species particularly susceptible to hunting (Isaac and Cowlishaw 2004). Concrete evidence of that susceptibility comes from the extraordinary countrywide foot survey of Gabon conducted twenty-five years ago by Caroline Tutin and Michel Fernandez. They estimated a 70% drop in gorilla numbers in heavily hunted areas (Tutin and Fernandez 1984).

Up to now, East African primates have suffered little from hunting for meat (Oates 1996a). For instance, none of the killed mountain gorillas that we know of were killed for their meat. However, one consequence of the ravages wrought by the war in eastern DRC has been an increase in the use of game meat, as hungry and out-of-control soldiers started to hunt wild animals, and political instability reduced the supply of domestic meat (Hall et al. 1998; Plumptre et al. 2003; Yamagiwa 1999b, 2003). Wild animals are, of course, not the only ones to suffer in this ongoing conflict in eastern DRC. This civil war has largely been ignored by the rest of the world. Yet in each six-month period over the five years before the December 2004 tsunami off Sumatra, Indonesia, probably as many people died in the conflict as were killed in total in that tsunami (Coghlan et al. 2006; Waltham 2005). And they are still being killed.

14.1.2. What to conserve?

14.1.2.1. The Red List classification of the gorilla

Perhaps the first source for anyone wondering what species need conserving globally is the Red List of Threatened Species of what is now called the World Conservation Union (still commonly referred to as IUCN, International Union for the Conservation of Nature and Natural Resources) (IUCN 2006). The Red List categorizes species, subspecies, and even populations into levels of threat. The three top levels for extant taxa are, in increasing order of degree of threat, Vulnerable, Endangered, and Critically Endangered. The categorizations are based on the size of populations; on degree of past, present, or projected future decline in population size; and on size and nature of the geographic range, for example, whether it is fragmented or not. Full details are at the Red List Web site (IUCN 2006).

According to the Red List, and other analyses of the data in it, the Order Primates might be one of the more threatened mammalian Orders (IUCN 2006; Mace and Balmford 2000). Across all 29 mammal Orders that appear in the IUCN Red List, the Primates has the seventh highest proportion of species classified as Threatened (i.e., Vulnerable, Endangered, Critically Endangered); across the 18 Orders with more than ten species, primates rank second (Perissodactyla [horses, tapirs, rhinos] are first); and across the 14 Orders with more than twenty species, the Primates has the highest proportion listed as Threatened (stats. 14.1). Primates have several attributes that might put them at risk. Their visibility (ch. 1.2.2) is surely one. Also, they breed slowly for their body size by comparison to other mammals (Read and Harvey 1989), which means that primate populations will recover relatively slowly from low numbers.

The gorilla is the largest primate, and large-bodied taxa, including of primates, are often more susceptible to extinction than are small-bodied species, especially when humans are the threat (Harcourt 1999; Johns and Skorupa 1987; Johnson 2002; Jolly 1986; Richard and Dewar 1991). Large-bodied species are often preferred by hunters (Brook and Bowman 2004; Caughley and Gunn 1996, ch. 2; de Thoisy, Renoux, and Julliot 2005; Isaac and Cowlishaw 2004; Peres 1990, 1999; Peres and Dolman 2000). They breed slowly (Johnson 2002). And they need more land, other things being equal (Clutton-Brock and Harvey 1977a; Milton and May 1976), which need correlates with large annual home range being a good predictor of susceptibility to extinction in primates (Harcourt 1998a; Harcourt and Schwartz 2001; Skorupa 1986).

The Red List recognizes two species of gorilla and four subspecies (ch. 3.1.2; box 3.1). It categorizes the western and the eastern lowland, or Grauer's, gorillas as Endangered on the basis of past, current, and projected future decline in population size. The Cross River gorillas and two mountain gorilla populations (Virunga and Bwindi) are classified as Critically Endangered on the basis of both past decline and projected future decline, exacerbated by already extremely small population size.

The *Cross River gorilla* is definitely Critically Endangered. Its population numbers probably no more than 250 animals, fragmented so much that probably no one population consists of more than 50 animals. The fragments are surrounded by dense populations of humans who used to hunt intensively and are still converting forest to farmland, despite the somewhat improved protection afforded the gorilla populations (Butynski 2001; Caldecott and Miles 2005, ch. 7; Harcourt et al. 1989b; Oates et al. 2003). However, while the 2004 Red List considers this population to be a subspecies, the recent discovery of other populations in the quite extensive Ebo forest between the Cross River and Sanaga River (Morgan et al. 2003) suggests that the Cross River gorillas might not be as isolated as previously thought, and therefore might not be a valid subspecies, if it ever was. The state of these newly discovered Ebo forest populations is unknown, but they are in several thousand square kilometers of forest only sparsely inhabited by humans, although heavily hunted (Morgan et al. 2003).

We agree that *western gorillas* are probably Endangered (declining population of perhaps 50,000–100,000, depending on the severity of the Ebola epidemic [sec. 14.1.1]). We might use different criteria than the Red List, but the outcome is the same: habitat destruction, hunting, and epidemics are hammering the population (Butynski 2001; Caldecott and Miles 2005, ch. 7, 13, 16; Huijbregts et al. 2003; Tutin 2001; Tutin and Vedder 2001; Walsh et al. 2003, 2005). Ten years ago, we argued that the western gorilla was Vulnerable (Harcourt 1996). The difference between now and then is the knowledge of the devastating impact of Ebola and of the huge increase in the bushmeat trade. Unlike for the other three types of gorillas, the majority of western gorillas live outside protected areas. Thus, assuming the other forms remain protected, the western gorilla is probably going to experience over the next century the greatest proportional drop in numbers, maybe over 80% (Harcourt 1996; Walsh et al. 2003), especially as much, even most, of the unprotected forest is already licensed as logging concession (Caldecott and Miles 2005, ch. 7, 13, 16; Harcourt 2000; Plumptre et al. 2003).

Such a huge drop in numbers could push the western gorillas into the Critically Endangered category, and the suggestion that they should be so classified has been publicly announced (Walsh et al. 2003). However, several very large (over 5,000 km^2) reserves contain a total of probably more than 30,000 gorillas (sec. 14.1.4). Such a number of gorillas in such large reserves is not sensibly considered Critically Endangered, particularly because the species is likely to receive more, not less, protection as numbers of gorillas outside reserves dwindle. The same point has been made for the chimpanzee (Oates 2006). It is true that even if the populations are well-protected, Ebola could devastate them. However, the calculations of the long-term effects of Ebola (Walsh et al. 2003) take no account of the likely rapid recovery of the populations as the few survivors find themselves in a habitat empty of competitors (Sibly et al. 2005).

As importantly, we suggest that the Critically Endangered category in the Red List would be diminished in significance if a species or subspecies of at least 30,000 animals in six widely separated reserves were to be classified as Critically Endangered, especially as currently the total population size might well be in excess of 50,000 animals. By comparison, the Critically Endangered kakapo (a New Zealand parrot) numbered less than 100 scattered individuals ten years ago (Elliott, Merton, and Jansen 2001). Broderick and colleagues (2006) make the same point about the IUCN's listing as Endangered an increasing population of perhaps over 2 million green turtles.

Eastern lowland (Grauer's) gorillas are certainly Endangered. Their population of perhaps 15,000 must be rapidly declining as a result of habitat destruction, hunting, and mining in war-ravaged eastern DRC (Butynski 2001; Caldecott and Miles 2005, Ch, 8, 13, 16; Hall et al. 1998; Plumptre et al. 2003; Tutin and Vedder 2001; Yamagiwa 2003).

IUCN categorization of the *mountain gorilla* is confusing. For most people studying gorillas, the mountain gorilla is *G. beringei beringei,* and consists of both the Virunga and Bwindi populations. Yet the 2006 Red List classifies only the Virunga population as *G.b.beringei,* listing the Bwindi gorillas as a subpopulation of *G. beringei* with no subspecies stated. The Red List classifies the Virunga population as Endangered, because it contains less than 250 mature animals. This classification is a change from the previous Critically Endangered status accorded to the population. The downgrading is warranted because the Virunga gorillas are now so well protected that their numbers have been increasing for two decades (Fig. 14.3) (Caldecott and Miles 2005, Ch. 8, 16; Kalpers et al. 2003).

By contrast, the 2006 Red List categorizes the Bwindi population as Critically Endangered because of its small size and a projected continuing

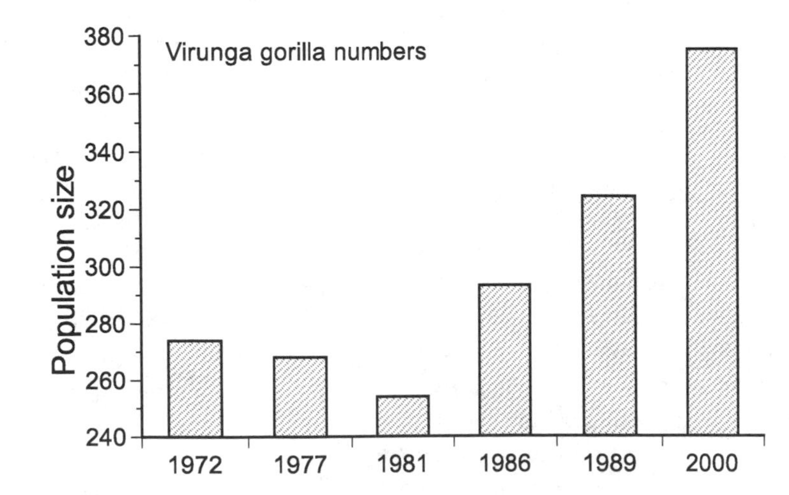

Fig. 14.3. Mountain (Virunga) gorilla numbers are increasing. *Details at end of chapter.* Source: Data from Kalpers et al. (2003).

decline. Yet the Bwindi population is now increasing (McNeilage et al. 2001). In fact, the Bwindi and Virunga populations are so similar in population size and trend (about 350 animals and rising), in the threats facing them, and in the level of protection afforded that they should have the same status. According to the Red List criteria, that status is Endangered, we argue. If the downgrading from Critically Endangered to Endangered of the two mountain gorilla populations were accompanied by the deserved fanfare and accolades to the fieldworkers and officials responsible for their protection, the two populations might even benefit from the fame.

14.1.2.2. Some reflections on the Red List

The Red List is deservedly and rightly the first major source for current estimates of the conservation status of species globally. Nevertheless, we would like to suggest some improvements.

The Red List employs four main criteria for categorization of conservation status. They are, in brief, population size and degree of fragmentation, decline in population size, the size and nature of geographic range size, and quantitative analyses of projected future decline. Thus the criteria, sensibly, consider the performance of the species. However, whether a species (or subspecies or population) is going to go extinct or not depends not only on what it itself is doing but also, even mainly, on what the threat is doing, or is going to do. If you want to know whether the houses below a dam are safe, you do not look at the strength of only the houses.

While analysis of threat is included as one means of estimating declines and fragmentation, we would like to suggest that adding an explicit index of threat to the Red List criteria, in other words, using five overt criteria, might give a better indication of the potential fate of species and their habitat (Harcourt and Parks 2003; see also Shi et al. 2005). Not only would more information be explicitly used in the categorizations, but there is often far more information available about the threat than about the species, especially if the species lives in tropical forest (Harcourt 1995; Harcourt and Parks 2003).

We have suggested that human density is a useful indicator of threat as it so often correlates with damage to the environment, and is far more often assessed, and probably assessed far more accurately than is population size of most other species, again especially tropical forest species (Harcourt 1995; Harcourt and Parks 2003). With respect to gorillas, the western gorillas are lucky in that most populations are sited in regions of current low human density (Harcourt and Parks 2003).

A major improvement to the Red List would be the provision of accessible, quantitative substantiation of the Red List classifications. Science is by definition an enterprise that produces testable hypotheses. However, most of the Red List substantiation is in the form of unpublished reports, and hence largely inaccessible reports, which means that neither the data nor its interpretation can be independently assessed.

For many species, there will in fact be no quantitative substantiation of the assessments, because the classification has to be based on the best guestimates of experts. Guesswork in the absence of hard data is properly involved in the case of the Red List. If IUCN could not rely on informed guestimates, a large proportion of the species would be classified as "data deficient," especially the rarest ones (Coppeto and Harcourt 2005), and lawmakers would almost certainly pay no attention to them.

But however a classification is arrived at, users of the Red List need to know the basis for it, especially as assessments of the status of the same species by different assessors can differ greatly—for some species, all the way from Least Concern to Critically Endangered (Keith et al. 2004; Regan et al. 2005). Informed and therefore effective management is difficult, even impossible, without knowledge of uncertainty (Dietz, Ostrom, and Stern 2003; Regan et al. 2005). Currently, the Red List does not supply information that will allow assessment of uncertainty, despite the fact that protocols for quantitatively incorporating uncertainty are available (Akçakaya et al. 2000). Consequently, both the Red List itself and users of the Red List act as if the List's classifications are certain (see, e.g., Cardillo, Mace, and Purvis 2005).

We strongly suspect that the charismatic gorilla engenders more concern than the data warrant, or than other less charismatic species would earn. If the same data were applied to, say, a tortoise (a relatively large-bodied, slowly breeding reptile), would it receive the same classification as the gorilla? We suspect not. For example, at the same time as some have proposed raising the status of the several tens of thousands of western gorillas from Endangered to Critically Endangered (Walsh et al. 2003), others are considering downgrading to Endangered the 500 Lear's macaws (Jamie Gilardi in litt. 2006). The contrast indicates that perhaps the Red List's objective criteria are not being applied equally to all species.

John Oates makes essentially the same point in his criticism of the application of the IUCN criteria to the chimpanzee (Oates 2006). He suggests that strict application of the criteria might not allow the current categorization of the chimpanzee as Endangered. Surprisingly enough, the Red List agrees: "Common chimpanzees are possibly rather than certainly Endangered under a strict application of the IUCN Red List Criteria; strictly, they might best be regarded as Vulnerable" (IUCN 2006).

The potential bias in favor of the gorilla (and other favored species) is not necessarily wrong. The wish to conserve species is in some senses a value judgment. Maybe taxa closely related to ourselves are more important than other taxa. The loss of the gorilla and chimpanzee, closely related to us as they are (ch. 3.1.2), would surely be a greater break in our link to the rest of the animal world than would be the loss of a galago or a guenon. But if loss of that link is a reason for greater concern, the potential bias needs to be openly admitted.

In fact, some classifications of conservation status explicitly incorporate uniqueness as a criterion by which to judge status of a species (Cowlishaw and Dunbar 2000, ch. 10; Oates 1996b). We should presumably be more worried about the loss of a species that is the only representative of its genus than by loss of a species that is just one of more than ten other species in the genus. The gorilla (in our view), the aye-aye, and the indri would be examples of the former, especially the aye-aye, which is the only representative of its family.

14.1.3. How much?

How many individuals are needed for a population or a species to persist? We immediately have to ask "persist for how long?" (Cowlishaw and Dunbar 2000, ch. 7, 10). A week is a long time in politics, as the saying goes. But for the purposes of conservation of species, there is presumably not much point in talking about persistence for any less than some millennia. Practically, the time period has to be shorter than that, but if we save a population that is large enough to survive for only a century, we can be fairly sure that it will be gone in a millennium—in which case why did we bother to save it for a century?

Calculations of millennial population sizes have been made theoretically, empirically, and a mixture of the two (Belovsky 1987; Cowlishaw and Dunbar 2000, ch. 7; Harcourt 2002; Reed et al. 2003). The results are hugely variable, depending as they do on the nature of the calculations and the biology of the species. Nevertheless, it looks as though for more or less large-bodied, long-lived, slowly reproducing taxa such as primates, and especially apes, a population of the order of 5,000–10,000 animals might be needed (Gurd, Nudds, and Rivard 2001; Harcourt 2002; Reed et al. 2003, 2004).

Let us take the more practical minimum lower value of 5,000. How much land do 5,000 gorillas need? While densities of 10 gorillas per km^2 have been recorded, the average over large areas is perhaps closer to 0.5–1 per km^2 (ch. 4.1.7). Let us say 1 per km^2 for areas that include more or less large proportions of suitable habitat (while accepting that censuses over large areas are often very inaccurate, and the vast majority of the gorilla's range is unsurveyed [see also Oates 2006]). In which case, we need 5,000 km^2 of well-protected land to save the gorilla. And just in case a catastrophe hits one 5,000 km^2 site (see Walsh et al. 2003), we should probably have a minimum of two or three populations of 5,000, that is, two or three well-protected areas of at least 5,000 km^2.

14.1.4. Where?

When doctors are treating the ill and injured in a major crisis, they leave aside those who might get better without treatment; they ignore those who will die even with treatment; they treat those who will die without it, but whom treatment might save. Conservationists have no such triage. Yes, there is the suggestion that we should save regions with many species, many endemic species, many endangered species—hotspots of various sorts (Mittermeier et al. 1998; Myers et al. 2000). And yes, there is the suggestion that we should save empty areas where conservation should be cheap (Mittermeier et al. 2003). But after that, it is difficult to see the application of any sort of triage rule—conservationists certainly don't give up on hopeless causes. Indeed, it is sometimes difficult to see the application of any rules (Cullen et al. 2005; Halpern et al. 2006; Margules and Pressey 2000).

While conservation does not have enough money or people to do all the conservation necessary, a sensible rule of thumb might be to concentrate at any one time on (a) those endangered species or areas that have the smallest numbers or size, and which at the same time are facing the greatest threat (providing the fight is not hopeless); and (b) those endangered species or areas that have the largest numbers or size, facing the least threat. The first are in the greatest need, while money and energy spent on the latter will probably bring the most efficient returns (Ayres, Bodmer, and Mittermeier 1991; Peres and Terborgh 1995).

"Providing the fight is not hopeless," we say. An area of 300 km^2 has a radius of 10 km, a distance quite easily penetrated by even a slow-moving hunter (Alvard 1998a). If conservation gave up on areas this small, it would abandon the Cross River populations, especially as they are so beset by threats (sec. 14.1.2). It would not abandon the two mountain gorilla populations (each in protected areas of little more than 300 km^2), because the reserves are now very well protected. Also, conservation is all about protecting diversity, if it is about anything, and the mountain gorillas and their habitat, especially the Virunga population, are an unusual example of their form, being as they are at the extreme altitudinal limit of the taxon. Furthermore, it looks as though common taxa are common, and hence safe, in part because of their greater variety of forms, that is, subtaxa (Harcourt, Coppeto, and Parks 2005). If so, the more subtaxa saved, the more likely it is that the taxon as a whole will be saved.

At the other extreme, we suggested that at least two populations of 5,000 animals, that is, two protected areas of at least 5,000 km^2, might be a minimum to ensure persistence of the gorilla for a millennium or more (sec. 14.1.3). It turns out that Africa does not just have the necessary two such reserves; it in fact might have seven of them, and if gorilla numbers recover, maybe eight (table 14.1). If these eight existing reserves can be adequately protected, the gorilla will have a good chance of surviving into the twenty-second century, and maybe even into the third millennium.

So our answer to where to conserve is the seven 5,000 km^2 protected areas in Africa (table 14.1), and the two mountain gorilla parks. That conclusion will probably annoy the several dedicated individuals fighting for gorilla conservation elsewhere. Happily, logic takes the world only so far.

Table 14.1. The seven protected areas in Africa that are 5,000 km^2 or more in size, and that might or do contain 5,000 gorillas or more. Latitude–longitude (lat–long) is N–E (except where S written) to nearest 30′. The number of gorillas might be the same as the size in km^2, because all the reserves probably contain relatively high densities of 1/km^2 or more. *Sources for reserve size (Res. size) and number of gorillas per reserve (Gor #s), and other details at end of chapter.*

Nation	Protected area	Size (km^2)	Lat–Long	Res. size	Gor#s
Western gorilla:					
Cameroon	Dja	6,200	3°–13°	1	A
	Boumba-Bek/Nki	6,100	2°30′– 14°30′	2	2, A
Cameroon/CAR/ Rep. Congo	Lac Lobéké-Dzanga-Sangha-Ndoki[1]	7,300/27,900	2°30′–16°30′	1, 2	A
Rep. Congo	Odzala-Koukoua/ Lossi/ Lékoli-Pandaka	13,400/41,900	0°30′–15°	1, 2	A
	Lac Télé/Likouala[2]	6,000/29,500	1°–17°	1, 2	2, B
Gabon	Gamba/Loango/ Moukalaba-Doudou	11,300/13,000	0°30′ S–11°30′	1, 2	2, A
Eastern (Grauer's) gorilla:					
DRC	Kahuzi-Biega	6,000	2°30′ S–28°	1	1, A

The world is not changed by sensible people. We are here arguing only what seems to us the logical solution, and we salute those who are ignoring us and are attempting to save the 200–250 Cross River gorilla (Oates et al. 2003), the 16 Mt. Tshiaberimu gorillas (Sarmiento 2003), and the several other small gorilla populations surrounded by dense human populations (Plumptre et al. 2003; Sarmiento 2003). Indeed, the logic of saving variety in order to save the whole (as mentioned earlier) could apply to these small populations.

14.1.5. How?

14.1.5.1. Africa is poor

Africa has little to no money to spend on conservation (Caldecott and Miles 2005, ch. 13). Wealthy individuals and governmental departments exist, of course, but in general its people are some of the poorest and therefore unhealthiest on the globe (fig. 14.4A, B).

Consequently, money for conservation must largely come from the rich countries of the world. Alternatively, the people and relevant government departments of the nations of Africa need to be made richer. Currently, about US$6 billion is spent globally on protected areas, a fraction of the $24 billion that has been calculated as the annual cost for conservation of 10% of the land surface of every nation or ecosystem on earth (James, Gaston, and Balmford 1999, 2001). Improving conservation by remediation in

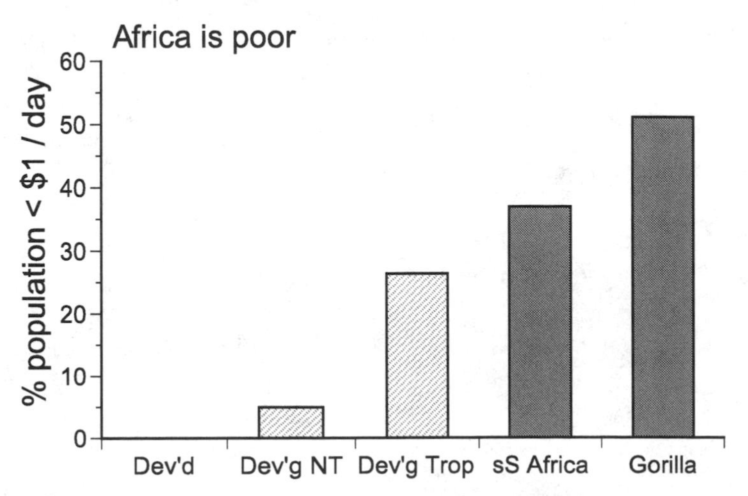

Fig. 14.4A. Africa is poor. Percent population earning , $1 per day, 1994–1999 in developed countries (Dev'd), developing non-tropical countries (Dev'g NT), developing tropical countries (Dev'g Trop), sub-Saharan Africa (sS Africa), and countries harboring gorillas (Gorilla). Source: Data from World Resources Institute (2004).

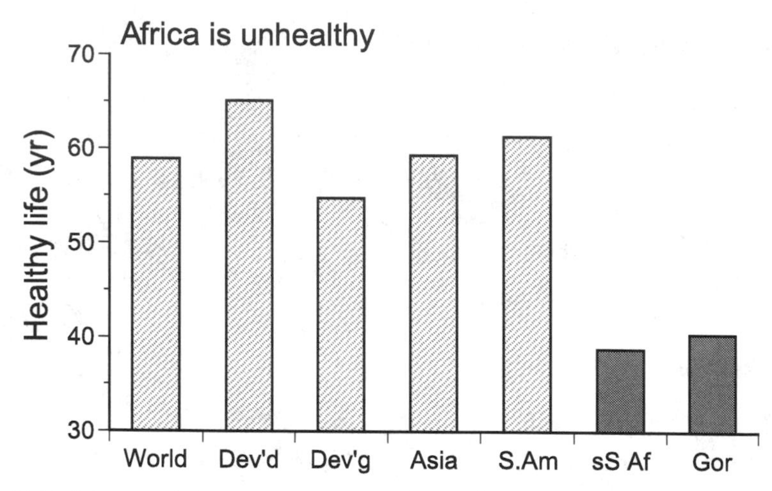

Fig. 14.4B. Expectancy of a healthy life is lowest in Africa, including gorilla countries. Abbreviations on the *x*-axis are as for figure 14.4A, with S Am = South America. Source: Data for 2002 from World Resources Institute (2004).

human-dominated landscapes might cost another US$290 billion. Call it US$315 billion in all. That sounds like an impossible figure. However, the baseless invasion of Iraq had by the end of 2005 already cost the United States over $200 billion (Rogers 2005). Globally, environmentally damaging subsidies might amount to US$1,000 billion per year (James et al. 1999, 2001). And globally again, well-protected wilderness might be worth US$33 trillion annually in various environmental benefits (Costanza et al. 1997). In other words, the money and benefits are clearly there.

How can the money be put into the hands of the people on the ground who do the conservation? The suggestion of trust funds for the salaries and infrastructure of the government departments responsible for conservation (Kramer and Sharma 1997; Oates 1999, ch. 9) seems to us to be eminently sensible, although we freely admit extreme financial naivety.

14.1.5.2. Protected areas and tourism?

Most wild gorillas are outside protected areas or reserves, suggesting that gorilla conservationists should echo calls from others, and advocate conservation programs outside as well as inside protected areas (Ceballos et al. 2005; Lindenmayer and Franklin 2002; Margules and Pressey 2000). However, for gorillas and Africa, protected areas seem to us the most useful means of conservation with, in some cases, associated tourism (Walsh et al. 2003). The difficulties of adequately caring for reserves are great enough without extending conservation beyond the edge of the reserve, except by means of conservation publicity and good community relations.

Zoos are important for publicity and education, but as places to conserve a population of gorillas (or indeed any other large-bodied mammal species), they are currently irrelevant (Balmford 2000). About 750 gorillas are now in captivity across the globe (International Species Information System [ISIS] 2005). Compare that to the well over 100,000 gorillas in the wild ten years ago (Harcourt 1996), and the more than 35,000 that could be conserved in the present seven protected areas, each of which contains by recent estimates at least 5,000 gorillas (table 14.1). Even if worst fears are realized (Walsh et al. 2003), and the total wild population drops or has dropped by 80% from 100,000, Africa will still have over twenty-five times as many in the wild as the globe does in captivity.

Furthermore, conservation in the wild is relatively cheap, especially in developing countries (Balmford, Leader-Williams, and Green 1995; Harcourt 1986; James et al. 1999, 2001). Thus we calculated that each wild Virunga gorilla cost $1,250 per year in the costly Virunga gorilla conservation program in the 1980s, compared to $1,500 for a zoo gorilla then (Harcourt 1986).

Protected areas, of course, have their problems (Blount 2001; Bookbinder et al. 1998; Borgerhoff Mulder and Coppolillo 2005, ch. 8; Lindsay 1987; Myers 1972; Newmark et al. 1994; Weber 1987; Western 1982). One is that they can cause local resentment among a poor populace prevented from using a needed nearby resource and not compensated for that deprivation (Weladji and Tchamba 2003). Nevertheless, the resentment and the deprivation caused by prevention of access to the wilderness can be mitigated by sensitively run programs that include overt benefit to the local people. Some conservation programs have thus become development programs.

The acronym ICDP (integrated conservation and development program) has become very much part of conservation writing, along with the accompanying question of whether programs for conservation of wildlife and wilderness must include programs for development of the local people (Borgerhoff Mulder and Coppolillo 2005; Caldecott 1996; Oates 1999; Sayer and Campbell 2004). If the integration should occur, it needs to be a two-way integration: development programs that do not also conserve the local environment on which rural people in tropical countries so heavily depend are as doomed to failure as conservation programs that ignore the needs of the local people.

A host of problems confront attempts to integrate conservation with development. Many result from the almost inescapable fact that because development almost always involves more consumption, development is fundamentally antithetical to conservation (Borgerhoff Mulder and Coppolillo 2005, ch. 10; Harcourt 2000; Oates 1999). Of course, development should not be stopped: the poverty of Africa is unacceptable (sec. 14.1.5.1). But as development programs generally have much more money and much more expertise in development than do conservation programs, conservation programs should probably not turn themselves into development programs. Instead, the conservation programs should work in association with development programs that themselves are sensitively run and overtly benefit the local people (Caldecott and Miles 2005, ch. 16; Harrison 1987, 1992; Sayer and Campbell 2004).

These coordinated programs will include, of course, ones run by the local people themselves (Borgerhoff Mulder and Coppolillo 2005, ch. 8; Caldecott and Miles 2005, ch. 16; Sayer and Campbell 2004). It is the height of imperialist arrogance to assume that conservation is imposed upon non-Western countries from outside. Conservation is not solely a Western idea (Adams and McShane 1992, ch. 12; Borgerhoff Mulder and Coppolillo 2005, ch. 4; Gadgil and Berkes 1991). At the same time, resentment about protected areas and conservation is never going to be eradicated. We doubt that there is a country in the world where those who wish to exploit do not resent those who prevent exploitation. Look at the vituperation poured on environmentalists in the United States by some of the world's already richest individuals, institutions, and companies (Ehrlich and Ehrlich 1996).

The alternative to fully protected areas might be multiple-use areas, that is, protected areas where some small-scale extraction is permitted. As protected areas are rarely in fact completely protected, especially in countries too poor to protect them properly, any protected area is effectively a multiple-use area anyway (Harcourt 1986). If these areas are still mostly wilderness despite the exploitation, as many are, then they are effectively multiple-use areas. The problem is that legal multiple use can be very difficult to manage, and surely normally leads to overuse (Sharma and Shaw 1993). Again, we are not talking only about tropical countries: the appearance on Japanese supermarket shelves in Osaka of whale meat from an animal caught by Icelandic whalers for "scientific study" is famous (Cipriano and Palumbi 1999).

Globally, the pejorative "paper park" has often been used to describe protected areas in the third world. John Terborgh (1999) in his alarming *Requiem for Nature* contrasts the poor achievements of developing nations with what he sees as the United States' dedication to conservation. However, as we have written elsewhere, "If you want to see a paper park, go to Yosemite National Park in the USA on a warm summer day. Look at the acres of tarmac inside the park, the luxury hotels, the thousands of visitors, and the smog haze on the valley floor from the hundreds of vehicles" (Harcourt 2000).

More concretely, adequately funded parks in the developing world can work just as well as they do in the United States. Protect the parks properly and numbers of endangered species will increase (Caro, Rejmánek, and Pelkey 2000; Kalpers et al. 2003; Leader-Williams and Albon 1988; Leader-Williams, Harrison, and Green 1990; McNeilage et al. 2001), even if there are interesting complications concerning differences between species in how they respond to different forms of protection (Caro et al. 2000). Globally, protected areas have less land cleared than do their surroundings, less logging, more wild animals, less land burned, and less grazing by domestic animals (Bruner et al. 2001).

Poor as they are, developing nations have on average gazetted as protected a greater proportion of their land area than have developed nations, 12.6% versus 8.9% in 2003 (World Resources Institute 2005). The poorest continent, sub-Saharan Africa, at 10.9%, does better than developed nations; and the nine countries in Africa that harbor gorillas average almost as much as developed nations, 8.2% (World Resources Institute 2005).

Not only that, but the average size of the set-aside land, the protected areas, is greater in developing than in developed countries (Leader-Williams et al. 1990). Thus, sub-Saharan Africa's median protected area is 205 km^2, compared to northern America's 50 km^2 and Europe's 6 km^2 (data from Iremonger, Ravilious, and Quinton 1997, for IUCN reserves I, II, IV).

And finally, several African countries spend (or spent) on their protected areas a greater proportion of their government's expenditure than do Western countries: in the 1990s, the United Kingdom spent just 0.03%, the United States spent 0.12%, but Botswana, Kenya, Namibia, Tanzania, and Zimbabwe each spent over 0.2% (Wilkie, Carpenter, and Zhang 2001).

The three-country Virunga Volcano protected region and the Bwindi Impenetrable National Park of Uganda are examples of effective protection in national parks in Africa (Caldecott and Miles 2005, ch. 8, 16; Harcourt 1986; Kalpers et al. 2003; McNeilage et al. 2001; Weber 1993). Surrounded by some of the densest human populations in Africa, these parks are fifty times the average size of a European park, the boundaries of the parks are effectively intact (fig. 14.5A, B), and as already shown, numbers of gorillas in both are increasing (fig. 14.3).

Both the mountain gorilla protected areas have benefited from large international conservation programs and from tourism (Caldecott and Miles 2005, ch. 16; Harcourt 1986; Weber 1993). Parks are not protected by goodwill and honesty alone—however hard the underpaid, even unpaid, guards of the DRC's Parc National de Virunga-Sud might try. Tourism can bring in the necessary cash (fig. 14.6), and that cash can make a difference

Fig. 14.5. Some developing nations' parks are well protected. (A) Border of Rwanda's Parc National des Volcans, © A. H. Harcourt; (B) Border of Uganda's Bwindi Impenetrable National Park. © C. S. Harcourt.

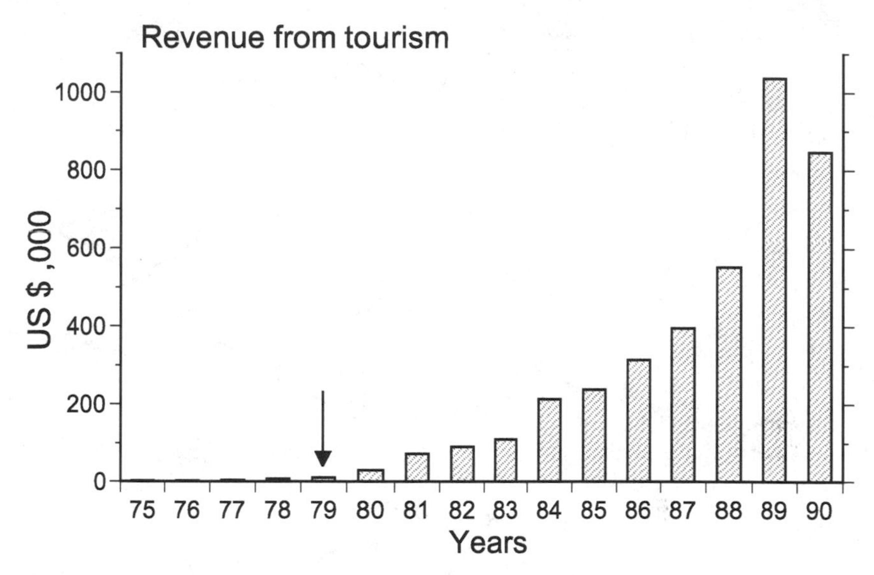

Fig. 14.6. Revenue from tourism to Parc National des Volcans, Rwanda, 1975–1990. Arrow marks start of Mountain Gorilla Project, which augmented protection, education, and tourism. Sources: Data from Harcourt (1986), Weber (1993), and Butynski and Kalina (1998).

to how important the government perceives the protected area to be (Harcourt 1986; Weber 1993).

Ecotourism is far from a panacea. A lot of tourist revenue does not go to the office of conservation, even not to the country (the safari is paid for outside the country, and the hotels are owned by expatriates). The tourists can cost more to support than they contribute (Borgerhoff Mulder and Coppolillo 2005, ch. 10; Weber 1987; Wilkie and Carpenter 1999). Damage can result both nationally and locally, both to the animals concerned and to the country and its people (Butynski and Kalina 1998; Ghimire and Pimbert 1997; Harcourt 1986; Kiss 2004; Myers 1972; Weber 1987, 1993). Not the least of the forms of damage is the exposure of the blatant disparity in wealth between the tourist and the local African village (Myers 1972). A very specific potential problem for gorilla tourism is transmission of disease (Butynski 2001; Butynski and Kalina 1998; Caldecott and Miles 2005, ch. 8, 16; Mudakikwa et al. 2001; Woodford, Butynski, and Karesh 2002). Probably as a result of Tom Butynski's emphasis of the potential problem, the Red List even has tourism as a "major threat" facing mountain gorillas (IUCN 2006).

Butynski and Kalina (1998) have stated that effectively no research has been conducted on the effects of tourism on gorillas. They are correct that far too little has been done, but they failed to mention that fifteen years previously, the effects were explicitly tested (Harcourt et al. 1983). We compared the performance of groups of gorillas visited by researchers and tourists with groups not legally visited by anyone. The result? Visited groups were faring significantly better as judged by the ratio of immatures to adults (stats. 14.2) (Harcourt et al. 1983).

Butynski and Kalina (1998) went on to argue that because so many other conservation activities are implemented at the same time as a tourism program, it is difficult, even impossible, to separate any benefits of tourism from these other activities. But that is precisely the point. Tourism necessarily involves other conservation activities, which benefit the protected area and the gorillas (Caldecott and Miles 2005, ch. 8; Weber 1993). Without them, the tourism program would not work. Without tourism, those other activities might not occur.

Tourism helps prevent illegal use of protected areas through the frequent monitoring of, in this case, the gorilla population; tourism helps prevent illegal use of protected areas through the extra guarding that normally accompanies ecotourism—no animals to watch, no ecotourism; and tourism helps prevent illegal use of protected areas because of the extra revenue that it brings, which both increases the importance of the protected area to a government with many other calls on its revenue, and provides the funds necessary to implement the increased protection (Caldecott and Miles 2005, ch. 8; Harcourt 1986; Weber 1993). No wonder that the two gorilla populations in Africa most visited by tourists are increasing in size (Kalpers et al. 2003; McNeilage et al. 2001), indeed they may be the only ones increasing in size.

That increase itself is yet more direct quantitative evidence that far from being a major threat to gorillas, properly run tourism programs can benefit them. Yes, there are dangers, but the crucial question that has to be asked is what the state of the gorilla populations (or any other wildlife areas) would be without tourism. Would they be as healthy as they are with it? Would they even exist if there was no tourism (Myers 1972; Weber 1987)? We think not.

14.2. Socioecology and Conservation

We have given some indication of the state of conservation of gorillas. We turn our attention now to the application of the science of socioecology to conservation, an understudied area of investigation (Janson 2000; Kappeler and van Schaik 2002). An immense amount of paper (and therefore trees) is used by the conservation biology industry. But how much of that biological science is practically applicable to conservation? If it is applicable, is it in fact applied? If it is applied, how relevant is it really?

14.2.1. Is biology necessary?

Whether a forest is conserved or not usually depends not on our understanding of its biology, nor on our understanding of the biology of its reaction to threat, but on the threats and on our ability to combat them. When an avalanche is bearing down on a house, survival depends not on the architect, but on diverting the avalanche. And an avalanche is bearing down on wilderness. A burgeoning human population and its ever-burgeoning desire for more is destroying wilderness faster than ever before (sec. 14.1.1.).

Where is the place for biology in this scenario? Where is the place for socioecology? Spend time in the offices of the people doing conservation, as opposed to writing about it, and you will not find them poring over books on conservation biology. They are getting on with the practice of conservation, not reading about the biology of the species or ecosystems that they are trying to conserve (Harcourt 2000).

We know why the gorilla needs conserving: its habitat is disappearing, and it is being hunted. We know which populations are the most threatened: Cross River and some of the tiny eastern DRC ones. We know the size of the populations needed to conserve the species in (relative) perpetuity—5,000 animals in 5,000 km^2. We know where to concentrate our conservation efforts—the seven or eight 5,000 km^2 protected areas, and the two mountain gorilla populations. We know how to go about conserving those sites—put a strong metaphorical fence around them. Where is the biology in all this knowledge? How much of that biology is socioecology? What more biology or socioecology is needed?

In the case of the mountain gorillas, surely a success story if ever there was one (sec. 14.1.2.1), what biology was needed in order for the conservation authorities to stop poaching and stop encroachment on the forests? What more was needed other than to leave the gorilla alone, and let them increase? What caused the mountain gorilla turnaround in the 1980s? Was it a massive increase in biological research or the application of that research? No, it was a massive increase in spending on protection, on the development of tourism to bring in funds to the offices of national parks, and on education and publicity about the gorillas in the host countries (Harcourt 1986; Weber 1993).

In a deeply pessimistic mood, Tony Whitten and colleagues (Whitten, Holmes, and Mackinnon 2001) suggested that most of conservation biology was, as they politely put it, displacement activity. For forty years they had been flying over the forests of Indonesia. For forty years they had been reading and themselves writing in the field of conservation biology. And despite all that science, the forests were disappearing at an ever faster rate. Conservation biology was not making a jot of difference.

The relevance of conservation biology is a different question from its usefulness (Harcourt 2000). As Margaret Kinnaird and Tim O'Brien (2001) replied to Whitten and colleagues (2001), the problem for Indonesian forests was not the scientific irrelevance of conservation biology, but the fact that the science was ignored by exploiters who were not interested in conservation.

In the case of the answers to the why, what, how much, where, and how of gorilla conservation, there is some substantive biology contained. Without it, we would not know, for example, the size of a population necessary to conserve a species in (relative) perpetuity (sec. 14.1.3). Is there more, though, than demography, than size of populations and their rate of change of size (IUCN 2006)? Does it matter how a population of 5,000 gorillas works as long as we manage to keep the population at or above 5,000? In a field necessarily dominated by population size of the species, or area of the ecosystem, and in a field to some extent dominated by outcome—numbers or hectares saved—does process matter?

Perhaps right now for protected areas of several thousand square kilometers, maybe not much biology is needed, not much understanding of process. By comparison to the politics and economics of conservation, by comparison to the massive increase in the size of the developing world's human population, and its aspirations, and by comparison to the scale of consumption of the developed world and the consequent pollution, biology is at the moment often a minor player. Nevertheless, biology does play a role, and it is going to play a far greater role (Chapman and Peres 2001; Cowlishaw and Dunbar 2000; Harcourt 2000; Sayer and Campbell 2004, ch. 11).

To say that maybe not much biology is needed is not to say that no biology is needed. For instance, animal epidemics, such as foot-and-mouth disease, are sometimes controlled by restricting movement of infected animals. However, the confirmation by biologists of the long-suspected possibility that fruit-bats harbor the Ebola virus (Leroy et al. 2005) would make nonsense of any attempts to restrict movement of potentially infected apes or humans.

As we have already said, there is no question but that most of the world's wilderness is going to disappear in the coming century. The remaining wilderness is going to be under even more intense pressure in an even more adverse landscape than is currently the case. The comparison of reserve sizes by continent that we have already described might give a portent of the future that we speak of. Currently, reserves in countries that harbor primates average 142 km^2; by contrast, Western Europe's wildlife has to survive in patches of 6 km^2 on average (data from Iremonger et al. 1997, for IUCN reserves classes I, II, IV).

Intense management is going to be increasingly necessary to preserve the diversity of these small remaining patches. Only the deepest understanding of the process of species' response to change of the environment, of predator-prey relationships, community ecology, food webs, social systems, and so on will enable us to manage the wilderness of the twenty-second century in such a way as to maintain even a fraction of its twenty-first century diversity (Beissinger 1997; Chapman and Peres 2001; Harcourt 1998a; Harcourt 2000; Thomas, Baguette, and Lewis 2000).

When reserves worldwide are of the order of only a few tens of square kilometers, then it will be highly relevant whether the species is territorial or not, polygynous or not, group-living or not (all of which affect potential stability of the population and its effective size) (Arcese, Keller, and Cary 1997; Creel 1998; Dobson and Poole 1998; Dobson and Lyles 1989; Dobson and Zinner 2003; Durant 2000; Durant and Mace 1994; Franklin 1980; Gilpin and Soulé 1986; Parker and Waite 1997). Distance dispersed and sex of disperser will profoundly affect responses to harvesting, the ability of populations to replenish each other by natural movement of animals between them, and the ability of populations to recover from low numbers (Clutton-Brock and Albon 1992; Clutton-Brock et al. 2002; Durant 2000; Van Vuren 1998). The socioecology of animals' responses to hunting and conversion of habitat, and of animals' use of the human habitat will be crucial (Chapman and Peres 2001; Cowlishaw and Dunbar 2000).

For biology to produce useful, reliable guidelines, far more data are needed than conservationists presently have the time or money to provide (Nicholls 1998). We barely begin to have the biological understanding required to implement successfully the sort of active conservation management that is going to be necessary in the next century (Wilson 2000), and we are especially lacking in our understanding of tropical biology (Coppeto and Harcourt 2005; Gaston and Blackburn 1999; Janzen 1986). Even less do we have the necessary integration of disciplines, including of biology with non-biological disciplines (Brown 2004; Ludwig, Mangel, and Haddad 2001).

14.2.2. How can socioecology help?

Steve Beissinger (1997) and Tim Caro (1998a) have separately and nicely summarized and reviewed the roles that behavioral ecology can play in conservation. Behavioral ecology, the science of the study of adaptive behavior of individuals, underlies socioecology, the science of how the interaction of an individual's adaptive strategies produces the observed society (ch. 1.1). Beissinger identified three main problems that conservationists face, and listed seven main methods that they currently use to solve those problems (table 14.2). Caro started with a different, though complementary, list of seven main methods used in conservation management, and asked which ones benefited from application of behavioral ecology (table 14.2, column B). Caro saw behavioral ecology as relevant to only four of conservation's methods (italicized in the table).

We see behavioral ecology and socioecology as relevant to all of Caro's management methods, and all of Beissinger's too. All forms of management on lists A and B in table 14.2 involve management of populations and can benefit from understanding of the demography of the populations. Therefore, all forms of management can benefit from application of socioecological understanding.

Preventing loss (table 14.2, no. 1), even if it is as crude or simple as building a fence, requires demographic knowledge, not least with regard to, for instance, the influence of dispersal of individuals on the structure of societies and therefore populations. We could not design reserves or manage them (table 14.2, no. 1.1), or design connections between reserves (table 14.2, no. 1.2), and certainly could not perform population viability analyses (table 14.2, no. 1.3) without an understanding of demographic process, and therefore of socioecology.

Table 14.2. Some main conservation management practices.

A.	Beissinger	B. Caro
1.	Preventing loss	
	1.1 Reserve design	Managing reserves
	1.2 Ecosystem management[1]	*Reserve connectivity*
	1.3 Population viability analyses	
2.	Conservation × Development	
	2.1 Sustainable development	Trade in wildlife; *Extractive reserves*
3.	Restoration	
	3.1 Managing reserves	
	3.2 Captive breeding; release	*Captive breeding; Reintroductions/ translocations, etc.*
	3.3 Ecosystem restoration	
4.		Prioritizing management plans

Sources: Beissinger (1997) and Caro (1998a).
[1]By "ecosystem management," Beissinger means management of areas outside reserves.

The linking of conservation to development (table 14.2, no. 2), particularly if it is going to be via some form of extraction (table 14.2, no. 2.1), requires deep understanding of process, and therefore of socioecology.

Reserve management (table 14.2, no. 1.1, 3.1) and captive breeding and release (table 14.2, no. 3.2) all require socioecological knowledge. Without it, how would anyone know, for example, which sex of gorilla to release and how (Harcourt 1987c)? If an ecosystem is to be restored, the effects on wildlife of the means of restoration and of the result need to be known, knowledge that requires an understanding of process, and hence of socioecology.

Finally, prioritizing of management plans (table 14.2, no. 4) must involve ability to predict what populations will do under different scenarios of management. The prioritization must therefore involve knowledge and understanding of how populations will respond as conditions change. That knowledge and understanding can come only from knowledge and understanding of the process of response, in other words, with contribution from socioecology.

Next, we describe some ways in which, we suggest, socioecology can and should be useful to conservation management of wild animal populations.

14.2.3. Socioecology and demography

Surely the most satisfying moment for any field biologist interested in the behavior of animals is the day when the biologist's arrival warrants the merest glance from the animals to check who he or she is before the animals go back to their daily routine. One adult female gorilla in one of our study groups at Dian Fossey's Karisoke Research Center continually frustrated us. None of us who tried to collect data on Fuddle will ever forget her bad-tempered stare at us as we approached, the sight of her getting up and disappearing into the thick vegetation, and the return to the Center with yet another day with no data on Fuddle.

Fuddle's bilious temper emphasizes that societies are always composed of different sorts of individuals doing different things even under apparently the same circumstances, and certainly as the circumstances differ. Details of those circumstances, of the proportions of individuals differing, and the influence of the differences on the structure of society are highly relevant to assessments of demographic stability of populations. The application to population management and conservation of understanding of individual behavior and societal behavior has been a topic of interest for decades (Chitty 1996; Cohen, Malpass, and Klein 1980; Sibly and Smith 1985) but now seems to be receiving increased attention (Akçakaya and

Ginzburg 1991; Clutton-Brock and Albon 1985; Clutton-Brock et al. 2002; Dittus 1977; Durant 2000; Durant and Mace 1994; Kendall and Fox 2002; Sutherland 1998). Books are being published on the subject (Caro 1998b; Clemmons and Buchholz 1997; Festa-Bianchet and Apollonio 2003; Gosling and Sutherland 2000; Lomnicki 1988; Sutherland 1996).

In simple brief, the more homogeneous the population, the less stable it is—because with all individuals doing the same thing at the same time, all can go extinct at the same time with the appropriate adverse change to their environment. If individuals do different things at different times in different places, populations are not nearly so sensitive to change in the environment. Models that incorporate the fact that individuals differ produce far more realistically stable populations than those that do not. And models that incorporate not just the fact that individuals differ, but knowledge of how they differ, and why they do so, and in what circumstances, are even better.

Sarah Durant's studies of gorilla demography are relevant in this context (Durant 2000; Durant and Mace 1994). Gorillas happen to have a society that makes their populations perhaps unusually resistant to environmental stochasticity. For instance, broken as the population is into scattered, quite stable groups, it is unlikely that all the population is going to be affected at once by any environmental catastrophe. Yet because of dispersal of both sexes, groups or areas that suffer can be rejuvenated with immigrants. The demographic models of Robbins and Robbins (2004) took even more account of the nature of gorilla society, incorporating, for instance, not just sex differences in dispersal, but the effect of death of the silverback on dispersal of females and hence on infanticide rates in the population. Comparison of the models' outcomes with the results of the much-censused Virunga population allowed detailed examination of the varying influence of different aspects of gorilla society on demography. The comparisons indicated higher male mortality than recorded and emphasized the influence of dispersal by males on the population's structure.

One general problem with dispersers is that they can carry disease (Hoeck 1982), which is why corridors between protected areas are not always necessarily a good idea (Hess 1994). Socioecologists have largely ignored the potential relation between the nature of an animal's society and transmission of disease within it (Janson 2000; Kappeler and van Schaik 2002). So also has demographic modeling, such as Durant's. As the threat of disease has only recently risen as an area of crucial study in conservation (sec. 14.1.1), detailed analysis of the relation between the nature of a species' society and its susceptibility to disease is only just beginning (Hess et al. 2002). For instance, socioecology was not mentioned in Chapman and colleagues' (2005) review of the ecology of disease in primates.

In the case of transmission of disease, the nature of gorilla's society could make it particularly susceptible to a contagious, lethal disease, such as Ebola (Nunn et al., submitted). Because female gorillas are attracted to a male (ch. 4.2.2.1; ch. 7.1), and usually only one male is in a group (ch. 4.2.1.1), if a disease enters a group and the male dies, females then disperse (ch. 4.2.1.2; ch. 8), carrying disease with them. By contrast, in chimpanzee society, the death of a male could have no effect at all on others.

While one report has suggested (though with no data) that gorillas might have suffered in Ebola outbreaks more severely than have chimpanzees (Leroy et al. 2004), others indicate that gorillas and chimpanzees suffer equally (Huijbregts et al. 2003; Walsh et al. 2003). Whichever turns out to be the case, any contradictions of the model's predictions should tell us something about the origins and transmission of Ebola in wild populations. Here, socioecological understanding, by informing us about the disease, could help not only the wildlife but also the local people, who are almost as susceptible as the apes are (Leroy et al. 2004).

14.2.4. Socioecology and reaction to the environment

Much conservation is necessarily based on our knowledge of what species and populations are doing now. For instance, conservationists pay a lot of attention to current hotspots of diversity (Mittermeier et al. 1998; Myers et al. 2000). That attention, that outlook is fine if the world does not change. But, of course, the very reason to act to conserve is the fact that the environment is changing. Unless we understand the process by which individuals, societies, populations, and species react to change, conservation management is going to be for only the short term (Brown 1995, ch. 12; Caughley 1994; Caughley and Gunn 1996).

If socioecology is concerned with anything, it is concerned with the effect on society of individuals' responses to changing environments. Understanding the process of interaction between behavior of individuals and the environment, between the nature of the society and the environment, and how both might change with a changing environment is going to be fundamental to conservation. For instance, an increased density of food or nest sites could change a monogamous system in which the male bird helps feed the offspring into a polygynous system in which he does not, with consequent decrease in nesting success per female (Davies 1992; Komdeur and Deerenberg 1997; Rubenstein 1998). The change might also result in a drop in effective population size (Komdeur and Deerenberg 1997), unless polygynous males breed for a shorter time than do monogamous ones. And with dispersal a response to competition, and dispersal fundamental to the nature of societies (ch. 2.4.2.6; ch. 8; ch. 9; ch. 10; ch. 11.2; ch. 11.3), understanding the socioecology of dispersal is of fundamental importance to conservation (Macdonald and Johnson 2001).

The southern Virunga Volcano area in Rwanda and DRC was cleared of cattle, people, and other disturbance in the mid-1970s (Harcourt 1986; Weber and Vedder 1983). The cleared area is a highly suitable habitat for gorillas, full of their main food plants, and relatively well protected. And yet, thirty years later, still no gorillas use the area (Kalpers et al. 2003). Members of normal mammalian societies, in which males and females range separately, and dispersing females move to suitable habitat where they are joined by males, would surely soon have colonized such an empty habitat (Pope 2000; Van Vuren 1998). But in many species, the attraction is not to suitable habitat, but to conspecifics (Stamps 1988). If each dispersing sex of gorillas goes where other gorillas are (ch. 4.2.1.2; ch. 8; ch. 10.1), not to where food is, and if there are no other gorillas there, no reason exists to go there. Are the Virunga gorillas ignoring prime habitat because of the nature of their society?

Territoriality has long been associated with strong influences on populations' reactions to change in the environment via competitive spacing of individuals. The study of territoriality of grouse, a highly lucrative game bird if its populations and habitat are well managed, has thus long been of interest (Mougeot et al. 2003; Watson et al. 1984). Chimpanzees are territorial; gorillas are not. Lee White and Caroline Tutin (2001) have suggested that this very difference is responsible for the contrast between the species in their response to logging. They found that chimpanzee densities dropped by over 80% after logging, whereas gorilla densities changed little. They argued that the territorial chimpanzees cannot escape the loggers by moving because they are surrounded by territorial neighbors, whereas because gorilla home ranges already overlap hugely (ch. 4.1.5), gorillas can easily shift their home range into another group's range.

The argument needs verification, and might not be general. Primate species that exploit widely distributed foods and have large home ranges, either absolutely or for their body mass, appear to be susceptible to alteration or loss of their habitat (Harcourt 1998a; Harcourt and Schwartz 2001; Skorupa 1986). Territorial species, which have small home ranges (see Mitani and Rodman 1979), should then be less susceptible (Skorupa 1986). However, neither Harcourt nor Skorupa found that this was the case. Correspondingly, Juichi Yamagiwa (1999a) suggested other social factors than territoriality explained differences between the two species in their responses to environmental change, such as differences between them in variability of group size and readiness of females to travel alone.

As a final example, consider the infanticide model of chapter 7.2. The model ignored predation and asked whether infanticide alone could cause gorilla females to associate with a male. If females associated with males for protection from only predation, we could see great changes in the nature of their society in response to disappearance of predators from fragmented forests (Michalski and Peres 2005; Terborgh et al. 2001), especially if the predators are hunted by humans (Harcourt, Parks, and Woodroffe 2001; Parks and Harcourt 2002; Woodroffe and Ginsberg 1998). Because of the great deal that primatology knows about the nature of gorilla (and chimpanzee and orangutan) society, we could model the changes caused by females no longer associating with males for protection from predation—and in this case show that the nature of the society probably would not change if predators disappeared, because females would continue to associate for protection from infanticide. What are the consequences for management? One is that lucrative photographic safaris to gorilla populations would still be viable.

However, in general, we know far too little about the socioecology of reaction to environmental change (Caughley 1994; Chapman and Peres 2001; Harcourt 1998a; Janson 2000). One obvious reason for our ignorance is that because of the need for large sample sizes for believable results, study sites are in areas of high density of our study species (Chapman and Peres 2001). In the case of long-lived species such as primates, that usually means that the study sites are in areas safe from rapid change. Conservation needs more study of species in marginal habitats (ch. 13.3.9.1).

14.2.5. Socioecology of human use of the environment

Humans are the main current threat to persistence of wilderness. Human need and greed are why we need to fight to conserve wilderness. What is behavioral ecology and socioecology all about if it is not about understanding the process of acquisition of resources, and its effect on both the resources and society, and society's effect on the means individuals use to acquire resources? So far behavioral ecological theory has been far more strongly applied than has socioecological theory to understanding of how and why humans use resources, and often overuse them (Alvard 1998b; Cowlishaw and Dunbar 2000, ch. 9; Milner-Gulland and Mace 1998; Wilson, Daly, and Gordon 1998; Winterhalder and Smith 2000). Common applications of behavioral ecological theory include quantitative understanding of overharvesting of various resources that arises because current return on effort is often greater than future return (Alvard 1998a, b; Clark 1989), and because of scramble competition for a communal resource (Borgerhoff Mulder and Ruttan 2000; Borgerhoff Mulder and Coppolillo 2005, ch. 6).

Behavioral ecology might be more applicable and useful as a discipline to apply to conservation than is socioecology, given the conservationist's interest in exploitation of resources. To a large extent, the nature of the society's response to the exploitation is irrelevant—all it took was one individual to shoot the last dodo, whatever sort of society they came from. However, we see a greater role for socioecology as conservationists and others think harder about integration of conservation with improvement of the livelihood of local people, and about bottom-up approaches to conservation, rather than top-down ones (Borgerhoff Mulder and Coppolillo 2005, ch. 4, 10; Gadgil and Berkes 1991).

14.2.6. Socioecology and active management

We have stressed so far the importance of socioecological understanding to broad appreciation of populations' responses to environmental change. Here we list a few examples of how socioecological understanding might be directly applicable to active management of populations.

The effects of culling and hunting are usually calculated according to the number of animals taken, with potential domino effects often ignored. If the socioecology of infanticide were not known, would an increase in infant mortality as a result of the hunting of male grizzly bears be even detected, let alone its cause understood (death of protective males) (Swenson et al. 1997). Similarly, while gorillas are rarely hunted for sport now, Robbins and Robbins's individual-based models show the influence of death of breeding male gorillas on infanticide rates, and therefore on population growth rates (Robbins and Robbins 2004).

Active management of populations can involve veterinary management, for example, the anesthetization of mothers in order to capture infants to treat them (Mudakikwa et al. 2001). If we did not understand the society of colobine monkeys, in which adult females often carry infants that are not their own (Mitani and Watts 1997), how often would we anesthetize the "mother" to treat an infant, only to discover the real mother arriving to retrieve the infant?

If we did not understand the socioecology of competition between the sexes for resources, would we understand why not culling female red deer causes a drop in numbers of stags as a result of the stags emigrating to poorer feeding areas in the face of increased competition from females, despite the fact that males are larger than females (Clutton-Brock et al. 2002)?

14.2.7. Socioecology and tourism: The relevance of socioecology

"[G]orilla tourism . . . is founded on a very weak research base, and presents a substantial threat to gorilla survival." This is Eleanor Milner-Gulland and Ruth Mace's (1998, p. 172) summary of Butysnki and Kalina's chapter in their edited book. We have discussed elsewhere why we disagree with the idea that tourism is dangerous (sec. 14.1.5.2). What Milner-Gulland and Mace mean by tourism founded on a weak research base is that the effects of tourism itself have, as Butynski and Kalina state, not received much study. That, though, is very different from gorilla tourism being implemented in a vacuum of knowledge and understanding about gorillas.

By the time a tourism program is started, it can be difficult to realize just how much knowledge and understanding has gone into the program. We are going to close this brief discussion of the relevance of socioecology to conservation by pointing out what a great deal we needed to know about gorillas in order to implement a tourism program, and did know about them (Harcourt 1986; Harcourt and Curry-Lindahl 1979; Tutin and Vedder 2001; Weber 1993).

It is surely no accident that arguably the most successful gorilla tourism program was initiated in the gorilla population with the longest-running, most successful field study of wild gorillas at the time, namely, Dian Fossey's Karisoke Research Center site in the Virunga mountain gorilla population, established in 1967 (ch. 3.1.3) (Fossey 1983; Robbins et al. 2001; Stewart et al. 2001). Besides the fact that gorillas are charismatic megafauna, why did conservationists (specifically the Mountain Gorilla Project [MGP] [Harcourt 1986; Harcourt and Curry-Lindahl 1979]) consider that a tourism program was likely to be successful? The answer is because an immense amount was known about the Virunga gorillas, thanks largely to the research conducted from Dian Fossey's Karisoke Research Center, which itself built on Schaller's (1963) studies started two decades earlier.

The MGP knew by the late 1970s that mountain gorillas lived in small home ranges. Consequently, it knew that once a group was found, the group would always be in the rough vicinity, within a few kilometers of where it was first found. The MGP knew, therefore, where to site tourist entry points to the park. It knew that those sites could be planned and used for the year. Long-term planning was possible, even easy.

The MGP could in fact do a lot better than that. Because the MGP knew that the daily distance traveled by a group was on average only half a kilometer, it knew that a tourism program was almost guaranteed to find the gorilla group easily every day. Indeed, the MGP knew that there could on occasion be time to find the group before tourists had finished their breakfast and return to lead the tourists at a leisurely pace (the park is over 3,500 m in altitude) straight to the group.

The MGP knew that gorilla groups had more or less stable membership for months, even years at a time. Consequently, once a group was habituated to visits by tourists, it would remain habituated.

Because the MGP knew that gorilla groups were stable for long periods, guides could know from the identity of the male what group had been found, and therefore the group's composition. The tourists' behavior could be managed accordingly, for instance, being extra quiet and careful right from the start of the visit if an infant had recently been born.

Because the MGP knew that gorillas lived in family groups of a male to which were attracted several females, the tourist guides knew that if they could find the male, they could find the rest of the group.

The MGP knew that gorilla group members were effectively subservient to the leading male, and consequently the MGP knew that habituation of the male was the key to habituation of all animals, so improving the efficiency of the process of habituation. This efficiency went both ways. If the male could not be habituated, as one in the Virunga population could not, time was not wasted trying to continue habituation on a hopeless cause. Indeed, guides could be warned if the unhabituated male was in the vicinity, so that they did not mistakenly take tourists to his group.

The MGP knew that gorillas lived in family groups of a male, several females, and their offspring of varying ages, and hence the MGP knew that if a group was found, several animals of a variety of sizes performing a variety of behaviors would be present. An amazing and varied tourist experience could be guaranteed. (By contrast, we have sat for three hours under a tree in Sumatra waiting for the single orangutan we saw that day to emerge from the canopy—he never emerged).

Because the MGP knew the gorillas' daily activity rhythm, and knew that the animals had a midday rest period of roughly an hour, the MGP knew it could implement the perfect schedule of gorilla viewing for the tourist. The gorillas' daily rhythm accorded very well with the tourists' daily rhythm: breakfast, a walk after breakfast to the gorilla, gorilla viewing, and a return to the hotel by mid-afternoon.

Because the MGP knew that most gorilla group members gathered for protection near the dominant silverback male, the MGP knew that if it could get tourists to a gorilla group around midday, then not only would the tourists be in the presence of several animals of a variety of sizes performing a variety of behaviors, but all those animals could quite easily be seen at once (fig. 14.7).

Fig. 14.7. Two tourists and a park guide in full view of resting gorillas. (Mountain gorilla.) © A. H. Harcourt

Because the gorillas were clumped, the tourists could be kept clumped—they did not need to spread out to see the gorillas—with two benefits. The clumping of the tourists meant that only one guide was needed to shepherd the tourists. Perhaps more importantly, the clumping of the tourists meant that the gorillas could see all the tourists at once, and therefore they could be relaxed despite the close presence of so many humans.

Other knowledge was beneficial too. Because the MGP knew that the gorilla was a female-emigrant species (unlike many Old World primates), the park authorities knew that disappearance of a female from the gorilla group, especially a nubile female, did not necessarily indicate trouble in the group, for example, a death.

Because the MGP knew how paternal the leading male could be, being the main sire of the group, the MGP knew that loss of an adult female did not necessarily mean loss of her infant, with the consequent relief that the orphan could probably quite safely be left with the group, instead of having to be captured and hand-reared.

Because the MGP knew of the crucial leadership role of the dominant male, it knew that if he died, the group would probably disperse in the very near future (unlike many Old World primates), and therefore the MGP could plan accordingly.

Finally, detailed knowledge of the socioecology of gorillas allows modeling of the nature of their society (ch. 7.2; ch. 10.1.3). As we described in section 14.2.4, one of the models allows us to predict what would happen to gorilla societies in small protected areas that had lost their predators. Were it not for the socioecological knowledge and the consequent ability to model gorilla society, one might have predicted that the absence of predators would lead to the dispersal of groups, as females no longer needed to remain with males for protection from predators. In fact, the modeling indicates that protection from infanticide alone will keep females with males (ch. 7.2). Socioecologists can therefore predict for offices of conservation managing tourist visits to gorilla groups that fragmentation need not affect the business.

In sum, we repeat our argument; deep knowledge of gorillas was what allowed such a successful tourism program. Was it socioecological knowledge or just natural history? What's the difference? The difference is in understanding. Natural history is knowing what the animal does; socioecology is knowing why. Did we need any "why" in the application of research on gorillas to implementation of a tourism program? Maybe not. However, socioecology gives understanding of process. Understanding of process is crucial when the situation changes, as it always does in conservation management. We do not need to understand much about the internal combustion engine to drive to and from work—as long as we do not have a breakdown. But when the car stops, the more we understand about the internal combustion engine, the better our chance of getting it going again, especially in the face of inadequate replacement parts and tools.

14.2.8. Socioecologists and conservation

We have argued that socioecology is relevant to successful conservation. Socioecologists might be even more important than socioecology.

It is the biologist—the socioecologist—who produces the data that the conservation manager uses. It might even be the socioecologist who is the most important, for socioecology requires a lot of time in the field and therefore produces a lot of long-term data from close observation of the animals. We will give just one example of the importance of the field biologist in this area. It is not in fact an illustration of their importance to conservation. It is an example of their importance to human health. Humans catch Ebola from eating infected wildlife (Leroy et al. 2004). Who studies the wildlife? Who sees the infection in the wildlife before it gets to humans? Who can therefore warn of an impending epidemic and help avert it if only the authorities would pay attention? Yes, the wildlife biologist (Leroy et al. 2004).

And perhaps most important of all, it is the biologists who are interested in conservation; it is the biologists who are in the field and who have noticed first that their species, their site is threatened; it is the biologists who more than any other are motivated to act to conserve. Most biologists did not go into the field initially to conserve. They went there to do biology. And they stayed there to save the species and the site that they came to love. That is why conservation is surely usually done where biologists work, whatever all the theoretical algorithms might say about where conservation needs to be done.

14.2.9. Conservation and socioecology

What is the relevance of conservation to socioecology? Simple—extinction means we have nothing to study (Janson 2000). If only for selfish reasons, conservation is vital to the discipline of socioecology. More generally, so also is a healthy environment, both natural and human (Dasgupta 1995, 2004; Homer-Dixon 1999; Homer-Dixon, Boutwell, and Rathjens 1993).

Conclusion

With animals, centuries-old trees, forests, and coral reefs dying at the hand of humans, we see no other metaphor than war for the enterprise that is conservation. Despite the occasional battle won for the moment, the Virunga and Bwindi Impenetrable national parks, for instance (Kalpers et al. 2003; McNeilage et al. 2001), the conservation war is overshadowed by the far greater number of battles lost. As an ivory woodpecker might have been found again in Arkansas bottomland forest (Fitzpatrick et al. 2005), so with far less fanfare did probably the last po'ouli die, only thirty-one years after this single-species genus of Hawaiian bird was first discovered, in 1973 (Thomas 2005).

We suspect that in our lifetime, the Sumatran orangutan will go extinct. It is already in serious trouble (Sodhi et al. 2004; van Schaik, Monk, and Robertson 2001). Locally, its habitat is fast being destroyed. And the destruction is not going to stop. Indonesia is already the fourth most populous country in the world, and it is due to increase its population by 20% over the next twenty years, adding 45 million people to its current 225 million (World Resources Institute 2005). If we have anything to learn from differences between the Sumatran and Bornean orangutans (Delgado and van Schaik 2000), we had better find it out now, for our current students' students might not have the chance.

Stories of the gorilla's demise in the wild in the next twenty-five years are grossly exaggerated. Gorillas are going to be around for a lot longer than is the Sumatran orangutan. Nevertheless, the Cross River gorilla population, first censused in 1988 (Harcourt et al. 1989a, b) could be gone in our lifetime, so serious are the threats it faces (Oates et al. 2003). Over the next one hundred years, we cannot see any alternative scenario to a huge crash in gorilla numbers as forest outside of reserves is destroyed (Harcourt 1996). But the dedication, even unto death in some cases, of some of Africa's park guards and other protectors of wildlife (Plumptre 2003), the willingness of many of Africa's leaders to establish as national parks huge swathes of their territory, their willingness to expend a greater proportion of their country's income on protection of wilderness than do many developed nations, and the remarkable story of the success of mountain gorilla conservation in eastern Africa are grounds for hope.

Table Details

Table 14.1. Reserves of 5,000 km^2 or more, with maybe 5,000 gorillas. Reserve sizes given to nearest lower 100 km^2. Where two reserve sizes given, smaller is from source 1: it is the core area of usually higher protected status, and therefore potentially with higher densities of gorillas. Sources: Size of reserves, 1 (Caldecott and Miles 2005); 2 (Tutin et al. 2005). Gorilla densities (Gor #s), A (Plumptre et al. 2003); B (Poulsen and Clark 2004).

The 7,500 km^2 Minkébé National Park in northeast Gabon does not now have a population of 5,000 gorillas, despite its size, because it suffered one of the earliest recorded devastating Ebola outbreaks (Huijbregts et al. 2003). However, it might have had in the early 1990s densities of $1/km^2$ (Huijbregts et al. 2003), and presumably could again if properly protected. The multi-organization Regional Action Plan (RAP) for western gorillas (Tutin et al. 2005) listed Lopé/Waka reserves as a priority area. It is not included here, because it probably does not have 5,000 gorillas. The RAP put it high on the list because of its size and because of its importance for other taxa, for ecotourism, and for research.

[1]No one country's core section of 7,300 km^2 is more than 4,600 km^2, but densities might be > 2 gorilla/km^2.

[2]Area from source 1 estimated from map.

Figure Details

Fig. 14.2. Percent deforestation by human density: total, $N = 26$, $r^2 = 0.28$; $F = 10.8$, $p < 0.01$; exc. 2 outliers, $N = 24$, $r^2 = 0.54$, $F = 28.4$, $p < 0.001$; gorilla countries (regression shown), $N = 6$, $r^2 = 0.98$, $F = 134.4$, $p < 0.001$.

Fig. 14.3. The numbers are based on near total counts of the population and could be some of the most accurate and precise for any wild population of animals. Most counts are quite imprecise. Further details on counting gorillas are in chapter 4.1.7, box 4.2.

Statistical Details

Stats. 14.1. IUCN categories: in Threatened, we do not include Low Risk, which is why our ranking differs slightly from Mace and Balmford's (2000).

Stats. 14.2. Tourism's effects: gorilla groups visited by researchers and tourists had 48.5% immatures; unvisited groups had 33% ($\chi^2 = 5.0$, $p < 0.03$) (Harcourt et al. 1983).

REFERENCES

Abacus Concepts, I. 1990–91. *Statview SE+*. Berkeley, CA: Abacus Concepts.

Abouheif, E., and D. J. Fairbairn. 1997. A comparative analysis of allometry for sexual size dimorphism: Assessing Rensch's rule. *Amer Nat 149:* 540–562.

Abrams, P. A. 2004. Mortality and lifespan. *Nature 431:* 1048–1049.

Acevedo-Whitehouse, K., F. Gulland, D. Greig, and W. Amos. 2003. Disease susceptibility in California sea lions. *Nature 422:* 35.

Achard, F., et al. 2002. Determination of deforestation rates of the world's humid tropical forests. *Science 297:* 999–1002.

Adams, E. S. 1990. Boundary disputes in the territorial ant *Azteca trigona:* Effects of asymmetries in colony size. *Anim Behav 39:* 321–328.

Adams, J. S., and T. O. McShane. 1992. *The myth of wild Africa: Conservation without illusion.* New York: W. W. Norton.

Agapow, P.-M., et al. 2004. The impact of the species concept on biodiversity studies. *Q Rev Biol 79:* 161–179.

Akçakaya, H. R., and L. R. Ginzburg. 1991. Ecological risk analysis for single and multiple populations. In *Species conservation: A population-biological approach,* eds. A. Seitz, and V. Loeschke, 73–87. Basel: Birkhaüser Verlag.

Akçakaya, H. R., et al. 2000. Making consistent IUCN classifications under uncertainty. *Conserv Biol 14:* 1001–1013.

Akeley, C. E. 1923. *In brightest Africa.* Garden City, NJ: Garden City Publishers.

Alcock, J. 2001. *Animal behavior: An evolutionary approach.* 7th ed. Sunderland, MA: Sinauer Associates.

Alexander, B. K., and J. M. Bowers. 1967. Social organization of a troop of Japanese monkeys in a two-acre enclosure. *Folia Primatol 10:* 230–242.

Alexander, R. D. 1974. The evolution of social behavior. *Ann Rev Ecol Syst 5:* 325–383.

Alexander, R. D., J. L. Hoogland, R. D. Howard, K. M. Noonan, and P. W. Sherman. 1979. Sexual dimorphisms and breeding systems in pinnipeds, ungulates, primates, and humans. In *Evolutionary biology and human social behavior,* eds. N. A. Chagnon, and W. D. Irons, 402–435. North Scituate, MA: Duxbury Press.

Allen, E., et al. 1976. Sociobiology—another biological determinism. *BioScience 26:* 182–186.

Altmann, S. A. 1974. Baboons, space, time, and energy. *Amer Zool 14:* 221–248.

———. 1998. *Foraging for survival: Yearling baboons in Africa.* Chicago: University of Chicago Press.

Altmann, S. A., and J. Altmann. 1970. *Baboon ecology.* Chicago: University of Chicago Press.

Alvard, M. 1998a. Indigenous hunting in the Neotropics: Conservation or optimal foraging? In *Behavioral ecology and conservation biology,* ed. T. Caro, 474–500. Oxford: Oxford University Press.

Alvard, M. S. 1998b. Evolutionary ecology and resource conservation. *Evol Anthropol 7:* 62–74.

Amos, B., C. Schlotterer, and D. Tautz. 1993. Social structure of pilot whales revealed by analytical DNA profiling. *Science 260:* 670–672.

Anadu, P. A., P. O. Elamah, and J. F. Oates. 1988. The bushmeat trade in southwestern Nigeria: A case study. *Human Ecol 16:* 199–208.

Andelman, S. J. 1986. Ecological and social determinants of cercopithecine mating patterns. In *Ecological aspects of social evolution,* eds. D. I. Rubenstein, and R. W. Wrangham, 201–216. Princeton, NJ: Princeton University Press.

Anderson, C. M. 1986. Predation and primate evolution. *Primates 27:* 15–39.

Anderson, D. P., E. V. Nordheim, C. Boesch, and T. C. Moermond. 2002. Factors influencing fission-fusion grouping in chimpanzees in the Taï National Park, Cote d'Ivoire. In *Behavioural diversity in chimpanzees and bonobos,* eds. C. Boesch, G. Hohmann, and L. F. Marchant, 90–101. Cambridge: Cambridge University Press.

Anderson, J. R. 1998. Sleep, sleeping sites, and sleep-related activities: Awakening to their significance. *Amer J Primatol 46:* 63–75.

Anderson, M. J., and A. F. Dixson. 2002. Motility and the midpiece in primates. *Nature 416:* 496.

Andersson, M. 1982. Female choice selects for extreme tail length in a widowbird. *Nature 299:* 818–820.

Andrade, M. C. B. 1996. Sexual selection for male sacrifice in the Australian redback spider. *Science 271:* 70–72.

Appleby, M. C. 1983. The probability of linearity in hierarchies. *Anim Behav 31:* 600–608.

Arcadi, A. C., and R. W. Wrangham. 1999. Infanticide in chimpanzees: Review of cases and a new within-group observation from the Kanyawara study group in Kibale National Park. *Primates 40:* 337–351.

Arcese, P., L. F. Keller, and J. R. Cary. 1997. Why hire a behaviorist into a conservation or management team? In *Behavioral approaches to conservation in the wild,* eds. J. R. Clemmons, and R. Buchholz, 48–71. Cambridge: Cambridge University Press.

Asibey, E. O. A. 1974. Wildlife as a source of protein in Africa south of the Sahara. *Biol Conserv 6:* 32–39.

Aureli, F., and F. B. M. de Waal, eds. 2000. *Natural conflict resolution.* Berkeley: University of California Press.

Ayres, J. M., R. E. Bodmer, and R. A. Mittermeier. 1991. Financial considerations of reserve design in countries with high primate diversity. *Conserv Biol 5:* 109–114.

Backwell, P. R. Y., and M. D. Jennions. 2004. Coalition among male fiddler crabs. *Nature 430:* 417.

Baglione, V., D. Canestrari, J. M. Marcos, and J. Ekman. 2003. Kin selection in cooperative alliances of carrion crows. *Science 300:* 1947–1949.

Balmford, A. 2000. Priorities for captive breeding—which mammals should board the ark? In *Priorities for the conservation of mammalian diversity: Has the panda had its day?* eds. A. Entwistle, and N. Dunstone, 291–307. Cambridge: Cambridge University Press.

Balmford, A., N. Leader-Williams, and M. J. B. Green. 1995. Parks or arks: Where to conserve threatened mammals? *Biodiv Conserv 4:* 595–607.

Barnes, R. F. W. 1982. Mate searching behaviour of elephant bulls in a semi-arid environment. *Anim Behav 30:* 1217–1223.

———. 1990. Deforestation trends in tropical Africa. *Afr J Ecol 28:* 161–173.

Barrett, L., and C. B. Lowen. 1998. Random walks and the gas model: Spacing behaviour of grey-cheeked mangabeys. *Func Ecol 12:* 857–865.

Barrett, L., and S. P. Henzi. 2005. Monkeys, markets, and minds: Biological markets and primate sociality. In *Cooperation in primates and humans: Mechanisms and evolution,* eds. P. M. Kappeler, and C. P. van Schaik, 209–232. Berlin: Springer-Verlag.

Bartlett, T. Q., R. W. Sussman, and J. M. Cheverud. 1993. Infant killing in primates: A review of observed cases with specific reference to the sexual selection hypothesis. *Amer Anthropol 95:* 958–990.

Barton, R. A. 2000a. Primate brain evolution: Cognitive demands of foraging or of social life? In *On the move: How and why animals travel in groups,* eds. S. Boinski, and P. A. Garber, 204–237. Chicago: University of Chicago Press.

———. 2000b. Socioecology of baboons: The interaction of male and female strategies. In *Primate males: Causes and consequences of variation in group composition,* ed. P. M. Kappeler, 97–107. Cambridge: Cambridge University Press.

Barton, R. A., R. W. Byrne, and A. Whiten. 1996. Ecology, feeding competition, and social structure in baboons. *Behav Ecol Sociobiol 38:* 321–329.

Basabose, A. K. 2002. Diet composition of chimpanzees inhabiting the montane forest of Kahuzi, Democratic Republic of Congo. *Amer J Primatol 58:* 1–21.

———. 2005. Ranging patterns of chimpanzees in a montane forest of Kahuzi, Democratic Republic of Congo. *Int J Primatol 26:* 33–54.

Bearder, S. K. 1987. Lorises, bushbabies, and tarsiers: Diverse societies in solitary foragers. In *Primate societies,* eds. B. B. Smuts, D. L. Cheney, R. M. Seyfarth, R. W. Wrangham, and T. T. Struhsaker, 11–24. Chicago: University of Chicago Press.

Beck, B. B., and M. L. Power. 1988. Correlates of sexual and maternal competence in captive gorillas. *Zoo Biol 7:* 339–350.

Beissinger, S. R. 1997. Integrating behavior into conservation biology: Potentials and limitations. In *Behavioral approaches to conservation in the wild,* eds. J. R. Clemmons, and R. Buchholz, 23–47. Cambridge: Cambridge University Press.

Belovsky, G. E. 1987. Extinction models and mammalian persistence. In *Viable populations for conservation,* ed. M. E. Soulé, 35–57. Cambridge: Cambridge University Press.

Bengtsson, B. O. 1978. Avoiding inbreeding: At what cost? *J Theor Biol 73:* 439–444.

Bennett, E. L., and A. G. Davies. 1994. The ecology of Asian colobines. In *Colobine monkeys: Their ecology, behaviour, and evolution,* eds. A. G. Davies, and J. F. Oates, 129–171. Cambridge: Cambridge University Press.

Bentley, G. R. 1999. Aping our ancestors: Comparative aspects of reproductive ecology. *Evol Anthropol 7:* 175–185.

Bergman, T. J., J. C. Beehner, D. L. Cheney, and R. M. Seyfarth. 2003. Hierarchical classification by rank and kinship in baboons. *Science 302:* 1234–1236.

Bermejo, M. 1997. Study of western lowland gorillas in the Lossi Forest of North Congo and a pilot gorilla tourism plan. *Gorilla Cons News 11:* 6–7.

———. 1999. Status and conservation of primates in Odzala National Park, Republic of the Congo. *Oryx 33:* 323–331.

———. 2004. Home-range use and intergroup encounters in western gorillas (*Gorilla g. gorilla*) at Lossi Forest, North Congo. *Amer J Primatol 64:* 223–232.

Bertram, B. C. R. 1975. Social factors influencing reproduction in wild lions. *J Zool, Lond 177:* 463–482.

———. 1978. Living in groups: Predators and prey. In *Behavioural ecology: An evolutionary approach,* eds. J. R. Krebs, and N. B. Davies, 64–96. Oxford: Blackwell Scientific.

Birkhead, T. R., and A. P. Møller, eds. 1998. *Sperm competition and sexual selection.* London: Academic Press.

Bischof, N. 1975. Comparative ethology of incest avoidance. In *Biosocial anthropology,* ed. R. Fox, 37–67. New York: John Wiley.

Blackburn, T. M., and K. J. Gaston. 1998. The distribution of mammal body masses. *Diversity and Distributions 4:* 121–133.

Bliege Bird, R. L., and D. W. Bird. 1997. Delayed reciprocity and tolerated theft: The behavioral ecology of food-sharing strategies. *Curr Anthrop 38:* 49–78.

Blom, A., C. Cipolletta, A. M. H. Brunsting, and H. H. T. Prins. 2004. Behavioral responses of gorillas to habituation in the Dzanga-Ndoki National Park, Central African Republic. *Int J Primatol 25:* 179–196.

Blount, B. G. 2001. Indigenous peoples and the uses and abuses of ecotourism. In *On biocultural diversity: Linking language, knowledge, and the environment,* ed. L. Maffi, 503–516. Washington, DC: Smithsonian Institution Press.

Boesch, C. 1991. The effects of leopard predation on grouping patterns in forest chimpanzees. *Behaviour 117:* 220–242.

———. 1996. Social grouping in Taï chimpanzees. In *Great ape societies,* eds. W. C. McGrew, L. F. Marchant, and T. Nishida, 101–113. Cambridge: Cambridge University Press.

Boesch, C., and H. Boesch-Achermann. 2000. *The chimpanzees of the Taï Forest: Behavioural ecology and evolution.* Oxford: Oxford University Press.

Boesch, C., G. Hohmann, and L. F. Marchant, eds. 2002. *Behavioural diversity in chimpanzees and bonobos.* Cambridge: Cambridge University Press.

Boinski, S. 1999. The social organization of squirrel monkeys: Implications for ecological models of social evolution. *Evol Anthropol 8:* 101–112.

Boinski, S., A. Treves, and C. A. Chapman. 2000. A critical evaluation of the influence of predators on primates: Effects of group travel. In *On the move: How and why animals travel in groups,* eds. S. Boinski, and P. A. Garber, 43–72. Chicago: University of Chicago Press.

Bookbinder, M. P., E. Dinerstein, A. Rijal, H. Cauley, and A. Rajarouia. 1998. Ecotourism's support of biodiversity conservation. *Conserv Biol 12:* 1399–1404.

Borgerhoff Mulder, M. 1991. Human behavioural ecology. In *Behavioural ecology: An evolutionary approach.* 3rd. ed., eds. J. R. Krebs, and N. B. Davies, 69–103. Oxford: Blackwell Scientific.

Borgerhoff Mulder, M., and P. Coppolillo. 2005. *Conservation: Linking ecology, economics, and culture.* Princeton, NJ: Princeton University Press.

Borgerhoff Mulder, M., and L. M. Ruttan. 2000. Grassland conservation and the pastoralist commons. In *Behaviour and conservation,* eds. L. M. Gosling, and W. J. Sutherland, 34–50. Cambridge: Cambridge University Press.

Borries, C., and A. Koenig. 2000. Infanticide in hanuman langurs: Social organization, male migration, and weaning age. In *Infanticide by males and its implications,* eds. C. P. van Schaik, and C. H. Janson, 99–122. Cambridge: Cambridge University Press.

Bowen-Jones, E., and S. Pendry. 1999. The threat to primates and other mammals from the bushmeat trade in Africa, and how this threat could be diminished. *Oryx 33:* 233–246.

Bowman, J., J. A. G. Jaeger, and L. Fahrig. 2002. Dispersal distance of mammals is proportional to home range size. *Ecology 83:* 2049–2055.

Boyd, R. 1992. The evolution of reciprocity when conditions vary. In *Coalitions and alliances in humans and other animals,* eds. A. H. Harcourt, and F. B. M. de Waal, 473–489. Oxford: Oxford University Press.

Boyd, R., and J. B. Silk. 1997. *How humans evolved.* New York: W. W. Norton.

Bradbury, J. W., and S. L. Vehrencamp. 1977. Social organization and foraging in emballonurid bats, III: Mating systems. *Behav Ecol Sociobiol 2:* 1–17.

Bradley, B. J., D. Doran-Sheehy, C. Boesch, and L.Vigilant. 2005. Related dyads of females are common in western gorilla groups despite routine female dispersal. *Amer J Phys Anthropol Suppl 40:* 77.

Bradley, B. J., D. M. Doran-Sheehy, D. Lukas, C. Boesch, and L. Vigilant. 2004. Dispersed male networks in western gorillas. *Current Biol 14:* 510–513.

Bradley, B. J., and L. Vigilant. 2002. The evolutionary genetics and molecular ecology of chimpanzees and bonobos. In *Behavioural diversity in chimpanzees and bonobos,* ed. P. T. Ellison, 259–276. Cambridge: Cambridge University Press.

Bradley, B. J., et al. 2005. Mountain gorilla tug-of-war: Silverbacks have limited control over reproduction in multimale groups. *Pro Nat Acad Sci USA 102:* 9418–9423.

Brandon-Jones, D., et al. 2004. Asian primate classification. *Int J Primatol 25:* 97–164.

Brashares, J. S., P. Arcese, and M. K. Sam. 2001. Human demography and reserve size predict wildlife extinction in West Africa. *Proc Roy Soc, Lond B 268:* 2473–2478.

Brashares, J. S., T. J. Garland, and P. Arcese. 2000. Phylogenetic analysis of coadaptation in behavior, diet, and body size in the African antelope. *Behav Ecol 11:* 452–463.

Broderick, A. C., et al. 2006. Are green turtles globally endangered? *Global Ecol Biogeog 15:* 21–26.

Brook, B. W., and D. M. J. S. Bowman. 2004. The uncertain blitzkrieg of Pleistocene megafauna. *J Biogeog 31:* 517–523.

Broom, M., C. Borries, and A. Koenig. 2004. Infanticide and infant defence by males—modelling the conditions in primate multi-male groups. *J Theor Biol 231:* 261–270.

Brown, J. H. 1995. *Macroecology.* Chicago: University of Chicago Press.

———. 2004. Concluding remarks. In *Frontiers of biogeography: New directions in the geography of nature,* eds. M. V. Lomolino, and L. R. Heaney, 361–368. Sunderland, MA: Sinauer Associates.

Brown, J. L. 1987. *Helping and communal breeding in birds: Ecology and evolution.* Princeton, NJ: Princeton University Press.

Bruner, A. G., R. E. Gullison, R. E. Rice, and G. A. B. da Foncesca. 2001. Effectiveness of parks in protecting tropical biodiversity. *Science 291:* 125–128.

Buchan, J. C., S. C. Alberts, J. B. Silk, and J. Altmann. 2003. True paternal care in a multi-male primate society. *Nature 425:* 179–181.

Buckland, S. T., D. R. Anderson, K. P. Burnham, and J. L. Laake. 1993. *Distance sampling: Estimating abundance of biological populations.* London: Chapman & Hall.

Burkhardt, F. 1996. *Charles Darwin's letters. A selection.* Cambridge: Cambridge University Press.

Butynski, T. M. 2001. Africa's great apes. In *Great apes and humans: The ethics of coexistence,* eds. B. Beck, et al., 3–56. Washington, DC: Smithsonian Institution Press.

Butynski, T. M., and J. Kalina. 1998. Gorilla tourism: A critical look. In *Conservation of biological resources,* eds. E. Milner-Gulland, and R. Mace, 294–314. Oxford: Blackwell Scientific.

Bygott, J. D., B. C. R. Bertram, and J. P. Hanby. 1979. Male lions in large coalitions gain reproductive advantages. *Nature 282:* 839–841.

Byrne, R. W., and A. Whiten, eds. 1988. *Machiavellian intelligence: Social expertise and the evolution of intellect in monkeys, apes, and humans.* Oxford: Oxford University Press.

Caldecott, J. 1996. *Designing conservation projects.* Cambridge: Cambridge University Press.

Caldecott, J., and L. Miles, eds. 2005. *World atlas of great apes and their conservation.* Berkeley: University of California Press.

Calder, W. A. 1984. *Size, function, and life history.* Cambridge, MA: Harvard University Press.

Calvert, J. J. 1985. Food selection by western gorillas in relation to food chemistry. *Oecologia 65.*

Campbell, C. 2004. Natural Worlds. Goliathus. C. Campbell. August 2004. http://www.naturalworlds.org/goliathus/

Campbell, C. J., A. Fuentes, K. C. MacKinnon, M. Panger, and S. K. Bearder, eds. 2007a. *Primates in perspective.* New York: Oxford University Press.

———. 2007b. Where we have been, where we are, and where we are going. In *Primates in perspective,* eds. C. J. Campbell, A. Fuentes, K. C. MacKinnon, M. Panger, and S. K. Bearder, 702–705. New York: Oxford University Press.

Cardillo, M., G. M. Mace, and A. Purvis. 2005. Problems of studying extinction risks. Response. *Science 310:* 1277–1278.

Cardillo, M., et al. 2004. Human population density and extinction risk in the world's carnivores. *Pub Lib Sci, Biol 2:* 909–914.

Carlson, A. A., and L. A. Isbell. 2001. Causes and consequences of single-male and multimale mating in free-ranging patas monkeys, *Erythrocebus patas. Anim Behav 62:* 1047–1058.

Caro, T. M. 1976. Observations on the ranging behaviour and daily activity of a lone silverback mountain gorilla. *Anim Behav 24:* 889–897.

Caro, T. 1998a. The significance of behavioral ecology for conservation biology. In *Behavioral ecology and conservation biology,* ed. T. Caro, 3–26. Oxford: Oxford University Press.

———, ed. 1998b. *Behavioral ecology and conservation biology.* New York: Oxford University Press.

———. 2005. *Antipredator defenses in birds and mammals.* Chicago: University of Chicago Press.

Caro, T. M., M. Rejmánek, and N. Pelkey. 2000. Which mammals benefit from protection in East Africa? In *Priorities for the conservation of mammalian diversity: Has the panda had its day?* eds. A. Entwistle, and N. Dunstone, 221–238. Cambridge: Cambridge University Press.

Carter, C. S., and B. S. Cushing. 2004. Proximate mechanisms regulating sociality and social monogamy, in the context of evolution. In *The origins and nature of sociality,* eds. R. W. Sussman, and A. R. Chapman, 99–121. New York: Aldine de Gruyter.

Carter, C. S., and L. L. Getz. 1993. Monogamy and the prairie vole. *Sci Amer 268:* 100–106.

Cashdan, E. 1992. Spatial organization and habitat use. In *Evolutionary ecology and human behavior,* eds. E. A. Smith, and B. Winterhalder, 237–266. New York: Aldine de Gruyter.

Casimir, M. J. 1975. Feeding ecology and nutrition of an eastern gorilla group in Mt. Kahuzi region (Republic of Zaire). *Folia Primatol 24:* 81–136.

Caughley, G. 1994. Directions in conservation biology. *J Anim Ecol 63:* 215–244.

Caughley, G., and A. Gunn. 1996. *Conservation biology in theory and practice.* Cambridge, MA: Blackwell Science.

Ceballos, G., P. R. Ehrlich, J. Soberón, I. Salazar, and J. P. Fay. 2005. Global mammal conservation: What must we manage? *Science 309:* 603–607.

Chapais, B. 1995. Alliances as a means of competition in primates: Evolutionary, developmental, and cognitive aspects. *Yrbk Phys Anthropol 38:* 115–136.

———. 2005. Kinship, competence, and cooperation in primates. In *Cooperation in primates and humans: Mechanisms and evolution,* eds. P. M. Kappeler, and C. P. van Schaik, 47–64. Berlin: Springer-Verlag.

Chapais, B., and P. Bélise. 2004. Constraints on kin selection in primate groups. In *Kinship and behavior in primates,* eds. B. Chapais, and C. Berman, 365–386. New York: Oxford University Press.

Chapman, C. A. 1990. Ecological constraints on group size in three species of neotropical primates. *Folia Primatol 55:* 1–9.

Chapman, C. A., and L. J. Chapman. 2000. Determinants of group size in primates: The importance of travel costs. In *On the move: How and why animals travel in groups,* eds. S. Boinski, and P. A. Garber, 24–42. Chicago: University of Chicago Press.

Chapman, C. A., L. J. Chapman, K. A. Bjorndal, and D. A. Onderdonk. 2002. Application of protein-to-fiber ratios to predict colobine abundance on different spatial scales. *Int J Primatol 23:* 283–310.

Chapman, C. A., T. R. Gillespie, and T. L. Goldberg. 2005. Primates and the ecology of their infectious diseases: How will anthropogenic change affect host-parasite interactions. *Evol Anthropol 14:* 134–144.

Chapman, C. A., and C. A. Peres. 2001. Primate conservation in the new millennium: The role of scientists. *Evol Anthropol 10:* 16–33.

Chapman, C., R. W. Wrangham, and L. Chapman. 1995. Ecological constraints on group size: An analysis of spider monkey and chimpanzee subgroups. *Behav Ecol Sociobiol 36:* 59–70.

Chapman, C. A., T. T. Struhsaker, and J. E. Lambert. 2005. Thirty years of research in Kibale National Park, Uganda, reveals a complex picture for conservation. *Int J Primatol 26:* 539–555.

Chapman, C. A., F. J. White, and R. W. Wrangham. 1994. Party size in chimpanzees and bonobos: A reevaluation of theory based on two similarly forested sites. In *Chimpanzee cultures,* eds. R. W. Wrangham, W. C. McGrew, F. B. M. de Waal, P. G. Heltne, and L. A. Marquardt, 41–58. Cambridge, MA: Harvard University Press.

Charnov, E. L. 1991. Evolution of life history variation among female mammals. *Proc Nat Acad Sci USA 88:* 1134–1137.

Charnov, E. L., and D. Berrigan. 1993. Why do female primates have such long lifespans and so few babies? *Evol Anthropol 1:* 191–194.

Chen, F. C., and W. H. Li. 2001. Genomic divergences between humans and other hominoids and the effective population size of the common ancestor of humans and chimpanzees. *Amer J Hum Genet 68:* 444–456.

Cheney, D. 1977. The acquisition of rank and the development of reciprocal alliances among free-ranging immature baboons. *Behav Ecol Sociobiol 2:* 303–318.

Cheney, D. L., and R. M. Seyfarth. 1983. Nonrandom dispersal in free-ranging vervet monkeys: Social and genetic consequences. *Amer Nat 122:* 392–412.

———. 1990. *How monkeys see the world: Inside the mind of another species.* Chicago: University of Chicago Press.

Cheney, D. L., R. M. Seyfarth, S. J. Andelman, and P. C. Lee. 1988. Reproductive success in vervet monkeys. In *Reproductive success,* ed. T. H. Clutton-Brock, 384–402. Chicago: University of Chicago Press.

Cheney, D. L., and R. W. Wrangham. 1987. Predation. In *Primate societies,* eds. B. B. Smuts, D. L. Cheney, R. M. Seyfarth, R. W. Wrangham, and T. T. Struhsaker, 227–239. Chicago: University of Chicago Press.

Cheney, D. L., et al. 2004. Factors affecting reproduction and mortality among baboons in the Okavango Delta, Botswana. *Int J Primatol 25:* 401–428.

Chitty, D. 1996. *Do lemmings commit suicide? Beautiful hypotheses and ugly facts.* New York: Oxford University Press.

Chivers, D. J., and C. M. Hladik. 1980. Morphology of the gastrointestinal tract in primates: Comparisons with other mammals in relation to diet. *J Morphol 166:* 337–386.

Cipolletta, C. 2003. Ranging patterns of a western gorilla group during habituation to humans in the Dzanga-Ndoki National Park, Central African Republic. *Int J Primatol 24:* 1207–1226.

———. 2004. Effects of group dynamics and diet on the ranging patterns of a western gorilla group (*Gorilla gorilla gorilla*) at Bai Hokou, Central African Republic. *Amer J Primatol 64:* 193–205.

Cipriano, F., and S. R. Palumbi. 1999. Genetic tracking of a protected whale. *Nature 397:* 307–308.

Clark, A. P., and R. W. Wrangham. 1994. Chimpanzee arrival pant-hoots: Do they signify food or status? *Int J Primatol 15:* 185–205.

Clark, C. W. 1989. Bioeconomics. In *Perspectives in ecological theory,* eds. J. Roughgarden, R. M. May, and S. A. Levin, 275–286. Princeton, NJ: Princeton University Press.

Clemmons, J. R., and R. Buchholz, eds. 1997. *Behavioral approaches to conservation in the wild.* Cambridge: Cambridge University Press.

Clutton-Brock, T. H., ed. 1977. *Primate ecology: Studies of feeding and ranging behaviour in lemurs, monkeys, and apes.* London: Academic Press.

———, ed. 1988. *Reproductive success.* Chicago: University of Chicago Press.

———. 1989a. Female transfer and inbreeding avoidance in social mammals. *Nature 337:* 70–72.

———. 1989b. Mammalian mating systems. *Proc Roy Soc, Lond B 236:* 339–372.

———. 2002. Breeding together: Kin selection and mutualism in cooperative vertebrates. *Science 296:* 69–72.

———. 2005. Cooperative breeding in mammals. In *Cooperation in primates and humans: Mechanisms and evolution,* eds. P. M. Kappeler, and C. P. van Schaik, 173–190. Berlin: Springer-Verlag.

Clutton-Brock, T. H., and S. D. Albon. 1985. Competition and population regulation in social mammals. In *Behavioural ecology: Ecological consequences of adaptive behaviour,* eds. R. M. Sibly, and R. H. Smith, 577–592. Oxford: Blackwell Scientific.

———. 1992. Trial and error in the Highlands. *Nature 358:* 11–12.

Clutton-Brock, T. H., T. N. Coulson, E. J. Milner-Gulland, D. Thomson, and H. M. Armstrong. 2002. Sex differences in emigration and mortality affect optimal management of deer populations. *Nature 415:* 633–637.

Clutton-Brock, T. H., F. E. Guinness, and S. D. Albon. 1982. *Red deer: Behavior and ecology of two sexes.* Edinburgh: Edinburgh University Press.

Clutton-Brock, T. H., and P. H. Harvey. 1976. Evolutionary rules and primate societies. In *Growing points in ethology,* eds. P. P. G. Bateson, and R. A. Hinde, 195–237. Cambridge: Cambridge University Press.

———. 1977a. Primate ecology and social organization. *J Zool, Lond 183:* 1–39.

———. 1977b. Species differences in feeding and ranging behaviour in primates. In *Primate ecology: Studies of feeding and ranging behaviour in lemurs, monkeys, and apes,* ed. T. H. Clutton-Brock, 557–584. London: Academic Press.

———. 1978. Antlers, body size, and breeding group size in the Cervidae. *Nature 285:* 565–567.

———. 1980. Primates, brains, and ecology. *J Zool, Lond 190:* 309–323.

———. 1983. The functional significance of variation in body size among mammals. In *Advances in the study of mammalian behavior,* eds. J. F. Eisenberg, and D. G. Kleiman, 632–663. Shippensburg, PA: American Society of Mammalogists.

Clutton-Brock, T. H., P. H. Harvey, and B. Rudder. 1977. Sexual dimorphism, socionomic sex ratio, and body weight in primates. *Nature 269:* 797–800.

Clutton-Brock, T. H., and G. A. Parker. 1992. Potential reproductive rates and the operation of sexual selection. *Q Rev Biol 67:* 437–456.

———. 1995. Sexual coercion in animal societies. *Anim Behav 49:* 1345–1365.

Clutton-Brock, T. H., and A. C. J. Vincent. 1991. Sexual selection and the potential reproductive rates of males and females. *Nature 351:* 58–60.

Cody, M. L. 1974. Optimization in ecology. *Science 183:* 1156–1164.

Coghlan, B., et al. 2006. Mortality in the Democratic Republic of Congo: A nationwide survey. *Lancet 367:* 44–51.

Cohen, M. N., R. S. Malpass, and H. G. Klein, eds. 1980. *Biosocial mechanisms of population regulation.* New Haven, CT: Yale University Press.

Collins, D. A., C. D. Busse, and J. Goodall. 1984. Infanticide in two populations of savanna baboons. In *Infanticide: Comparative and evolutionary aspects,* eds. G. Hausfater, and S. B. Hrdy, 193–215. Hawthorne, NY: Aldine Publishing.

Coltman, D. W., J. G. Pilkington, J. A. Smith, and J. M. Pemberton. 1999. Parasite-mediated selection against inbred Soay sheep in a free-living, island population. *Evolution 53:* 1259–1267.

Constable, J., M. Ashley, J. Goodall, and A. E. Pusey. 2001. Noninvasive paternity assignment in Gombe chimpanzees. *Mol Ecol 10:* 1279–1300.

Coppeto, S. A., and A. H. Harcourt. 2005. Is a biology of rarity in primates yet possible? *Biodiv Conserv 14:* 1017–1022.

Cords, M. 2000. The number of males in guenon groups. In *Primate males: Causes and consequences of variation in group composition,* ed. P. M. Kappeler, 84–96. Cambridge: Cambridge University Press.

Corlett, R. T., and R. B. Primack. 2006. Tropical rainforests and the need for cross-continental comparisons. *Trends Ecol Evol 21:* 104–110.

Costanza, R., et al. 1997. The value of the world's ecosystem services and natural capital. *Nature 387:* 253–260.

Courtenay, J. 1987. Post-partum amenorrhoea, birth intervals, and reproductive potential in captive chimpanzees. *Primates 28:* 543–546.

Cowlishaw, G. 1994. Vulnerability to predation in baboon populations. *Behaviour 131:* 293–304.

———. 1997. Refuge use and predation risk in a desert baboon population. *Anim Behav 54:* 241–253.

Cowlishaw, G., and R. Dunbar. 2000. *Primate conservation biology.* Chicago: Chicago University Press.

Creel, S. 1998. Social organization and effective population size in carnivores. In *Behavioral ecology and conservation biology,* ed. T. Caro, 246–265. Oxford: Oxford University Press.

Crockett, C. M., and C. H. Janson. 2000. Infanticide in red howlers: Female group size, male membership, and a possible link to folivory. In *Infanticide by males and its implications,* eds. C. van Schaik, and C. H. Janson, 75–98. Cambridge: Cambridge University Press.

Crook, J. H. 1965. The adaptive significance of avian social organisations. *Symp Zool Soc Lond 14:* 181–218.

———. 1972. Sexual selection, dimorphism and social organization in the primates. In *Sexual selection and the descent of man,* ed. B. Campbell, 231–281. London: Heineman.

Crook, J. H., J. E. Ellis, and J. D. Goss-Custard. 1976. Mammalian social systems: Structure and function. *Anim Behav 24:* 261–274.

Crook, J. H., and J. S. Gartlan. 1966. Evolution of primate societies. *Nature 210:* 1200–1203.

Cullen, R., K. F. D. Hughey, G. Fairburn, and E. Moran. 2005. Economic analyses to aid nature conservation decision making. *Oryx 39:* 327–334.

Dagg, A. I. 1999. Infanticide by male lions hypothesis: A fallacy influencing research into human behavior. *Amer Anthropol 100:* 940–950.

Daly, M., and M. Wilson. 1988. *Homicide.* New York: Aldine de Gruyter.

Darwin, C. 1859. *On the origin of species.* London: John Murray.

———. 1871. *The descent of man, and selection in relation to sex.* London: John Murray.

———. 1872. *The expression of the emotions in man and animals.* London: John Murray.

Dasgupta, P. S. 1995. Population, poverty, and the local environment. *Sci Amer* February: 40–45.

———. 2004. *Human well-being and the natural environment.* Oxford: Oxford University Press.

Daszak, P., A. A. Cunningham, and A. D. Hyatt. 2000. Emerging infectious diseases of wildlife—threats to biodiversity and human health. *Science 287:* 443–449.

Datta, S. B. 1992. Effects of availability of allies on female dominance structure. In *Coalitions and alliances in humans and other animals,* eds. A. H. Harcourt, and F. B. M. de Waal, 61–82. Oxford: Oxford University Press.

Davies, A. G. 1994. Colobine populations. In *Colobine monkeys: Their ecology, behaviour, and evolution,* eds. A. G. Davies, and J. F. Oates, 285–310. Cambridge: Cambridge University Press.

Davies, N. B. 1991. Mating systems. In *Behavioural ecology: An evolutionary approach.* 3rd. ed., eds. J. R. Krebs, and N. B. Davies, 263–294. Oxford: Blackwell Scientific.

———. 1992. *Dunnock behaviour and social evolution.* Oxford: Oxford University Press.

Davies, N. B., and A. I. Houston. 1981. Owners and satellites: The economics of territory defence in the pied wagtail, *Motacilla alba. J Anim Ecol 50:* 157–180.

———. 1984. Territory economics. In *Behavioural ecology: An evolutionary approach.* 2nd. ed., eds. J. R. Krebs, and N. B. Davies, 148–169. Oxford: Blackwell Scientific.

De Lathouwers, M., and L. Van Elsacker. 2005. Reproductive parameters of female *Pan paniscus* and *P. troglodytes:* Quality versus quantity. *Int J Primatol 26:* 55–71.

de Mérode, E., M. Bermejo, and G. Illera. 2001. Aire protégée et tourisme. *Canopée 20:* 15–16.

de Thoisy, B., F. Renoux, and C. Julliot. 2005. Hunting in northern French Guiana and its impact on primate communities. *Oryx 39:* 149–157.

de Waal, F. B. M. 1989a. *Peacemaking among primates.* Cambridge, MA: Harvard University Press.

———. 1989b. Dominance "style" and primate social organization. In *Comparative socioecology: The behavioral ecology of humans and other mammals,* eds. V. Standen, and R. A. Foley, 243–263. Oxford: Blackwell Scientific.

———. 2000. Attitudinal reciprocity in food sharing among brown capuchin monkeys. *Anim Behav 60:* 253–261.

Deaner, R. O., R. A. Barton, and C. P. van Schaik. 2003. Primate brains and life histories: Renewing the connection. In *Primate life histories and socioecology,* eds. P. M. Kappeler, and M. E. Pereira, 233–265. Chicago: University of Chicago Press.

Deblauwe, I., J. Dupain, G. M. Nguenang, D. Werdenich, and L. Van Elsacker. 2003. Insectivory by *Gorilla gorilla gorilla* in southeast Cameroon. *Int J Primatol 24:* 493–502.

Deem, S. L., W. B. Karesh, and W. Weisman. 2001. Putting theory into practice: Wildlife health in conservation. *Conserv Biol 15:* 1224–1233.

Delgado, R. A., and C. P. van Schaik. 2000. The behavioral ecology and conservation of the orangutan (*Pongo pygmaeus*): A tale of two islands. *Evol Anthropol 9:* 201–218.

DeVore, I., and K. R. L. Hall. 1965. Baboon ecology. In *Primate behavior: Field studies of monkeys and apes,* ed. I. DeVore. New York: Holt, Rinehart and Winston.

DFGFI 2005. The return of Amahoro, *DFGFI Field News,* June 2005. http://www.gorillafund.org.

Di Fiore, A. 2003. Molecular genetic approaches to the study of primate behavior, social organization, and reproduction. *Yrbk Phys Anthropol 46:* 62–99.

Di Fiore, A., and D. Rendall. 1994. Evolution of social organization: A reappraisal for primates by using phylogenetic methods. *Proc Nat Acad Sci USA 91:* 9941–9945.

Dietz, T., E. Ostrom, and P. C. Stern. 2003. The struggle to govern the commons. *Science 302:* 1907–1912.

Digby, L. J. 1995. Infant care, infanticide, and female reproductive strategies in polygynous groups of common marmosets (*Callithrix jacchus*). *Behav Ecol Sociobiol 37:* 51–61.

———. 2000. Infanticide by female mammals: Implications for the evolution of social systems. In *Infanticide by males and its implications,* eds. C. P. van Schaik, and C. H. Janson, 423–446. Cambridge: Cambridge University Press.

Digby, L. J., S. F. Ferrari, and W. Saltzman. 2007. Callitrichines: The role of competition in cooperatively breeding species. In *Primates in perspective,* eds. C. J. Campbell, A. Fuentes, K. C. MacKinnon, M. Panger, and S. K. Bearder, 85–106. New York: Oxford University Press.

Dittus, W. P. J. 1977. The social regulation of population density and age-sex distribution in the toque monkey. *Behaviour 63:* 281–322.

Dixson, A. F. 1987. Observations on the evolution of the genitalia and copulatory behaviour in male primates. *J Zool, Lond 213:* 423–443.

———. 1998. *Primate sexuality.* Oxford: Oxford University Press.

Dobson, A. P., and A. M. Lyles. 1989. The population dynamics and conservation of primate populations. *Conserv Biol 3:* 362–380.

Dobson, A., and J. Poole. 1998. Conspecific aggregation and conservation biology. In *Behavioral ecology and conservation biology,* ed. T. Caro, 193–208. Oxford: Oxford University Press.

Dobson, F. S. 1982. Competition for mates and predominant juvenile male dispersal in mammals. *Anim Behav 30:* 1183–1192.

Dobson, F. S., and B. Zinner. 2003. Social groups, genetic structure, and conservation. In *Animal behavior and wildlife conservation,* eds. M. Festa-Bianchet, and M. Apollonio, 211–228. Washington, DC: Island Press.

Doran, D. M. 1996. Comparative positional behavior of the African apes. In *Great ape societies,* eds. W. C. McGrew, L. F. Marchant, and T. Nishida, 213–224. Cambridge: Cambridge University Press.

———. 1997. Influence of seasonality on activity patterns, feeding behavior, ranging, and grouping patterns in Taï chimpanzees. *Int J Primatol 18:* 183–206.

Doran, D. M., and A. McNeilage. 1998. Gorilla ecology and behavior. *Evol Anthropol 6:* 120–131.

———. 2001. Subspecific variation in gorilla behavior: The influence of ecological and social factors. In *Mountain gorillas: Three decades of research at Karisoke,* eds. M. M. Robbins, P. Sicotte, and K. J. Stewart, 123–149. Cambridge: Cambridge University Press.

Doran, D. M., W. L. Jungers, Y. Sugiyama, J. G. Fleagle, and C. P. Heesy. 2002a. Multivariate and phylogenetic approaches to understanding chimpanzee and bonobo behavioral diversity. In *Behavioural diversity in chimpanzees and bonobos,* eds. C. Boesch, G. Hohmann, and L. F. Marchant, 14–34. Cambridge: Cambridge University Press.

Doran, D. M., et al. 2002b. Western lowland gorilla diet and resource availability: New evidence, cross-site comparisons, and reflections on indirect sampling methods. *Amer J Primatol 58:* 91–116.

Doran-Sheehy, D. M., D. Greer, P. Mongo, and D. Schwindt. 2004. Impact of ecological and social factors on ranging in western gorillas. *Amer J Primatol 64:* 207–222.

Doran-Sheehy, D. M., N. F. Shah, and L. A. Heimbauer. 2006. Sympatric western gorilla and mangabey diet: Reexamination of ape and monkey foraging strategies. In *Feeding ecology in apes and other primates. Ecological, physical, and behavioral aspects,* eds. G. Hohmann, M. M. Robbins, and C. Boesch, 49–72. Cambridge: Cambridge University Press.

Du Chaillu, P. B. 1861. *Explorations and adventures in equatorial Africa; with accounts of the manners and customs of the people, and of the chase of the gorilla, the crocodile, leopard, elephant, hippopotamus, and other animals.* New York: Harper.

Dugatkin, L. A. 1997. *Cooperation among animals: An evolutionary perspective.* New York: Oxford University Press.

Dunbar, R. 2003. Why are apes so smart? In *Primate life histories and socioecology,* eds. P. M. Kappeler, and M. E. Pereira, 285–298. Chicago: University of Chicago Press.

Dunbar, R. I. M. 1979. Population demography, social organization, and mating strategies. In *Primate ecology and human origins,* eds. I. S. Bernstein, and E. O. Smith, 65–88. New York: Garland STPM Press.

———. 1984. *Reproductive decisions: An economic analysis of gelada baboon social strategies.* Princeton, NJ: Princeton University Press.

———. 1988. *Primate social systems.* London: Croom Helm.

———. 1996. Determinants of group size in primates: A general model. In *Evolution of social behaviour patterns in primates and man,* eds. W. G. Runciman, J. Maynard Smith, and R. I. M. Dunbar, 33–57. Oxford: Oxford University Press.

———. 2000. Male mating strategies: A modeling approach. In *Primate males: Causes and consequences of variation in group composition,* ed. P. M. Kappeler, 259–268. Cambridge: Cambridge University Press.

———. 2002. Modelling primate behavioral ecology. *Int J Primatol 23:* 785–819.

Durant, S. 2000. Dispersal patterns, social organization, and population viability. In *Behaviour and conservation,* eds. L. M. Gosling, and W. J. Sutherland, 172–197. Cambridge: Cambridge University Press.

Durant, S. M., and G. M. Mace. 1994. Species differences and population structure in population viability analyses. In *Creative conservation: Interactive management of wild and captive animals,* eds. P. G. S. Olney, G. M. Mace, and A. T. C. Feistner, 67–91. London: Chapman & Hall.

Ebensperger, L. A. 1998. Strategies and counterstrategies to infanticide in mammals. *Biol Rev 73:* 321–346.

Ehrlich, P. R., and A. H. Ehrlich. 1996. *Betrayal of science and reason: How anti-environmental rhetoric threatens our future.* Washington, DC: Island Press.

Ehrlich, P., and A. Ehrlich. 2004. *One with Nineveh: Politics, consumption, and the human future.* Washington, DC: Island Press.

Eisenberg, J. F., N. A. Muckenhirn, and R. Rudran. 1972. The relation between ecology and social structure in primates. *Science 176:* 863–874.

Elgar, M. 1986. House sparrows establish foraging flocks by giving chirrup calls if the resources are divisible. *Anim Behav 34:* 169–174.

Elgar, M. A., and D. Clode. 2001. Inbreeding and extinction in island populations: A cautionary note. *Conserv Biol 15:* 284–286.

Elliott, G. P., D. V. Merton, and P. W. Jansen. 2001. Intensive management of a critically endangered species: The kakapo. *Biol Conserv 99:* 121–133.

Elliott, R. C. 1976. Observations on a small group of mountain gorillas (*Gorilla gorilla beringei*). *Folia Primatol 25:* 12–24.

Emlen, S. T. 1991. Evolution of cooperative breeding in birds and mammals. In *Behavioural ecology: An evolutionary approach.* 3rd. ed., eds. J. R. Krebs, and N. B. Davies, 301–337. Oxford: Blackwell Scientific.

———. 1995. An evolutionary theory of the family. *Proc Nat Acad Sci USA 92:* 8092–8099.

———. 1997. Predicting family dynamics in social vertebrates. In *Behavioural ecology: An evolutionary approach.* 4th. ed., eds. J. R. Krebs, and N. B. Davies, 228–253. Oxford: Blackwell Scientific.

Emlen, S. T., N. J. Demon, and D. J. Emlen. 1989. Experimental induction of infanticide in female wattled jacanas. *The Auk 106:* 1–7.

Emlen, S. T., and L. W. Oring. 1977. Ecology, sexual selection, and the evolution of mating systems. *Science 197:* 215–223.

Emmons, L. H. 1999. Of mice and monkeys: Primates as predictors of mammal community richness. In *Primate communities,* eds. J. G. Fleagle, C. H. Janson, and K. E. Reed, 171–188. Cambridge: Cambridge University Press.

Enisco, A. E., J. M. Calcagno, and K. C. Gold. 1999. Social interactions between captive adult male and infant lowland gorillas: Implications regarding kin selection and zoo management. *Zoo Biol 18:* 53–62.

Enstam, K. L., and L. A. Isbell. 2007. The guenons (Genus *Cercopithecus*) and their allies. Behavioral ecology of polyspecific associations. In *Primates in perspective,* eds. C. J. Campbell, A. Fuentes, K. C. MacKinnon, M. Panger, and S. K. Bearder, 252–274. New York: Oxford University Press.

Evans, C. S., L. Evans, and P. Marler. 1993. On the meaning of alarm calls: Functional reference in an avian vocal system. *Anim Behav 46:* 23–38.

Fa, J. E., C. A. Peres, and J. Meeuwig. 2002. Bushmeat exploitation in tropical forests: An intercontinental comparison. *Conserv Biol 16:* 232–237.

Fa, J. E., S. F. Ryan, and D. J. Bell. 2005. Hunting vulnerability, ecological characteristics, and harvest rates of bushmeat species in afrotropical forests. *Biol Conserv 121:* 167–176.

Fa, J. E., et al. 2006. Getting to grips with the magnitude of exploitation: Bushmeat in the Cross-Sanaga rivers region, Nigeria and Cameroon. *Biol Conserv 129:* 497–510.

Fashing, P. J. 2001. Male and female strategies during intergroup encounters in guerezas (*Colobus guereza*): Evidence for resource defense mediated through males and a comparison with other primates. *Behav Ecol Sociobiol 50:* 219–230.

Fay, J. M. 1989. Partial completion of a census of the western lowland gorilla (*Gorilla g. gorilla* [Savage and Wyman]) in southwestern Central African Republic. *Mammalia 53:* 203–215.

Fay, J. M., and M. Agnagna. 1992. Census of gorillas in northern Republic of Congo. *Amer J Primatol 27:* 275–284.

Fay, J. M., M. Agnagna, J. Moore, and R. Oko. 1989. Gorillas (*Gorilla gorilla gorilla*) in the Likouala swamp forests of North Central Congo: Preliminary data on populations and ecology. *Int J Primatol 10:* 477–486.

Fay, J. M., R. Carroll, J. C. K. Peterhans, and D. Harris. 1995. Leopard attack on and consumption of gorillas in the Central African Republic. *J Hum Evol 29:* 93–99.

Festa-Bianchet, M., and M. Apollonio, eds. 2003. *Animal behavior and wildlife conservation.* London: Island Press.

Fichtel, C., and P. M. Kappeler. 2002. Anti-predator behavior of group-living Malagasy primates: Mixed evidence for a referential alarm call system. *Behav Ecol Sociobiol 51:* 262–275.

Fitzpatrick, J. W., et al. 2005. Ivory-billed woodpecker (*Campephilus principalis*) persists in continental North America. *Science 308:* 1460–1462.

Fleagle, J. C. 1999. *Primate adaptation and evolution.* 2nd. ed. New York: Academic Press.

Fletcher, A. 2001. Development of infant independence from the mother in wild mountain gorillas. In *Mountain gorillas: Three decades of research at Karisoke,* eds. M. M. Robbins, P. Sicotte, and K. J. Stewart, 153–182. Cambridge: Cambridge University Press.

Foerster, K., K. Delhey, A. Johensen, J. T. Lifjeld, and B. Kempenaers. 2003. Female increase offspring heterozygosity and fitness through extra-pair matings. *Nature 425:* 714–717.

Foley, R. 1987. *Another unique species.* Harlow, Essex: Longman Scientific & Technical.

Foley, R. A., and P. C. Lee. 1989. Finite social space, evolutionary pathways, and reconstructing hominid behavior. *Science 243:* 901–906.

Fossey, D. 1972. Vocalizations of the mountain gorilla (*Gorilla gorilla beringei*). *Anim Behav 20:* 36–53.

———. 1979. Development of the mountain gorilla (*Gorilla gorilla beringei*): The first thirty-six months. In *The great apes,* eds. D. A. Hamburg, and E. R. McCown, 138–184. Menlo Park, CA: Benjamin/Cummings.

———. 1983. *Gorillas in the mist.* London: Hodder and Stoughton.

———. 1984. Infanticide in mountain gorillas (*Gorilla gorilla beringei*) with comparative notes on chimpanzees. In *Infanticide: Comparative and evolutionary perspectives,* eds. G. Hausfater, and S. B. Hrdy, 217–236. Hawthorne, NY: Aldine Publishing.

Fossey, D., and A. H. Harcourt. 1977. Feeding ecology of free ranging mountain gorilla (*Gorilla gorilla beringei*). In *Primate ecology,* ed. T. H. Clutton-Brock, 415–447. London: Academic Press.

Fox, E. A. 2002. Female tactics to reduce sexual harassment in the Sumatran orangutan (*Pongo pygmaeus*). *Behav Ecol Sociobiol 52:* 93–101.

Frankham, R., J. D. Ballou, and D. A. Briscoe. 2002. *Introduction to conservation genetics.* Cambridge: Cambridge University Press.

Franklin, I. R. 1980. Evolutionary change in small populations. In *Conservation biology: An evolutionary-ecological perspective,* eds. M. E. Soulé, and B. A. Wilcox, 135–149. Sunderland, MA: Sinauer Associates.

Fruth, B., and G. Hohmann. 2002. How bonobos handle hunts and harvests: Why share food? In *Behavioural diversity in chimpanzees and bonobos,* eds. C. Boesch, G. Hohmann, and L. F. Marchant, 231–243. Cambridge: Cambridge University Press.

Fuentes, A. 1999. Re-evaluating primate monogamy. *Amer Anthropol 100:* 890–907.

Furuichi, T. 1987. Sexual swelling, receptivity, and grouping of wild pygmy chimpanzee females at Wamba, Zaïre. *Primates 28:* 309–318.

———. 1989. Social interactions and the life history of female *Pan paniscus* in Wamba, Zaire. *Int J Primatol 10:* 173–197.

———. 1997. Agnostic interactions and matrifocal dominance rank of wild bonobos (*Pan paniscus*) at Wamba. *Int J Primatol 18:* 855–875.

Furuichi, T., and C. Hashimoto. 2002. Why female bonobos have a lower copulation rate during estrus than chimpanzees. In *Behavioural diversity in chimpanzees and bonobos,* eds. C. Boesch, G. Hohmann, and L. F. Marchant, 156–167. Cambridge: Cambridge University Press.

Furuichi, T., et al. 1998. Population dynamics of wild bonobos (*Pan paniscus*) at Wamba. *Int J Primatol 19:* 1029–1044.

Gadgil, M., and F. Berkes. 1991. Traditional resource management systems. *Res Manage Optim 18:* 127–141.

Galdikas, B. M. F. 1979. Orangutan adaptation at Tanjung Puting Reserve: Mating and ecology. In *The great apes,* eds. D. A. Hamburg, and E. R. McCown, 195–233. Menlo Park, CA: Benjamin/Cummings.

———. 1981. Orangutan reproduction in the wild. In *Reproductive biology of the great apes,* ed. C. E. Graham, 281–300. New York: Academic Press.

———. 1985a. Subadult male orangutan sociality and reproductive behavior at Tanjung Puting. *Amer J Primatol 8:* 87–99.

———. 1985b. Orangutan socialty at Tanjung Puting. *Amer J Primatol 9:* 101–119.

———. 1988. Orangutan diet, range, and activity at Tanjung Puting, Central Borneo. *Int J Primatol 9:* 1–35.

———. 1995. Social and reproductive behavior of wild adolescent female orangutans. In *The neglected ape,* eds. R. D. Nadler, B. M. F. Galdikas, L. K. Sheeran, and N. Rosen, 163–182. New York: Plenum Press.

Ganas, J., and M. M. Robbins. 2004. Intrapopulation differences in ant eating in the mountain gorillas of Bwindi Impenetrable National Park, Uganda. *Primates 45:* 275–278.

———. 2005. Ranging behavior of the mountain gorillas (*Gorilla beringei beringei*) in Bwindi Impenetrable National Park, Uganda: A test of the ecological constraints model. *Behav Ecol Sociobiol 58:* 277–288.

Ganas, J., M. M. Robbins, J. B. Nkurunungi, B. A. Kaplin, and A. McNeilage. 2004. Dietary variability of mountain gorillas in Bwindi Impenetrable National Park, Uganda. *Int J Primatol 25:* 1043–1072.

Garber, P. A. 2000. Evidence for the use of spatial, temporal, and social information by some primate foragers. In *On the move: How and why animals travel in groups,* eds. S. Boinski, and P. A. Garber, 261–298. Chicago: University of Chicago Press.

Garner, K. J., and O. A. Ryder. 1992. Some applications of PCR to studies in wildlife genetics. *Symp Zool Soc Lond 64:* 167–181.

Garner, R. L. 1896. *Gorillas & chimpanzees.* London: Osgood, McIlvaine.

Gaston, K. J., and T. M. Blackburn. 1999. A critique for macroecology. *Oikos 84:* 353–368.

Gatti, S., F. Levréro, N. Ménard, and A. Gautier-Hion. 2004. Population and group structure of western lowland gorillas (*Gorilla gorilla gorilla*) at Lokoué, Republic of Congo. *Amer J Primatol 63:* 111–123.

Gatti, S., F. Levréro, N. Ménard, E. Petit, and A. Gautier-Hion. 2003. Bachelor groups of western lowland gorillas (*Gorilla gorilla gorilla*) at Lokue Clearing, Odzala National Park, Republic of Congo. *Folia Primatol 74:* 195–196.

Gautier-Hion, A. 1979. Food niches and coexistence in sympatric primates in Gabon. In *Recent advances in primatology. Vol. 1. Behaviour,* eds. D. J. Chivers, and J. Herbert, 269–286. London: Academic Press.

Gerald, N. 1995. *Demography of the Virunga mountain gorilla* (*Gorilla gorilla beringei*). M.Sc. diss. Pp. 81. Princeton University, Princeton, NJ.

Gerald Steklis, N., and H. D. Steklis. In press. The value of long-term research: The mountain gorilla as a case study. In *Gorilla conservation: Challenges of the 21st century,* eds. T. Stoinski, H. D. Steklis, and P. T. Mehlman. New York: Springer.

Gerloff, U., B. Hartung, B. Fruth, G. Hohmann, and D. Tautz. 1999. Intracommunity relationships, dispersal pattern, and paternity success in a wild living community of bonobos (*Pan paniscus*) determined from DNA analysis of faecal samples. *Proc R Soc, Lond B 266:* 1189–1195.

Ghiglieri, M. 1984. *The chimpanzees of Kibale Forest.* New York: Columbia University Press.

Ghimire, K. B., and M. P. Pimbert, eds. 1997. *Social change and conservation: Environmental politics and impacts of national parks and protected areas.* London: Earthscan Publications.

Gilg, O., I. Hanski, and B. Sittler. 2003. Cyclic dynamics in a simple vertebrate predator-prey community. *Science 302:* 866–868.

Gill, F. B., and L. L. Wolf. 1975. Foraging strategies and energetics of East African sunbirds at mistletoe flowers. *Amer Nat 109:* 491–510.

Gilpin, M. E., and M. E. Soulé. 1986. Minimum viable populations: Processes of species extinction. In *Conservation biology: The science of scarcity and diversity,* ed. M. E. Soulé, 19–34. Sunderland, MA: Sinauer Associates.

Goldizen, A. W. 1987. Tamarins and marmosets: Communal care of offspring. In *Primate societies,* eds. B. B. Smuts, D. L. Cheney, R. M. Seyfarth, R. W. Wrangham, and T. T. Struhsaker, 34–43. Chicago: University of Chicago Press.

Goldsmith, M. L. 1999. Ecological constraints on the foraging effort of western gorillas (*Gorilla gorilla gorilla*) at Bai Hokōu, Central African Republic. *Int J Primatol 20:* 1–23.

———. 2003. Comparative behavioral ecology of a lowland and highland gorilla population: Where do Bwindi gorillas fit? In *Gorilla biology. A multidisciplinary perspective,* eds. A. B. Taylor, and M. L. Goldsmith, 358–384. Cambridge: Cambridge University Press.

Gomendio, M., A. H. Harcourt, and E. Roldan. 1998. Sperm competition in mammals. In *Sperm competition,* eds. T. M. Birkhead, and A. P. Møller, 667–755. London: Academic Press.

Goodall, A. G. 1977. Feeding and ranging behaviour of a mountain gorilla group (*Gorilla gorilla beringei*) in the Tshibinda-Kahuzi region (Zaïre). In *Primate ecology,* ed. T. H. Clutton-Brock, 449–479. London: Academic Press.

Goodall, J. 1968. The behaviour of free-living chimpanzees in the Gombe Stream Reserve. *Anim Behav Monog 1:* 161–311.

———. 1986. *The chimpanzees of Gombe.* Cambridge, MA: Belknap Press.

Goodman, M., et al. 1998. Toward a phylogenetic classification of primates based on DNA evidence complemented by fossil evidence. *Mol Phylog Evol 9:* 585–598.

Goodman, S. M., S. O'Connor, and O. Langrand. 1993. A review of predation on lemurs: Implications for the evolution of social behavior in small, nocturnal primates. In *Lemur social systems and their ecological basis,* eds. P. M. Kappeler, and J. U. Ganzhorn, 51–66. New York: Plenum Press.

Gosling, L. M., and W. J. Sutherland, eds. 2000. *Behaviour and conservation.* Cambridge: Cambridge University Press.

Goss-Custard, J. D., and W. J. Sutherland. 1997. Individual behaviour, populations, and conservation. In *Behavioural ecology: An evolutionary approach.* 4th. ed., eds. J. R. Krebs, and N. B. Davies, 373–395. Oxford: Blackwell Scientific.

Goudie, A. 1993. *The human impact on the natural environment,* 4th. ed. Oxford: Blackwell Scientific.

Gould, L., R. W. Sussman, and M. Sauther. 2003. Demographic and life-history patterns in a population of ring-tailed lemurs (*Lemur catta*) at Beza Mahafaly Reserve, Madagascar: A 15-year perspective. *Amer J Phys Anthropol 120:* 182–194.

Grainger, A. 1993. Rates of deforestation in the humid tropics: Estimates and measurements. *Geog J 159:* 33–44.

Grant, J. W. A., C. A. Chapman, and K. S. Richardson. 1992. Defended versus undefended home range size of carnivores, ungulates, and primates. *Behav Ecol Sociobiol 31:* 149–161.

Grant, P. R. 1999. *Ecology and evolution of Darwin's finches.* Princeton, NJ: Princeton University Press.

Gray, M. et al. 2005. Virunga Volcanoes range mountain gorilla census, 2003. Joint Organisers' report, UWA/ORTPN/ICCN.

Greene, H. W. 2005. Organisms in nature as a central focus for biology. *Trends Ecol Evol 20:* 23–27.

Greenwood, P. J. 1980. Mating systems, philopatry, and dispersal in birds and mammals. *Anim Behav 28:* 1140–1162.

Groves, C., and A. Meder. 2001. A model of gorilla life history. *Australasian Primatol 15:* 2–15.

Groves, C. P. 1970. Population systematics of gorilla. *J Zool, Lond 161:* 287–300.

———. 1971. Distribution and place of origin of the gorilla. *Man 6:* 44–51.

———. 1993. Order Primates. In *Mammal species of the world: A taxonomic and geographic reference,* eds. D. E. Wilson, and D. M. Reeder, 243–277. Washington, DC: Smithsonian Institution Press.

———. 2001a. *Primate taxonomy.* Washington, DC: Smithsonian Institution Press.

———. 2001b. Why taxonomic stability is a bad idea, or why are there so few species of primates (or are there?). *Evol Anthropol 10:* 192–198.

———. 2003. A history of gorilla taxonomy. In *Gorilla biology. A multidisciplinary perspective,* eds. A. B. Taylor, and M. L. Goldsmith, 15–34. Cambridge: Cambridge University Press.

Grubb, P., et al. 2003. Assessment of the diversity of African primates. *Int J Primatol 24:* 1301–1357.

Grubb, P. 2006. Geospecies and superspecies in the African primate fauna. *Primate Conserv 20:* 75–78.

Gurd, D. B., T. D. Nudds, and D. H. Rivard. 2001. Conservation of mammals in eastern North American wildlife reserves: How small is too small? *Conserv Biol 15:* 1355–1363.

Hall, J. S., et al. 1998. Survey of Grauer's gorillas (*Gorilla gorilla graueri*) and eastern chimpanzees (*Pan troglodytes schweinfurthi*) in the Kahuzi-Biega National Park lowland sector and adjacent forest in eastern Democratic Republic of Congo. *Int J Primatol 19:* 207–235.

Hallé, N. 1987. *Cola lizae* N. Hallé (Sterculiaecea) Nouvelle espèce du Moyen Ogooue (Gabon). *Adansonia 3:* 229–237.

Halperin, S. 1979. Temporary association patterns in free ranging chimpanzees: An assessment of individual grouping preferences. In *The great apes,* eds. D. Hamburg, and E. McCown, 491–499. Menlo Park, CA: Benjamin/Cummings.

Halpern, B. S., et al. 2006. Gaps and mismatches between global conservation priorities and spending. *Conserv Biol 20:* 56–64.

Hamai, M., T. Nishida, H. Takasaki, and L. A. Turner. 1992. New records of within-group infanticide and cannibalism in wild chimpanzees. *Primates 33:* 151–162.

Hannah, L., D. Lohse, C. Hutchinson, J. L. Carr, and A. Lankerani. 1994. A preliminary inventory of human disturbance of world ecosystems. *Ambio 23:* 246–250.

Hannon, S. J., R. L. Mumme, W. D. Koenig, and F. A. Pitelka. 1985. Replacement of breeders and within-group conflict in the cooperatively breeding acorn woodpecker. *Behav Ecol Sociobiol 17:* 303–312.

Harcourt, A. H. 1978a. Strategies of emigration and transfer by primates with particular reference to gorillas. *Z Tierpsychol 48:* 401–420.

———. 1978b. Activity periods and patterns of social interaction: A neglected problem. *Behaviour 66:* 121–135.

———. 1979a. Social relationships between adult male and female mountain gorillas. *Anim Behav 27:* 325–342.

———. 1979b. Social relationships among adult female mountain gorillas. *Anim Behav 27:* 251–264.

———. 1979c. Contrasts between male relationships in wild gorilla groups. *Behav Ecol Sociobiol 5:* 39–49.

———. 1986. Gorilla conservation: Anatomy of a campaign. In *Primates: The road to self-sustaining populations,* ed. K. Benirschke, 31–46. New York: Springer-Verlag.

———. 1987a. Dominance and fertility in female primates. *J Zool, Lond 213:* 471–487.

———. 1987b. Behaviour of wild gorillas *Gorilla gorilla* and their management in captivity. *Int Zoo Yrbk 26:* 248–255.

———. 1987c. Options for unwanted or confiscated primates. *Prim Conserv 8:* 111–113.

———. 1991a. Sperm competition and the evolution of non-fertilizing sperm in mammals. *Evolution 45:* 314–328.

———. 1991b. Help, cooperation, and trust in animals. In *Cooperation and prosocial behaviour,* eds. R. A. Hinde, and J. Groebel, 15–26. Cambridge: Cambridge University Press.

———. 1992. Coalitions and alliances: Are primates more complex than non-primates? In *Coalitions and alliances in humans and other animals,* eds. A. H. Harcourt, and F. B. M. de Waal, 445–472. Oxford: Oxford University Press.

———. 1995. Population viability estimates: Theory and practice for a wild gorilla population. *Conserv Biol 9:* 134–142.

———. 1996. Is the gorilla a threatened species? How should we judge? *Biol Conserv 75:* 165–176.

———. 1997. Sperm competition in primates. *Amer Nat 149:* 189–194.

———. 1998a. Ecological indicators of risk for primates, as judged by susceptibility to logging. In *Behavioral ecology and conservation biology,* ed. T. M. Caro, 56–79. New York: Oxford University Press.

———. 1998b. Does primate socio-ecology need non-primate socio-ecology? *Evol Anthropol 7:* 3–7.

———. 1999. Biogeographic relationships of primates on south-east Asian islands. *Global Ecol Biogeog 8:* 55–61.

———. 2000. Conservation in practice. *Evol Anthropol 15:* 258–265.

———. 2001. Gorilla socio-ecology: Conflict and compromise between the sexes. In *Model systems in behavioral ecology,* ed. L. Dugatkin, 491–511. Princeton, NJ: Princeton University Press.

———. 2002. Empirical estimates of minimum viable population sizes for primates: Tens to tens of thousands? *Anim Conserv 5:* 237–244.

Harcourt, A. H., S. A. Coppeto, and S. A. Parks. 2005. The distribution-abundance (i.e. density) relationship: Its form and causes in a tropical mammal order, Primates. *J Biogeog 32:* 565–579.

Harcourt, A. H., and K. Curry-Lindahl. 1979. Conservation of the mountain gorilla and its habitat in Rwanda. *Environ Conserv 6:* 143–147.

Harcourt, A. H., and F. B. M. de Waal, eds. 1992a. *Coalitions and alliances in humans and other animals.* Oxford: Oxford University Press.

———. 1992b. Cooperation in conflict: From ants to anthropoids. In *Coalitions and alliances in humans and other animals,* eds. A. H. Harcourt, and F. B. M. de Waal, 493–510. Oxford: Oxford University Press.

Harcourt, A. H., D. Fossey, and J. Sabater Pi. 1981a. Demography of *Gorilla gorilla. J Zool, Lond 195:* 215–233.

Harcourt, A. H., D. Fossey, K. J. Stewart, and D. P. Watts. 1980. Reproduction in wild gorillas and some comparisons with chimpanzees. *J Reprod Fertil Suppl 28:* 59–70.

Harcourt, A. H., and J. Gardiner. 1994. Sexual selection and genital anatomy of male primates. *Proc Roy Soc, Lond B 255:* 47–53.

Harcourt, A. H., and J. Greenberg. 2001. Do gorilla females join males to avoid infanticide? A quantitative model. *Anim Behav 62:* 905–915.

Harcourt, A. H., and A. F. G. Groom. 1972. Gorilla census. *Oryx 11:* 355–363.

Harcourt, A. H., and S. A. Harcourt. 1984. Insectivory by gorillas. *Folia Primatol 43:* 229–233.

Harcourt, A. H., P. H. Harvey, S. G. Larson, and R. V. Short. 1981c. Testis weight, body weight, and breeding system in primates. *Nature 293:* 55–57.

Harcourt, A. H., J. Kineman, G. Campbell, J. Yamagiwa, and I. Redmond. 1983. Conservation and the Virunga gorilla population. *Afr J Ecol 21:* 139–142.

Harcourt, A. H., and S. A. Parks. 2003. Threatened primate taxa experience high human densities: Adding an index of threat to the IUCN Red List criteria. *Biol Conserv 109:* 137–149.

Harcourt, A. H., S. A. Parks, and R. Woodroffe. 2001. Human density as an influence on species/area relationships: Double jeopardy for small African reserves? *Biodiv Conserv 10:* 1011–1026.

Harcourt, A. H., A. Purvis, and L. Liles. 1995. Sperm competition: Mating system, not breeding season, affects testes size of primates. *Func Ecol 9:* 468–476.

Harcourt, A. H., and M. W. Schwartz. 2001. Primate evolution: A biology of Holocene extinction and survival on the south-east Asian Sunda Shelf islands. *Amer J Phys Anthropol 114:* 4–17.

Harcourt, A. H., and K. J. Stewart. 1981. Gorilla male relationships: Can differences during immaturity lead to contrasting reproductive tactics in adulthood. *Anim Behav 29:* 206–210.

———. 1984. Gorillas' time feeding: Aspects of methodology, body size, competition, and diet. *Afr J Ecol 22:* 207–215.

———. 1987. The influence of help in contests on dominance rank in primates: Hints from gorillas. *Anim Behav 35:* 182–190.

———. 1989. Functions of alliances in contests within wild gorilla groups. *Behaviour 109:* 176–190.

———. 2001. Vocal relationships of wild mountain gorillas. In *Mountain gorillas: Three decades of research at Karisoke,* eds. M. M. Robbins, P. Sicotte, and K. J. Stewart, 241–262. Cambridge: Cambridge University Press.

Harcourt, A. H., and K. J. Stewart. Unpublished. Unpublished data.

Harcourt, A. H., K. J. Stewart, and D. Fossey. 1976. Male emigration and female transfer in wild mountain gorilla. *Nature 263:* 226–227.

———. 1981b. Gorilla reproduction in the wild. In *Reproductive biology of the great apes,* ed. C. E. Graham, 265–279. New York: Academic Press.

Harcourt, A. H., K. J. Stewart, and M. Hauser. 1993. Functions of wild gorilla "close" calls. I. Repertoire, context, and interspecific comparison. *Behaviour 124:* 89–122.

Harcourt, A. H., K. J. Stewart, and I. M. Inahoro. 1989a. Nigeria's gorillas: A survey and recommendations. *Prim Conserv 10:* 73–76.

———. 1989b. Gorilla quest in Nigeria. *Oryx 23:* 7–13.

Harrison, P. 1987. *The greening of Africa.* London: Paladin.

———. 1992. *The third revolution: Environment, population, and a sustainable world.* London: I. B. Tauris, Penguin Books.

Harvell, C. D., et al. 2002. Climate warming and disease risks for terrestrial and marine biota. *Science 296:* 2158–2162.

Harvey, P. H., and A. H. Harcourt. 1984. Sperm competition, testes size, and breeding systems in primates. In *Sperm competition and the evolution of animal mating systems,* ed. R. L. Smith, 589–600. San Diego, CA: Academic Press.

Harvey, P. H., M. Kavanagh, and T. H. Clutton-Brock. 1978. Sexual dimorphism in primate teeth. *J Zool 186:* 175–185.

Harvey, P. H., R. D. Martin, and T. H. Clutton-Brock. 1987. Life histories in comparative perspective. In *Primate societies,* eds. B. B. Smuts, D. L. Cheney, R. M. Seyfarth, R. W. Wrangham, and T. T. Struhsaker, 181–196. Chicago: University of Chicago Press.

Harvey, P. H., and M. D. Pagel. 1989. Comparative studies in evolutionary ecology: Using the data base. In *Toward a more exact ecology,* eds. P. J. Grubb, and J. B. Whittaker, 209–227. Oxford: Blackwell Scientific.

———. 1991. *The comparative method in evolutionary biology.* Oxford: Oxford University Press.

Harvey, P. H., D. E. L. Promislow, and A. F. Read. 1989. Causes and correlates of life history differences among mammals. In *Comparative socioecology,* eds. V. Standen, and R. A. Foley, 305–318. Oxford: Blackwell Scientific.

Hasegawa, T., and M. Hiraiwa-Hasegawa. 1990. Sperm competition and mating behavior. In *The chimpanzees of the Mahale Mountains: Sexual and life history strategies,* ed. T. Nishida, 115–132. Tokyo: University of Tokyo Press.

Hassell, M. P., J. Latto, and R. M. May. 1989. Seeing the wood for the trees: Detecting density dependence from existing life-table studies. *J Anim Ecol 58:* 883–892.

Hauser, M. 1992. Costs of deception: Cheaters are punished in rhesus monkeys (*Macaca mulatta*). *Proc Nat Acad Sci USA 89:* 12137–12139.

———. 1993. Primatology: Some lessons from and for related disciplines. *Evol Anthropol 2:* 182–186.

Hausfater, G. 1982. Long-term consistency of dominance relations among female baboons (*Papio cynocpehalus*). *Science 217:* 752–755.

Hausfater, G., and S. B. Hrdy, eds. 1984. *Infanticide: Comparative and evolutionary aspects.* Hawthorne, NY: Aldine Publishing.

Hawkes, K., J. F. O'Connell, N. G. Blurton-Jones, H. Alvarez, and E. L. Charnov. 1998. Grandmothering, menopause, and the evolution of human life histories. *Proc Nat Acad Sci USA 95:* 1336–1339.

Hebert, P. D. N., E. H. Penton, J. M. Burns, D. H. Janzen, and W. Hallwachs. 2004. Ten species in one: DNA barcoding reveals cryptic species in the neotropical skipper butterfly *Astraptes fulgerator. Proc Nat Acad Sci USA 101:* 14812–14817.

Hedrick, A. V., and E. J. Temeles. 1989. The evolution of sexual dimorphism in animals: Hypotheses and tests. *Trends Ecol Evol 4:* 136–138.

Hess, G. R. 1994. Conservation corridors and contagious disease: A cautionary note. *Conserv Biol 8:* 256–262.

Hess, G. R., et al. 2002. Spatial aspects of disease dynamics. In *The ecology of wildlife diseases,* eds. P. J. Hudson, A. Rizzoli, B. T. Grenfell, H. Heesterbeek, and A. P. Dobson, 102–118. Oxford: Oxford University Press.

Hey, J., R. S. Waples, M. L. Arnold, R. K. Butlin, and R. G. Harrison. 2003. Understanding and confronting species uncertainty in biology and conservation. *Trends Ecol Evol 18:* 597–603.

Hill, D. A. 2004. The effects of demographic variation on kinship structure and behavior in cercopithecines. In *Kinship and behavior in primates,* eds. B. Chapais, and C. Berman, 132–150. New York: Oxford University Press.

Hill, K., et al. 2001. Mortality rates among wild chimpanzees. *J Hum Evol 40:* 437–450.

Hill, R. A., and R. I. M. Dunbar. 1998. An evaluation of the roles of predation rate and predation risk as selective pressures on primate grouping behavior. *Behaviour 135:* 411–430.

Hill, R. A., and P. C. Lee. 1998. Predation risk as an influence on group size in cercopithecoid primates: Implications for social structure. *J Zool, Lond 245:* 447–456.

Hinde, R. A. 1956. The biological significance of the territories of birds. *Ibis 98:* 340–369.

———. 1974. *Biological bases of human social behaviour.* New York: McGraw-Hill.

———. 1976. Interactions, relationships, and social structure. *Man 11:* 1–17.

———. 1982. *Ethology: Its nature and relations with other sciences.* Oxford: Oxford University Press.

———, ed. 1983. *Primate social relationships: An integrated approach.* Oxford: Blackwell Scientific.

Hiraiwa-Hasegawa, M., R. W. Byrne, H. Takasaki, and J. M. E. Byrne. 1986. Aggression toward large carnivores by wild chimpanzees of Mahale Mountains National Park, Tanzania. *Folia Primatol 47:* 8–13.

Hiraiwa-Hasegawa, M., and T. Hasegawa. 1994. Infanticide in nonhuman primates: Sexual selection and local resource competition. In *Infanticide and parental care,* eds. S. Parmigiani, and F. S. vom Saal, 137–154. Chur, Switzerland: Harwood Academic Publishers.

Hoare, D. J., I. D. Couzin, J.-G. J. Godin, and J. Krause. 2004. Context-dependent group size choice in fish. *Anim Behav 67:* 155–164.

Hoeck, H. N. 1982. Population dynamics, dispersal and genetic isolation in two species of hyrax (*Heterohyrax brucei* and *Procavia johnstoni*) on habitat islands in the Serengeti. *Z Tierpsychol 59:* 177–210.

Hoelzer, G. A., J. C. Morales, and D. J. Melnick. 2004. Dispersal and the population genetics of primate species. In *Kinship and behavior in primates,* eds. B. Chapais, and C. M. Berman, 109–131. Oxford: Oxford University Press.

Hoff, M. P., R. D. Nadler, and T. L. Maple. 1981. The development of infant play in a captive group of lowland gorillas (*Gorilla gorilla gorilla*). *Amer J Primatol 1:* 65–72.

———. 1983. Maternal transport and infant motor development in a captive group of lowland gorillas. *Primates 24:* 77–85.

Hoffman, K. A., S. P. Mendoza, M. B. Hennessy, and W. A. Mason. 1995. Responses of infant titi monkeys, *Callicebus moloch,* to removal of one or both parents: Evidence for parental attachment. *Devel Psychobiol 28:* 399–407.

Hohmann, G., and B. Fruth. 2002. Dynamics in social organization of bonobos (*Pan paniscus*). In *Behavioural diversity in chimpanzees and bonobos,* eds. C. Boesch, G. Hohmann, and L. F. Marchant, 138–150. Cambridge: Cambridge University Press.

Hohmann, G., U. Gerloff, D. Tautz, and B. Fruth. 1999. Social bonds and genetic ties: Kinship, association, and affiliation in a community of bonobos (*Pan paniscus*). *Behaviour 136:* 1219–1235.

Homer-Dixon, T. F. 1999. *Environment, scarcity, and violence.* Princeton, NJ: Princeton University Press.

Homer-Dixon, T. F., J. H. Boutwell, and G. W. Rathjens. 1993. Environmental change and violent conflict. *Sci Amer* February: 38–45.

Horn, H. S. 1968. The adaptive significance of colonial nesting in the Brewer's blackbird (*Euphagus cyanocephalus*). *Ecology 49:* 682–694.

Horr, D. A. 1975. The Borneo orang-utan: Population structure and dynamics in relationship to ecology and reproductive strategy. In *Primate behavior,* ed. L. A. Rosenblum. New York: Academic Press.

Hrdy, S. B. 1977. *The langurs of Abu: Female and male strategies of reproduction.* Cambridge, MA: Harvard University Press.

———. 1979. Infanticide among animals: A review, classification, and examination of the implications for the reproductive strategies of females. *Ethol Sociobiol 1:* 13–40.

———. 1999. *Mother nature.* New York: Pantheon Books.

Hudson, P. J., A. Rizzoli, B. T. Grenfell, H. Heesterbeek, and A. P. Dobson, eds. 2002. *The ecology of wildlife diseases.* Oxford: Oxford University Press.

Huijbregts, B., P. De Wachter, L. Sosth, N. Obiang, and M. E. Akou. 2003. Ebola and the decline of gorilla *Gorilla gorilla* and chimpanzee *Pan troglodytes* populations in Minkebe Forest, north-eastern Gabon. *Oryx 37:* 437–443.

Idani, G. 1991. Social relationships between immigrant and resident bonobo (*Pan paniscus*) females at Wamba. *Folia Primatol 57:* 83–95.

Ingold, T. 1989. The social and environmental relations of human being and other animals. In *Comparative socioecology: The behavioural ecology of humans and other animals,* eds. V. Standen, and R. A. Foley, 495–512. Oxford: Blackwell Scientific.

International Species Information System (ISIS). 2005. Species holdings—gorilla. International Species Information System (ISIS). 4 June 2005. https://www.isis.org/CMSHOME/.

Iremonger, S., C. Ravilious, and T. Quinton, eds. 1997. *A global overview of forest conservation CD-ROM.* Cambridge, UK: WCMC and CIFOR.

Isaac, N. J. B., and G. Cowlishaw. 2004. How species respond to multiple extinction threats. *Proc Roy Soc, Lond B 271:* 1135–1141.

Isaac, N. J. B., and A. Purvis. 2004. The "species problem" and testing macroevolutionary hypotheses. *Diversity Distrib 10:* 275.

Isabirye-Basuta, G. 1988. Food competition among individuals in a free-ranging chimpanzee community in Kibale Forest, Uganda. *Behaviour 105:* 135–147.

Isbell, L. A. 1991. Contest and scramble competition: Patterns of female aggression and ranging behavior among primates. *Behav Ecol 2:* 143–155.

———. 1994. Predation on primates: Ecological patterns and evolutionary consequences. *Evol Anthropol 3:* 61–71.

———. 2004. Is there no place like home? Ecological bases of female dispersal and philopatry and their consequences for the formation of kin groups. In *Kinship and behavior in primates,* eds. B. Chapais, and C. Berman, 71–108. New York: Oxford University Press.

Isbell, L. A., J. D. Pruetz, and T. P. Young. 1998. Movements of adult female vervets (*Cercopithecus aethiops*) and patas monkeys (*Erythrocebus patas*) as estimators of food resource size, density, and distribution. *Behav Ecol Sociobiol 42:* 123–133.

Isbell, L. A., and D. Van Vuren. 1996. Differential costs of locational and social dispersal and the consequences for female group-living primates. *Behaviour 133:* 1–36.

Isbell, L. A., and T. P. Young. 1993. Human presence reduces predation in a free-ranging vervet monkey population in Kenya. *Anim Behav 45:* 1233–1235.

———. 2002. Ecological models of female social relationships in primates: Similarities, disparities, and some directions for future clarity. *Behaviour 139:* 177–202.

IUCN 2006. 2006 IUCN Red List of Threatened Species. IUCN. The World Conservation Union. http://www.iucnredlist.org.

Jack, K. M., and L. Fedigan. 2004. Male dispersal patterns in white-faced capuchins, *Cebus capucinus.* Part 2: Patterns and causes of secondary dispersal. *Anim Behav 67:* 771–782.

James, A. N., K. J. Gaston, and A. Balmford. 1999. Balancing the earth's accounts. *Nature 401:* 323–324.

———. 2001. Can we afford to conserve biodiversity? *BioScience 51:* 43–52.

Janson, C. 1985. Aggressive competition and individual food consumption in wild brown capuchin monkeys *Cebus apella. Behav Ecol Sociobiol 18:* 125–138.

Janson, C. H. 1988. Food competition in brown capuchin monkeys (*Cebus apella*): Quantitative effects of group size and tree productivity. *Behaviour 105:* 53–76.

———. 1998a. Testing the predation hypothesis for vertebrate sociality: Prospects and pitfalls. *Behaviour 135:* 389–410.

———. 1998b. Experimental evidence for spatial memory in foraging wild capuchin monkeys, *Cebus apella. Anim Behav 55:* 1229–1243.

———. 2000. Primate socio-ecology: The end of a golden age. *Evol Anthropol 9:* 83–86.

———. 2003. Puzzles, predation, and primates: Using life history to understand selection pressures. In *Primate life histories and socioecology,* eds. P. M. Kappeler, and M. E. Pereira, 103–131. Chicago: University of Chicago Press.

Janson, C. H., and C. A. Chapman. 1999. Resources and primate community structure. In *Primate Communities,* eds. J. G. Fleagle, C. H. Janson, and K. E. Reed, 237–267. Cambridge: Cambridge University Press.

Janson, C. H., and M. L. Goldsmith. 1995. Predicting group size in primates: Foraging costs and predation risks. *Behav Ecol 6:* 326–336.

Janson, C. H., and C. P. van Schaik, eds. 1988. Food competition in primates. *Behaviour 105* (1, 2): 1–186.

———. 2000a. The behavioral ecology of infanticide by males. In *Infanticide by males and its implications,* eds. C. H. Janson, and C. P. van Schaik, 469–494. Cambridge: Cambridge University Press.

———, eds. 2000b. *Infanticide by males and its implications.* Cambridge: Cambridge University Press.

Janzen, D. H. 1986. The future of tropical biology. *Ann Rev Ecol Syst 17:* 305–324.

Jarman, P. J. 1974. The social organisation of antelope in relation to their ecology. *Behaviour 48:* 215–267.

Jennions, M. D., and M. Petrie. 2000. Why do females mate multiply? A review of the genetic benefits. *Biol Rev 75:* 21–64.

Jensen-Seaman, M. I., A. S. Deinard, and K. K. Kidd. 2003. Mitochondrial and nuclear DNA estimates of divergence between western and eastern gorillas. In *Gorilla biology: A multidisciplinary perspective,* eds. A. B. Taylor, and M. L. Goldsmith, 247–268. Cambridge: Cambridge University Press.

Johns, A. D., and J. P. Skorupa. 1987. Responses of rain-forest primates to habitat disturbance: A review. *Int J Primatol 8:* 157–191.

Johnson, C. N. 1986. Sex-biased philopatry and dispersal in mammals. *Oecologia 69:* 626–627.

———. 2002. Determinants of loss of mammal species during the Late Quaternary "megafauna" extinctions: Life history and ecology, but not body size. *Proc Roy Soc, Lond B 269:* 2221–2227.

Johnson, M. L., and M. S. Gaines. 1990. Evolution of dispersal: Theoretical models and empirical tests using birds and mammals. *Ann Rev Ecol Syst 21:* 449–480.

Jolly, A. 1966. Lemur social behavior and primate intelligence. *Science 153:* 501–506.

———. 1972. *The evolution of primate behavior.* New York: Macmillan.

———. 1986. Lemur survival. In *Primates: The road to self-sustaining populations,* ed. K. Benirschke, 71–98. New York: Springer-Verlag.

Jolly, A., et al. 2000. Infant killing, wounding, and predation in *Eulemur* and *Lemur. Int J Primatol 21:* 21–40.

Jones, C. B., and J. Sabater Pi. 1971. Comparative ecology of *Gorilla gorilla* (Savage and Wyman) and *Pan troglodytes* (Blumenbach) in Rio Muni, West Africa. *Bibl Primatol 13:* 1–96.

Judge, D. S., and J. R. Carey. 2000. Postreproductive life predicted by primate patterns. *J Gerontol 55A:* B201–B209.

Jungers, W. L., ed. 1985. *Size and scaling in primate biology.* New York: Plenum Press.

Kalpers, J., et al. 2003. Gorillas in the crossfire: Population dynamics of the Virunga mountain gorillas over the past three decades. *Oryx 37:* 326–337.

Kano, T. 1992. *The last ape: Pygmy chimpanzee behavior and ecology.* Palo Alto, CA: Stanford University Press.

Kano, T., and R. Asato. 1994. Hunting pressure on chimpanzees and gorillas in the Motaba River area, northeastern Congo. *Afr St Monog 15:* 143–162.

Kaplan, K., K. Hill, J. Lancaster, and M. A. Hurtado. 2000. A theory of human life history evolution: Diet, intelligence, and longevity. *Evol Anthropol 9:* 156–184.

Kappeler, P. M. 1993. Sexual selection and lemur social systems. In *Lemur social systems and their ecological basis,* eds. P. M. Kappeler, and J. U. Ganzhorn, 225–242. New York: Plenum Press.

———. 1999. Lemur social structure and convergence in primate socioecology. In *Comparative primate socioecology,* ed. P. C. Lee, 273–299. Cambridge: Cambridge University Press.

Kappeler, P. M., ed. 2000. *Primate males: Causes and consequences of variation in group composition.* Cambridge: Cambridge University Press.

Kappeler, P. M., and J. U. Ganzhorn. 1993. The evolution of primate communities and societies in Madagascar. *Evol Anthropol 2:* 159–171.

Kappeler, P. M., and E. W. Heymann. 1996. Nonconvergence in the evolution of primate life history and socio-ecology. *Biol J Linn Soc 59:* 297–326.

Kappeler, P. M., M. E. Pereira, and C. P. van Schaik. 2003. Primate life histories and socioecology. In *Primate life histories and socioecology,* eds. P. M. Kappeler, and M. E. Pereira, 1–23. Chicago: University of Chicago Press.

Kappeler, P. M., and C. P. van Schaik. 2002. Evolution of primate social systems. *Int J Primatol 23:* 707–740.

———, eds. 2004. *Sexual selection in primates: New and comparative perspectives.* Cambridge: Cambridge University Press.

———, eds. 2005. *Cooperation in primates and humans: Mechanisms and evolution.* Berlin: Springer-Verlag.

Kapsalis, E. 2004. Matrilineal kinship and primate behavior. In *Kinship and behavior in primates,* eds. B. Chapais, and C. Berman, 153–176. New York: Oxford University Press.

Karpanty, S. M. 2006. Direct and indirect impacts of raptor predation on lemurs in southeastern Madagascar. *Int J Primatol 27:* 239–261.

Kates, R. W., and T. M. Parris. 2003. Long-term trends and a sustainability transition. *Proc Nat Acad Sci USA 100:* 8062–8067.

Kawai, M. 1958/1965. On the system of social ranks in a natural troop of Japanese monkeys. I. Basic rank and dependent rank, *Primates,* 1–2: 111–130. In *Japanese monkeys: A collection of translations,* eds. K. Imanishi, and S. A. Altmann. Atlanta, GA: S. A. Altmann.

Kawamura, S. 1958/1965. Matriarchal social ranks in the Minoo B troop: A study of the rank system of Japanese monkeys. *Primates,* 1–2: 149–156. In *Japanese monkeys: A collection of translations,* eds. S. A. Altmann, and K. Imanishi. Atlanta, GA: S. A. Altmann.

Keith, D. A., et al. 2004. Protocols for listing threatened species can forecast extinction. *Ecology Letters 7:* 1101–1108.

Keller, L., and H. K. Reeve. 1995. Why do females mate with multiple males? The sexually selected sperm hypothesis. *Adv St Behav 24:* 291–315.

Keller, L. F., and D. M. Waller. 2002. Inbreeding effects in wild populations. *Trends Ecol Evol 17:* 230–241.

Kendall, B. E., and G. A. Fox. 2002. Variation among individuals and reduced demographic stochasticity. *Conserv Biol 16:* 109–116.

Kerr, J. T., and D. J. Currie. 1995. Effects of human activity on global extinction risk. *Conserv Biol 9:* 1528–1538.

Kinnaird, M. F., and T. G. O'Brien. 2001. Who's scratching whom? Reply to Whitten et al. *Conserv Biol 15:* 1459.

Kipling, R. 1892. *Ballads and barrack room ballads.* London: MacMillan.

Kiss, A. 2004. Is community-based ecotourism a good use of biodiversity conservation funds? *Trends Ecol Evol 19:* 232–237.

Kleiman, D. G. 1977. Monogamy in mammals. *Q Rev Biol 52:* 39–69.

Knott, C. D. 1998a. Changes in orangutan caloric intake, energy balance, and ketones in response to fluctuating fruit availability. *Int J Primatol 19:* 1061–1079.

———. 1998b. Social system dynamics, ranging patterns, and male and female strategies in wild Bornean orangutans (*Pongo pygmaeus*). *Amer J Phys Anthropol Suppl 26:* 140.

———. 1999. Orangutan behavior and ecology. In *The nonhuman primates,* eds. P. Dolhinow, and A. Fuentes, 50–57. Mountain View, CA: Mayfield Publishing.

———. 2001. Female reproductive ecology of apes. In *Reproductive ecology and human evolution,* ed. P. Ellison, 429–463. New York: Aldine de Gruyter.

Knott, C. D., and S. M. Kahlenberg. 2007. Orangutans in perspective. In *Primates in perspective,* eds. C. J. Campbell, A. Fuentes, K. C. MacKinnon, M. Panger, and S. K. Bearder, 290–305. New York: Oxford University Press.

Koenig, A. 1995. Group size, composition, and reproductive success in wild common marmosets (*Callithrix jacchus*). *Amer J Primatol 35:* 311–317.

———. 2002. Competition for resources and its behavioral consequences among female primates. *Int J Primatol 23:* 759–783.

Koenig, A., J. Beise, M. K. Chalise, and J. U. Ganzhorn. 1998. When females should contest for food—testing hypotheses about resource density, distribution, size, and quality with Hanuman langurs (*Presbytis entellus*). *Behav Ecol Sociobiol 42:* 225–237.

Koenig, A., and C. Borries. 2001. Socioecology of Hanuman langurs: The story of their success. *Evol Anthropol 10:* 122–137.

Koenig, W. D. 1981. Reproductive success, group size, and the evolution of cooperative breeding in the acorn woodpecker. *Amer Nat 117:* 421–443.

Koenig, W. D., D. Van Vuren, and P. N. Hooge. 1996. Detectability, philopatry, and the distribution of dispersal distances in vertebrates. *Trends Ecol Evol 11:* 514–517.

Komdeur, J. 1992. Importance of habitat saturation and territory quality for evolution of cooperative breeding in the Seychelles warbler. *Nature 358:* 493–495.

Komdeur, J., and C. Deerenberg. 1997. The importance of social behavior studies for conservation of natural populations. In *Behavioral approaches to conservation in the wild,* eds. J. R. Clemmons, and R. Buchholz, 262–276. Cambridge: Cambridge University Press.

Kortlandt, A. 1962. Chimpanzees in the wild. *Sci Amer 206:* 128–138.

Kramer, R. A., and N. Sharma. 1997. Tropical forest biodiversity protection: Who pays and why. In *Last stand: Protected areas and the defense of tropical biodiversity,* eds. R. Kramer, C. van Schaik, and J. Johnson, 162–186. New York: Oxford University Press.

Krause, J., and G. D. Ruxton. 2002. *Living in groups.* Oxford: Oxford University Press.

Krebs, J. R., and N. B. Davies, eds. 1993. *An introduction to behavioural ecology.* Oxford: Blackwell Scientific.

Kumar, S., A. Filipski, V. Swarna, A. Walker, and S. B. Hedges. 2005. Placing confidence limits on the molecular age of the human-chimpanzee divergence. *Proc Nat Acad Sci USA 102:* 18842–18847.

Kummer, H. 1968. *Social organization of hamadryas baboons.* Chicago: University of Chicago Press.

Kuroda, S. 1989. Developmental retardation and behavioral characteristics of pygmy chimpanzees. In *Understanding chimpanzees,* eds. P. G. Heltne, and L. A. Marquardt, 184–193. Cambridge, MA: Harvard University Press.

———. 1992. Ecological interspecies relationships between gorillas and chimpanzees in the Ndoki-Nouabale Reserve, Northern Congo. In *Topics in primatology. 2: Behavior, ecology, and conservation,* eds. N. Itoigawa, Y. Sugiyama, G. P. Sackett, and R. K. R. Thompson, 385–394. Tokyo: University of Tokyo Press.

Kuroda, S., T. Nishihara, S. Suzuki, and R. A. Oko. 1996. Sympatric chimpanzees and gorillas in the Ndoki Forest, Congo. In *Great Ape Societies,* eds. W. C. McGrew, L. F. Marchant, and T. Nishida, 71–81. Cambridge: Cambridge University Press.

LaBarbera, M. 1989. Analyzing body size as a factor in ecology and evolution. *Ann Rev Ecol Syst 20:* 97–117.

Lack, D. L. 1939. The behaviour of the robin: I and II. *Proc R Soc Lond, A 109:* 169–178.

Lambert, J. E. 1998. Primate digestion: Interactions among anatomy, physiology, and feeding ecology. *Evol Anthropol 7:* 8–20.

Laurance, W. F., et al. 2002. Predictors of deforestation in the Brazilian Amazon. *J Biogeog 29:* 737–748.

Layton, R. H. 1989. Are sociobiology and social anthropology compatible? The significance of sociocultural resources in human evolution. In *Comparative socioecology: The behavioural ecology of humans and other animals,* eds. V. Standen, and R. A. Foley, 433–455. Oxford: Blackwell Scientific.

Leader-Williams, N., and S. D. Albon. 1988. Allocation of resources for conservation. *Nature 336:* 533–535.

Leader-Williams, N., J. Harrison, and M. J. B. Green. 1990. Designing protected areas to conserve natural resources. *Sci Progress Oxford 74:* 189–204.

Lee, P. C. 1996. The meanings of weaning: Growth, lactation, and life history. *Evol Anthropol 5:* 87–96.

Lee, P. C., and J. A. Johnson. 1992. Sex differences in alliances, and the acquisition and maintenance of dominance status among immature primates. In *Coalitions and alliances in humans and other animals,* eds. A. H. Harcourt, and F. de Waal, 391–414. Oxford: Oxford University Press.

Lee, P. C., and P. M. Kappeler. 2003. Socioecological correlates of phenotypic plasticity of primate life histories. In *Primate life histories and socioecology,* eds. P. M. Kappeler, and M. E. Pereira, 41–65. Chicago: University of Chicago Press.

Lee, P. C., J. Thornback, and E. L. Bennett. 1988. *Threatened primates of Africa. The IUCN Red Data Book.* Gland, Switzerland: IUCN.

Leendertz, F. H., et al. 2004. Anthrax kills wild chimpanzees in a tropical rainforest. *Nature 430:* 451–452.

Lefebvre, L., P. Whittle, and E. Lascaris. 1997. Feeding innovations and forebrain size in birds. *Anim Behav 53:* 549–560.

Lehmann, L., and N. Perrin. 2003. Inbreeding avoidance through kin recognition: Choosy females boost male dispersal. *Amer Nat 162:* 638–652.

Leigh, S. R., and G. E. Blomquist. 2007. Life history. In *Primates in perspective,* eds. C. J. Campbell, A. Fuentes, K. C. MacKinnon, M. Panger, and S. K. Bearder, 396–407. New York: Oxford University Press.

Leigh, S. R., and B. T. Shea. 1995. Ontogeny and the evolution of adult body size dimorphism in apes. *Amer J Primatol 36:* 37–60.

Leighton, D. R. 1987. Gibbons: Territoriality and monogamy. In *Primate societies,* eds. B. B. Smuts, D. L. Cheney, R. M. Seyfarth, R. W. Wrangham, and T. T. Struhsaker, 135–145. Chicago: University of Chicago Press.

Leighton, M., et al. 1995. Orangutan life history and vortex analysis. In *The neglected ape,* eds. R. D. Nadler, B. M. F. Galdikas, L. K. Sheeran, and N. Rosen, 97–107. New York: Plenum Press.

Leroy, E. M., et al. 2004. Multiple Ebola virus transmission events and rapid decline of central African wildlife. *Science 303:* 387–390.

———. 2005. Fruit bats as reservoirs of Ebola virus. *Nature 438:* 575–576.

Leutenegger, W. 1979. Evolution of litter size in primates. *Amer Nat 114:* 525–531.

Levréro, F. 2001. *Rencontres inter-groupes chez le gorille de plaine (Gorilla gorilla gorilla): Essai sur leur rôle dans la dynamique sociale.* Diplome d'études approfondies eco-ethologie evolutive. Pp. 31. Université de Rennes I, Paimpont.

Levréro, F., et al. 2006. Living in non-breeding groups: An alternative strategy for maturing gorillas. *Amer J Primatol 68:* 275–291.

Lindenfors, P., and B. S. Tullberg. 1998. Phylogenetic analyses of primate size evolution: The consequences of sexual selection. *Biol J Linn Soc 64:* 413–447.

Lindenmayer, B. D., and J. F. Franklin. 2002. *Conserving forest biodiversity: A comprehensive multiscaled approach.* Covelo, CA: Island Press.

Lindsay, W. K. 1987. Integrating parks and pastoralists: Some lessons from Amboseli. In *Conservation in Africa: People, politics, and practice,* eds. D. Anderson, and R. Grove, 149–168. Cambridge: Cambridge University Press.

Liu, J., G. C. Daily, P. R. Ehrlich, and G. W. Luck. 2003. Effects of household dynamics on resource consumption and biodiversity. *Nature 421:* 530–533.

Lomborg, B. 2001. *The skeptical environmentalist: Measuring the real state of the world.* Cambridge: Cambridge University Press.

Lomnicki, A. 1988. *Population ecology of individuals.* Princeton, NJ: Princeton University Press.

Ludwig, D., M. Mangel, and B. Haddad. 2001. Ecology, conservation, and public policy. *Ann Rev Ecol Syst 32:* 481–517.

Macdonald, D. W., and D. D. P. Johnson. 2001. Dispersal in theory and practice: Consequences for conservation biology. In *Dispersal,* eds. J. Clobert, E. Danchin, A. A. Dhondt, and J. D. Nichols, 358–372. Oxford: Oxford University Press.

Mace, G. M., and A. Balmford. 2000. Patterns and processes in contemporary mammalian extinction. In *Priorities for the conservation of mammalian diversity: Has the panda had its day?* eds. A. Entwistle, and N. Dunstone, 27–52. Cambridge: Cambridge University Press.

MacKinnon, J. 1974. The behaviour and ecology of wild orang-utans (*Pongo pygmaeus*). *Anim Behav 22:* 3–74.

Madsen, T., R. Shine, J. Loman, and T. Hakansson. 1992. Why do female adders copulate so frequently? *Nature 355:* 440–441.

Maggioncalda, A. N., N. M. Czekala, and R. M. Sapolsky. 2000. Growth hormone and thyroid stimulating hormone concentrations in captive male orangutans: Implications for understanding developmental arrest. *Amer J Primatol 50:* 67–76.

Magliocca, F., and A. Gautier-Hion. 2002. Mineral content as a basis for food selection by western lowland gorillas in a forest clearing. *Amer J Primatol 57:* 67–77.

Magliocca, F., S. Querouil, and A. Gautier-Hion. 1999. Population structure and group composition of western lowland gorillas in north-western Republic of Congo. *Amer J Primatol 48:* 1–14.

Malthus, T. R. 1798. *An essay on the principle of population.* London: Macmillan.

Mangel, M. 1990. Resource divisibility, predation, and group formation. *Anim Behav 39:* 1163–1172.

Manson, J. H. 2007. Mate choice. In *Primates in perspective,* eds. C. J. Campbell, A. Fuentes, K. C. MacKinnon, M. Panger, and S. K. Bearder, 447–463. New York: Oxford University Press.

Margules, C. R., and R. L. Pressey. 2000. Systematic conservation planning. *Nature 405:* 243–253.

Marler, P. 1996. Social cognition. Are primates smarter than birds? In *Current ornithology, Vol. 13,* eds. V. Nolan, and E. D. Ketterson, 1–32. New York: Plenum Press.

Martin, R. D., D. J. Chivers, A. M. Maclarnon, and C. M. Hladik. 1985. Gastrointestinal allometry in primates and other mammals. In *Size and scaling in nonhuman primates,* ed. W. L. Jungers, 61–89. New York: Plenum Press.

Martin, R. D., M. Genoud, and C. K. Hemelrijk. 2005. Problems of allometric scaling analysis: Examples from mammalian reproductive biology. *J Exp Biol 208:* 1731–1747.

Matsubara, M., and D. S. Sprague. 2004. Mating tactics in response to costs incurred by mating with multiple males in wild female Japanese macaques. *Int J Primatol 25:* 901–918.

Matsumoto-Oda, A. 2002. Social relationships between cycling females and adult males in Mahale chimpanzees. In *Behavioural diversity in chimpanzees and bonobos,* eds. C. Boesch, G. Hohmann, and L. F. Marchant, 168–180. Cambridge: Cambridge University Press.

Matsumoto-Oda, A., K. Hosaka, M. Huffman, and K. Kawanaka. 1998. Factors affecting party size in chimpanzees of the Mahale Mountains. *Int J Primatol 19:* 999–1011.

Matsumoto-Oda, A., and R. Oda. 1998. Changes in the activity budget of cycling female chimpanzees. *Amer J Primatol 46:* 157–166.

Matsumura, S. 1999. The evolution of "egalitarian" and "despotic" social systems among macaques. *Primates 40:* 23–31.

McComb, K., C. Packer, and A. E. Pusey. 1994. Roaring and numerical assessment in contests between groups of female lions, *Panthera leo. Anim Behav 47:* 379–387.

McFarland Symington, M. 1988. Food competition and foraging party size in the black spider monkey (*Ateles paniscus chamek*). *Behaviour 105:* 117–134.

McKinney, M. L. 2001. Role of human population size in raising bird and mammal threat among nations. *Anim Conserv 4:* 45–57.

McMahon, T. A., and J. T. Bonner. 1983. *On size and life.* New York: Scientific American Library.

McNab, B. K. 1963. Bioenergetics and the determination of home range size. *Amer Nat 97:* 133–140.

———. 1999. On the comparative ecological and evolutionary significance of total and mass-specific rates of metabolism. *Physiol Biochem Zool 72:* 642–644.

McNeely, J. A., M. Gadgil, C. Levèque, C. Padoch, and K. Redford. 1995. Human influences on biodiversity. In *Global biodiversity assessment,* eds. V. H. Heywood, and R. T. Watson, 711–821. Cambridge: Cambridge University Press.

McNeilage, A. 2001. Diet and habitat use of two mountain gorillas groups in contrasting habitats in the Virungas. In *Mountain gorillas: Three decades of research at*

Karisoke, eds. M. M. Robbins, P. Sicotte, and K. J. Stewart, 263–292. Cambridge: Cambridge University Press.

McNeilage, A., A. J. Plumptre, A. Brock-Doyle, and A. Vedder. 2001. Bwindi Impenetrable National Park, Uganda: Gorilla census 1997. *Oryx 35:* 39–47.

Meder, A. 1990. Sex differences in the behaviour of immature captive lowland gorillas. *Primates 31:* 51–63.

———. 2005. Integration of hand-reared gorillas into breeding groups. *Zoo Biol 9:* 157–164.

Ménard, N. 2004. Do ecological factors explain variation in social organization? In *Macaque societies: A model for the study of social organization,* eds. B. Thierry, M. Singh, and W. Kaumanns, 237–262. Cambridge: Cambridge University Press.

Mendoza, S. P. 1991. Behavioural and physiological indices of social relationships: Comparative studies of New World monkeys. In *Primate responses to environmental change,* ed. H. O. Box, 311–335. London: Chapman & Hall.

Mendoza, S. P., and W. A. Mason. 1986. Parental division of labour and differentiation of attachments in a monogamous primate (*Callicebus moloch*). *Anim Behav 34:* 1336–1347.

Menzel, C. R. 1997. Primates' knowledge of their natural habitat: As indicated in foraging. In *Machiavellian intelligence II,* eds. A. Whitem, and R. W. Byrne, 207–239. Cambridge: Cambridge University Press.

Mercador, J. 2002. Forest people: The role of African rainforests in human evolution and dispersal. *Evol Anthropol 11:* 117–124.

Merfield, F. G., and H. Miller. 1956. *Gorilla hunter.* New York: Farrar, Straus and Chuday.

Messier, W., and C.-B. Stewart. 1997. Episodic adaptive evolution of primate lysozymes. *Nature 385:* 151–154.

Michalski, F., and C. A. Peres. 2005. Anthropogenic determinants of primate and carnivore local extinctions in a fragmented forest landscape of southern Amazonia. *Biol Conserv 124:* 383–396.

Milinski, M. 2005. Reputation, personal identity, and cooperation in a social dilemma. In *Cooperation in primates and humans: Mechanisms and evolution,* eds. P. M. Kappeler, and C. P. van Schaik, 263–278. Berlin: Springer-Verlag.

Millennium Ecosystem Assessment 2005. Millennium ecosystem assessment synthesis reports. Millennium Ecosystem Assessment. http://www.millenniumassessment.org//en/Products.Synthesis.aspx.

Miller, L. E., and A. Treves. 2007. Predation on primates: Past studies, current challenges, and directions for the future. In *Primates in perspective,* eds. C. J. Campbell, A. Fuentes, K. C. MacKinnon, M. Panger, and S. K. Bearder, 525–543. New York: Oxford University Press.

Milner-Gulland, E. J., and R. Mace. 1998. *Conservation of biological resources.* Malden, MA: Blackwell Science.

Milton, K. 1984. The role of food-processing factors in primate food choice. In *Adaptations for foraging in nonhuman primates,* eds. P. S. Rodman, and J. G. H. Cant, 249–279. New York: Columbia University Press.

———. 1988. Foraging behaviour and the evolution of primate intelligence. In *Machiavellian intelligence,* eds. R. W. Byrne, and A. Whiten, 285–305. Oxford: Oxford University Press.

Milton, K., and M. L. May. 1976. Body weight, diet, and home range area in primates. *Nature 259:* 459–462.

Mitani, J. C. 1985a. Sexual selection and adult male orangutan (*Pongo pygmaeus*) long calls. *Anim Behav 33:* 272–283.

———. 1985b. Mating behaviour of male orangutans in the Kutai Game Reserve, Indonesia. *Anim Behav 33:* 392–402.

———. 1992. Preliminary results of the studies on wild western lowland gorillas and other sympatric diurnal primates in the Ndoki Forest, Northern Congo. In *Topics in primatology 2: Behavior, ecology, and conservation,* eds. N. Itoigawa, Y. Sugiyama, G. P. Sackett, and R. K. R. Thompson, 215–224. Tokyo: University of Tokyo Press.

Mitani, J. C., G. F. Grether, P. S. Rodman, and D. Priatna. 1991. Associations among wild orang-utans: Sociality, passive aggregations, or chance. *Anim Behav 42:* 33–46.

Mitani, J. C., J. Gros-Louis, and J. H. Manson. 1996a. Number of males in primate groups: Comparative tests of competing hypotheses. *Amer J Primatol 38:* 315–332.

Mitani, J. C., J. Gros-Louis, and A. F. Richards. 1996b. Sexual dimorphism, the operational sex ratio, and the intensity of male competition in polygynous primates. *Amer Nat 147:* 966–980.

Mitani, J. C., and P. S. Rodman. 1979. Territoriality: The relation of ranging pattern and home range size to defendability, with an analysis of territoriality among primate species. *Behav Ecol Sociobiol 5:* 241–251.

Mitani, J. C., and D. P. Watts. 1997. The evolution of non-maternal caretaking among anthropoid primates: Do helpers help? *Behav Ecol Sociobiol 40:* 213–220.

———. 2001. Why do chimpanzees hunt and share meat? *Anim Behav 61:* 915–924.

Mitani, J. C., D. P. Watts, and J. S. Lwanga. 2002a. Ecological and social correlates of chimpanzee party size and composition. In *Behavioural diversity in chimpanzees and bonobos,* eds. C. Boesch, G. Hohmann, and L. F. Marchant, 102–111. Cambridge: Cambridge University Press.

Mitani, J. C., D. P. Watts, and M. N. Muller. 2002b. Recent developments in the study of wild chimpanzee behavior. *Evol Anthropol 11:* 9–25.

Mitchell, R. W. 1989. Functions and consequences of infant-adult male interactions in a captive group of lowland gorillas (*Gorilla gorilla gorilla*). *Zoo Biol 8:* 125–137.

Mittermeier, R. A., N. Myers, J. R. Thomsen, G. A. B. da Fonseca, and S. Olivieri. 1998. Biodiversity hotspots and major tropical wilderness areas: Approaches to setting conservation priorities. *Conserv Biol 12:* 516–520.

Mittermeier, R. A., et al. 2003. Wilderness and biodiversity conservation. *Proc Nat Acad Sci USA 100:* 10309–10313.

Møller, A. P. 1988. Ejaculate quality, testes size, and sperm competition in primates. *J Hum Evol 17:* 479–488.

Moore, B. D., and W. J. Foley. 2005. Tree use by koalas in a chemically complex landscape. *Nature 435:* 488–490.

Moore, J., and R. Ali. 1984. Are dispersal and inbreeding avoidance related? *Anim Behav 32:* 94–112.

Morgan, B. J., C. Wild, and A. Ekobo. 2003. Newly discovered gorilla population in the Ebo Forest, Littoral Province, Cameroon. *Int J Primatol 24:* 1129–1137.

Mougeot, F., S. M. Redpath, F. Leckie, and P. J. Hudson. 2003. The effect of aggressiveness on the population dynamics of a territorial bird. *Nature 421:* 737–739.

Moura, A. C. d. A., and P. C. Lee. 2004. Capuchin stone tool use in Caatinga dry forest. *Science 306:* 1909.

Muchaal, P. K., and G. Ngandjui. 1999. Impact of village hunting on wildlife populations in the western Dja Reserve, Cameroon. *Conserv Biol 13:* 385–396.

Mudakikwa, A. B., M. R. Cranfield, J. M. Sleeman, and U. Eilenberger. 2001. Clinical medicine, preventive healthcare and research on mountain gorillas in the Virunga Volcanoes region. In *Mountain gorillas: Three decades of research at Karisoke,* eds. M. M. Robbins, P. Sicotte, and K. J. Stewart, 341–360. Cambridge: Cambridge University Press.

Muller, M. N. 2002. Agonistic relations among Kanyawara chimpanzees. In *Behavioural diversity in chimpanzees and bonobos,* eds. C. Boesch, G. Hohmann, and L. F. Marchant, 112–123. Cambridge: Cambridge University Press.

Muller, M. N., and R. W. Wrangham. 2001. The reproductive ecology of male hominoids. In *Reproductive ecology and human evolution,* ed. P. T. Ellison, 397–427. New York: Aldine de Gruyter.

———. 2004. Dominance, aggression, and testosterone in wild chimpanzees: A test of the "challenge hypothesis." *Anim Behav 67:* 113–123.

Muniz, L., S. Perry, J. H. Manson, J. Gros-Louis, and L. Vigilant. 2006. Father-daughter inbreeding avoidance in a wild primate population. *Curr Biol 16:* R156–R157.

Myers, N. 1972. National parks in savannah Africa. *Science 178:* 1255–1263.

———. 1984. *The primary source: Tropical forests and our future.* New York: W.W. Norton.

Myers, N., and J. Kent. 2003. New consumers: The influence of affluence on the environment. *Proc Nat Acad Sci USA 100:* 4963–4968.

Myers, N., R. A. Mittermeier, C. G. Mittermeier, G. A. B. da Fonseca, and J. Kent. 2000. Biodiversity hotspots for conservation priorities. *Nature 403:* 853–858.

Nadler, R. D. 1975a. Sexual cyclicity in captive lowland gorillas. *Science 189:* 813–814.

———. 1975b. Determinants of variability in maternal behavior of captive female gorillas. In *Proceedings of the symposium of the 5th congress of the International Primatological Society,* eds., S. Kondo, M. Kawai, A. Ehara, and S. Kawamura, 207–216. Japan Science Press, Tokyo.

———. 1981. Laboratory research on sexual behavior of the great apes. In *Reproductive biology of the great apes,* ed. C. E. Graham, 192–238. New York: Academic Press.

Newmark, W. D., D. N. Manyanza, D. M. Gamassa, and H. I. Sariko. 1994. The conflict between wildlife and local people living adjacent to protected areas in Tanzania: Human density as a predictor. *Conserv Biol 8:* 249–255.

Newton-Fisher, N. E. 2002. Relationships of male chimpanzees in the Budongo Forest, Uganda. In *Behavioural diversity in chimpanzees and bonobos,* eds. C. Boesch, G. Hohmann, and L. F. Marchant, 124–137. Cambridge: Cambridge University Press.

Nicholls, A. O. 1998. Integrating population abundance, dynamics, and distribution into broad-scale priority setting. In *Conservation in a changing world,* eds. G. M. Mace, A. Balmford, and J. R. Ginsberg, 251–272. Cambridge: Cambridge University Press.

Niinemets, U. 1995. Distribution of foliar carbon and nitrogen across the canopy of *Fagus sylvatica:* Adaptation to a vertical light gradient. *Acta Oecol 16:* 525–541.

Nishida, T. 1979. The social structure of chimpanzees of the Mahale Mountains. In *The great apes,* eds. D. A. Hamburg, and E. R. McCown, 73–121. Menlo Park, CA: Benjamin/Cummings.

———. 1990. *The chimpanzees of the Mahale Mountains: Sexual and life history strategies.* Tokyo: University of Tokyo Press.

Nishida, T., M. Hiraiwa-Hasegawa, T. Hasegawa, and Y. Takahata. 1985. Group extinction and female transfer in wild chimpanzees in the Mahale National Park, Tanzania. *Z Tierpsychol 67:* 284–301.

Nishida, T., and K. Hosaka. 1996. Coalition strategies among adult male chimpanzees of the Mahale Mountains, Tanzania. In *Great ape societies,* eds. W. C. McGrew, L. F. Marchant, and T. Nishida, 114–134. Cambridge: Cambridge University Press.

Nishida, T., H. Takasaki, and Y. Takahata. 1990. Demography and reproductive profiles. In *The chimpanzees of the Mahale Mountains. Sexual and life history strategies,* ed. T. Nishida, 63–97. Tokyo: University of Tokyo Press.

Nishida, T., et al. 2003. Demography, female life history, and reproductive profiles among the chimpanzees of Mahale. *Amer J Primatol 59:* 99–121.

Nishihara, T. 1994. Population density and group organization of gorillas (*Gorilla gorilla gorilla*) in the Nouabalé-Ndoki National Park. *J African Stud 44:* 29–45.

———. 1995. Feeding ecology of western lowland gorillas in the Nouabalé-Ndoki National Park, northern Congo. *Primates 35:* 151–168.

Nkurunungi, J. B., J. Ganas, M. M. Robbins, and C. B. Stanford. 2004. A comparison of two mountain gorilla habitats in Bwindi Impenetrable National Park, Uganda. *Afr J Ecol 42:* 289–297.

Noë, R. 1992. Alliance formation among male baboons: Shopping for profitable partners. In *Coalitions and alliances in humans and other animals,* eds. A. H. Harcourt, and F. B. M. de Waal, 285–322. Oxford: Oxford University Press.

———. 1994. Biological markets: Supply and demand determine the effect of partner choice in cooperation, mutualism, and mating. *Behav Ecol Sociobiol 35:* 1–11.

Noë, R., and R. Bshary. 1997. The formation of red colobus-diana monkey associations under predation pressure from chimpanzees. *Proc Roy Soc, Lond B 264:* 253–259.

Norconk, M. A. 1990. Ecological and behavioral correlates of polyspecific primate troops. *Amer J Primatol 21:* 81–85.

Nowak, M. A., and K. Sigmund. 2005. Evolution of indirect reciprocity. *Nature 437:* 1291–1298.

Nowak, R. M. 1999. *Walker's primates of the world.* Baltimore: Johns Hopkins University Press.

Nsubuga, A. M. 2005. *Genetic analysis of the social structure in wild mountain gorillas (Gorilla beringei beringei) of Bwindi Impenetrable National Park, Uganda.* Ph.D. diss. University of Leipzig.

Nunn, C. L. 1999a. The evolution of exaggerated sexual swellings in primates and the graded-signal hypothesis. *Anim Behav 58:* 229–246.

———. 1999b. The number of males in primate social groups: A comparative test of the socioecological model. *Behav Ecol Sociobiol 46:* 1–13.

Nunn, C. L., and R. A. Barton. 2001. Comparative methods for studying primate adaptation and allometry. *Evol Anthropol 10:* 81–98.

Nunn, C. L., J. L. Gittleman, and J. Antonovics. 2000. Promiscuity and the primate immune system. *Science 290:* 1168–1170.

Nunn, C. L., P. H. Thrall, K. J. Stewart, and A. H. Harcourt. Submitted. Emerging infectious diseases and animal mating and social systems. *Evol Ecol.*

Nunn, C. L., and C. P. van Schaik. 2000. Social evolution in primates: The relative roles of ecology and intersexual conflict. In *Infanticide by males and its implications,* eds. C. P. van Schaik, and C. H. Janson, 388–419. Cambridge: Cambridge University Press.

Oates, J. F. 1994. The natural history of African colobines. In *Colobine monkeys: Their ecology, behaviour, and evolution,* eds. A. G. Davies, and J. F. Oates, 75–128. Cambridge: Cambridge University Press.

———. 1996a. Habitat alteration, hunting, and the conservation of folivorous primates in African forests. *Aus J Ecol 21:* 1–9.

———. 1996b. *African primates: Status survey and conservation action plan.* Gland, Switzerland: IUCN.

———. 1999. *Myth and reality in the rain forest.* Berkeley: University of California Press.

———. 2006. Is the chimpanzee, *Pan troglodytes,* an endangered species? It depends on what "endangered" means. *Primates 47:* 102–112.

Oates, J. F., et al. 1990. Determinants of variation in tropical forest primate biomass: New evidence from West Africa. *Ecology 71:* 328–343.

———. 2003. The Cross River gorilla: Natural history and status of a neglected and critically endangered subspecies. In *Gorilla biology: A multidisciplinary perspective,* eds. A. B. Taylor, and M. L. Goldsmith, 472–497. Cambridge: Cambridge University Press.

Olsson, M., A. Gullberg, H. Tegelström, T. Madsen, and R. Shine. 1994. Can female adders multiply? *Nature 369:* 528.

Olupot, W., and P. M. Waser. 2001. Activity patterns, habitat use, and mortality risks of mangabey males living outside social groups. *Anim Behav 61:* 1227–1235.

Owen-Smith, R. N. 1988. *Megaherbivores: The influence of very large body size on ecology.* Cambridge: Cambridge University Press.

Packer, C. 1979. Inter-troop transfer and inbreeding avoidance in *Papio anubis. Anim Behav 27:* 1–36.

———. 2001. Infanticide is no fallacy. *Amer Anthropol 102:* 829–831.

Packer, C., D. A. Gilbert, A. E. Pusey, and S.J. O'Brien. 1991. A molecular genetic analysis of kinship and cooperation in African lions. *Nature 351:* 562–565.

Packer, C., and A. E. Pusey. 1982. Cooperation and competition within coalitions of male lions. *Nature 296:* 740–742.

———. 1984. Infanticide in carnivores. In *Infanticide: Comparative and evolutionary aspects,* eds. G. Hausfater, and S. B. Hrdy, 31–42. Hawthorne, NY: Aldine Publishing.

Packer, C., D. Scheel, and A. E. Pusey. 1990. Why lions form groups: Food is not enough. *Amer Nat 136:* 1–19.

Packer, C., et al. 1988. Reproductive success of lions. In *Reproductive success,* ed. T. H. Clutton-Brock, 363–383. Chicago: University of Chicago Press.

Pagel, M. D., and P. H. Harvey. 1993. Evolution of the juvenile period in mammals. In *Juvenile primates: Life history, development, and behavior,* eds. M. E. Pereira, and L. A. Fairbanks, 28–37. New York: Oxford University Press.

Palombit, R. A. 1999. Infanticide and the evolution of pair bonds in nonhuman primates. *Evol Anthropol 7:* 117–129.

———. 2000. Infanticide and the evolution of male-female bonds in animals. In *Infanticide by males and its implications,* eds. C. P. van Schaik, and C. H. Janson, 239–268. Cambridge: Cambridge University Press.

———. 2003. Male infanticide in savanna baboons: Adaptive significance and intraspecific variation. In *Sexual selection and reproductive competition in primates: New perspectives and directions,* ed. C. B. Jones, 367–411. Norman, OK: American Society of Primatologists.

Palombit, R. A., D. L. Cheney, and R. M. Seyfarth. 2001. Female-female competition for male "friends" in wild chacma baboons, *Papio cynocephalus ursinus. Anim Behav 61:* 1159–1171.

Palombit, R. A., R. M. Seyfarth, and D. L. Cheney. 1997. The adaptive value of "friendships" to female baboons: Experimental observational evidence. *Anim Behav 54:* 599–614.

Palombit, R. A., et al. 2000. Male infanticide and defense of infants in chacma baboons. In *Infanticide by males and its implications,* eds. C. P. van Schaik, and C. H. Janson, 123–152. Cambridge: Cambridge University Press.

Parish, A. R. 1994. Female relationships in bonobos (*Pan paniscus*): Evidence for bonding, cooperation, and female dominance in a male-philopatric species. *Human Nat 7:* 61–96.

Parker, G. A. 1970. Sperm competition and its evolutionary consequences in the insects. *Biol Rev 45:* 525–567.

Parker, P. G., and T. A. Waite. 1997. Mating systems, effective population size, and conservation of natural populations. In *Behavioral approaches to conservation in the wild,* eds. J. R. Clemmons, and R. Buchholz, 243–261. Cambridge: Cambridge University Press.

Parks, S. A., and A. H. Harcourt. 2002. Reserve size, local human density, and mammalian extinctions in U.S. protected areas. *Conserv Biol 16:* 800–808.

Parmesan, C., and G. Yohe. 2003. A globally coherent fingerprint of climate change impacts across natural systems. *Nature 421:* 37–42.

Parmigiani, S., P. Palanza, D. Mainardi, and P. F. Brain. 1994. Infanticide and protection of young in house mice (*Mus domesticus*): Female and male strategies. In *Infanticide & parental care,* eds. S. Parmigiani, and F. S. vom Saal, 341–363. Chur, Switzerland: Harwood Academic Publishers.

Parmigiani, S., and F. S. vom Saal, eds. 1994. *Infanticide & parental care.* Chur, Switzerland: Harwood Academic Publishers.

Parnell, R. J. 2002. Group size and structure in western lowland gorillas (*Gorilla gorilla gorilla*) at Mbeli Bai, Republic of Congo. *Amer J Primatol 56:* 193–206.

Parnell, R. J., and H. M. Buchanan-Smith. 2001. An unusual social display by gorillas. *Nature 412:* 294.

Paul, A. 2002. Sexual selection and mate choice. *Int J Primatol 23:* 877–904.

Paul, A., and J. Kuester. 2004. The impact of kinship in mating and reproduction. In *Kinship and behavior in primates,* eds. B. Chapais, and C. Berman, 271–291. New York: Oxford University Press.

Paz-y-Mino, C. G., A. B. Bond, A. C. Kamil, and R. P. Balda. 2004. Pinyon jays use transitive inference to predict social dominance. *Nature 430:* 778–781.

Peres, C. A. 1990. Effects of hunting on western Amazonian primate communities. *Biol Conserv 54:* 47–59.

———. 1999. Effects of subsistence hunting and forest types on the structure of Amazonian primate communities. In *Primate communities,* eds. J. G. Fleagle, C. H. Janson, and K. E. Reed, 268–283. Cambridge: Cambridge University Press.

Peres, C. A., and P. M. Dolman. 2000. Density compensation in neotropical primate communities: Evidence from 56 hunted and nonhunted Amazonian forests of varying productivity. *Oecologia 122:* 175–189.

Peres, C. A., and J. W. Terborgh. 1995. Amazonian nature reserves: An analysis of the defensibility status of existing conservation units and design criteria for the future. *Conserv Biol 9:* 34–46.

Pérez-Barbería, F. J., I. J. Gordon, and M. Page. 2002. The origins of sexual dimorphism in body size in ungulates. *Evolution 56:* 1276–1285.

Perrigo, G., and F. S. vom Saal. 1994. Behavioral cycles and the neural timing of infanticide and parental behavior in male house mice. In *Infanticide & Parental Care,* eds. S. Parmigiani, and F. S. vom Saal, 365–396. Chur, Switzerland: Harwood Academic Publishers.

Peters, R. H. 1983. *The ecological implications of body size.* Cambridge: Cambridge University Press.

Peterson, D. 2003. *Eating apes.* Berkeley: University of California Press.

Pimm, S. L., G. J. Russell, J. L. Gittleman, and T. M. Brooks. 1995. The future of biodiversity. *Science 269:* 347–350.

Plavcan, J. M. 1999. Mating systems, intrasexual competition, and sexual dimorphism in primates. In *Comparative primate socioecology,* ed. P. C. Lee, 241–260. Cambridge: Cambridge University Press.

———. 2004. Sexual selection, measures of sexual selection, and sexual dimorphism in primates. In *Sexual selection in primates: New and comparative perspectives,* eds. P. M. Kappeler, and C. P. van Schaik, 230–252. Cambridge: Cambridge University Press.

Plumptre, A. J. 2000. Monitoring mammal populations with line transect techniques in African forests. *J Appl Ecol 37:* 356–368.

———. 2003. Lessons learned from on-the-ground conservation in Rwanda and Democratic Republic of the Congo. In *War and tropical forests: Conservation in areas of armed conflict,* ed. S. V. Price, 71–91. New York: Food Products Press.

Plumptre, A. J., A. McNeilage, J. S. Hall, and E. A. Williamson. 2003. The current status of gorillas and threats to their existence at the beginning of a new millennium. In *Gorilla biology: A multidisciplinary perspective,* eds. A. B. Taylor, and M. L. Goldsmith, 414–431. Cambridge: Cambridge University Press.

Plumptre, A. J., and E. A. Williamson. 2001. Conservation-oriented research in the Virunga region. In *Mountain gorillas. Three decades of research at Karisoke,* eds. M. M. Robbins, P. Sicotte, and K. J. Stewart, 361–389. Cambridge: Cambridge University Press.

Pope, T. R. 2000. The evolution of male philopatry in neotropical monkeys. In *Primate males: Causes and consequences of variation in group composition,* ed. P. M. Kappeler, 219–235. Cambridge: Cambridge University Press.

Poulsen, J. R., and C. J. Clark. 2004. Densities, distributions, and seasonal movements of gorillas and chimpanzees in swamp forest in northern Congo. *Int J Primatol 25:* 285–306.

Promislow, D. E. L., and P. H. Harvey. 1990. Living fast and dying young: A comparative analysis of life-history variation among mammals. *J Zool 220:* 417–437.

Pulliam, H. R., and T. Caraco. 1984. Living in groups: Is there an optimal group size? In *Behavioural ecology: An evolutionary approach.* 2nd. ed., eds. J. R. Krebs, and N. B. Davies, 122–147. Oxford: Blackwell Scientific.

Purvis, A. 1995. A composite estimate of primate phylogeny. *Phil Trans Roy Soc Lond Ser B 348:* 405–421.

———. 2004. Comparative analysis by independent contrasts. http://www.bio.ic.ac.uk/evolve/software/caic/.

Purvis, A., and A. Rambaut. 1995. Comparative analysis by independent contrasts (CAIC): An Apple Macintosh application for analysing comparative data. *Computer Applic Biosci 11:* 247–251.

Purvis, A., and A. J. Webster. 1999. Phylogenetically independent comparisons and primate phylogeny. In *Comparative primate socioecology,* ed. P. C. Lee, 44–70. Cambridge: Cambridge University Press.

Purvis, A., A. J. Webster, P. Agapow, K. E. Jones, and N. J. B. Isaac. 2003. Primate life histories and phylogeny. In *Primate life histories and socioecology,* eds. P. M. Kappeler, and M. E. Pereira, 25–40. Chicago: University of Chicago Press.

Pusey, A. E. 1983. Mother-offspring relationships in chimpanzees after weaning. *Anim Behav 31:* 363–377.

———. 1987. Sex-biased dispersal and inbreeding avoidance in birds and mammals. *Trends Ecol Evol 2:* 295–299.

———. 1988. Primate dispersal: Reply. *Trends Ecol Evol 3:* 145–146.

———. 1992. The primate perspective on dispersal. In *Animal dispersal: Small mammals as a model,* eds. N. C. Stenseth, and W. Z. Lidicker, 243–259. London: Chapman & Hall.

Pusey, A. E., G. W. Oehlert, J. M. Williams, and J. Goodall. 2005. Influence of ecological and social factors on body mass of wild chimpanzees. *Int J Primatol 26:* 3–31.

Pusey, A. E., and C. Packer. 1987. Dispersal and philopatry. In *Primate societies,* eds. B. B. Smuts, D. L. Cheney, R. M. Seyfarth, R. W. Wrangham, and T. T. Struhsaker, 250–266. Chicago: University of Chicago Press.

———. 1997. The ecology of relationships. In *Behavioural ecology: An evolutionary approach.* 4th. ed., eds. J. R. Krebs, and N. B. Davies, 254–283. Oxford: Blackwell Scientific.

Pusey, A., J. Williams, and J. Goodall. 1997. The influence of dominance rank on the reproductive success of female chimpanzees. *Science 277:* 828–831.

Pusey, A., and M. Wolf. 1996. Inbreeding avoidance in animals. *Trends Ecol Evol 11:* 201–206.

Quinn, J. L., and Y. Kokorev. 2002. Trading-off risks from predators and from aggressive hosts. *Behav Ecol Sociobiol 51:* 455–460.

Ralls, K., and J. Ballou. 1983. Extinction: Lessons from zoos. In *Genetics and conservation,* eds. C. M. Schonewald-Cox, S. M. Chambers, B. MacBryde, and W. L. Thomas, 164–184. Menlo Park, CA: Benjamin/Cummings.

Ralls, K., P. H. Harvey, and M. E. Soulé. 1986. Inbreeding in natural populations of birds and mammals. In *Conservation biology: The science of scarcity and diversity,* ed. M. E. Soulé, 35–56. Sunderland, MA: Sinauer Associates.

Read, A. F., and P. H. Harvey. 1989. Life history differences among the eutherian radiations. *J Zool, Lond 219:* 329–353.

Reader, S. M., and K. M. Laland. 2005. Social intelligence, innovation, and enhanced brain size in primates. *Proc Nat Acad Sci USA 99:* 4436–4441.

Reed, D. H., J. J. O'Grady, B. W. Brook, J. D. Ballou, and R. Frankham. 2003. Estimates of minimum viable population sizes for vertebrates and factors influencing those estimates. *Biol Conserv 113:* 23–34.

———. 2004. Large estimates of minimum viable population sizes. *Conserv Biol 18:* 1179.

Refisch, J., and I. Koné. 2005a. Market hunting in the Taï region, Côte d'Ivoire and implications for monkey populations. *Int J Primatol 26:* 621–629.

———. 2005b. Impact of commercial hunting on monkey populations in the Taï region, Côte d'Ivoire. *Biotropica 37:* 136–144.

Regan, T. J., et al. 2005. The consistency of extinction risk classification protocols. *Conserv Biol 19:* 1969–1977.

Remis, M. 1995. Effects of body size and social context on the arboreal activities of lowland gorillas in the Central African Republic. *Amer J Phys Anthropol 97:* 413–433.

Remis, M. J. 1997a. Ranging and grouping patterns of a western lowland gorilla group at Bai Hokou, Central African Republic. *Amer J Primatol 43:* 111–133.

———. 1997b. Western lowland gorillas (*Gorilla gorilla gorilla*) as seasonal frugivores: Use of variable resources. *Amer J Primatol 43:* 87–109.

———. 1999. Tree structure and sex differences in arboreality among western lowland gorillas (*Gorilla gorilla gorilla*) at Bai Hokou, Central African Republic. *Primates 40:* 383–396.

———. 2003. Are gorillas vacuum cleaners of the forest floor? The roles of body size, habitat, and food preferences on dietary flexibility and nutrition. In *Gorilla biology: A multidisciplinary perspective,* eds. A. B. Taylor, and M. L. Goldsmith, 385–404. Cambridge: Cambridge University Press.

Rendall, D. 2004. "Recognizing kin": Mechanisms, media, minds, modules, and muddles. In *Kinship and behavior in primates,* eds. B. Chapais, and C. Berman, 295–316. New York: Oxford University Press.

Reynolds, V. 1965. Some behavioral comparisons between the chimpanzee and the mountain gorilla in the wild. *Amer Anthropol 67:* 691–706.

Richard, A. F., and R. E. Dewar. 1991. Lemur ecology. *Ann Rev Ecol Syst 22:* 145–175.

Rijksen, H. D. 1978. *A field study on Sumatran orangutans (Pongo pygmaeus abelii Lesson 1827): Ecology, behaviour, and conservation.* Wageningen: H. Veenman & Zonen, B.V.

Rijksen, H. D., et al. 1995. Estimates of orangutan distribution and status in Borneo. In *The neglected ape,* eds. R. D. Nadler, B. M. F. Galdikas, L. K. Sheeran, and N. Rosen, 117–128. New York: Plenum Press.

Robbins, A. M., and M. M. Robbins. 2005. Fitness consequences of dispersal decisions for male mountain gorillas (*Gorilla beringei beringei*). *Behav Ecol Sociobiol 58:* 295–309.

Robbins, M. M. 1995. A demographic analysis of male life history and social structure of mountain gorillas. *Behaviour 132:* 21–47.

———. 1996. Male-male interactions in heterosexual and all-male wild mountain gorilla groups. *Ethology 102:* 942–965.

———. 1999. Male mating patterns in wild multimale mountain gorilla groups. *Anim Behav 57:* 1013–1020.

———. 2001. Variation in the social system of mountain gorillas: The male perspective. In *Mountain gorillas: Three decades of research at Karisoke,* eds. M. M. Robbins, P. Sicotte, and K. J. Stewart, 29–58. Cambridge: Cambridge University Press.

———. 2003. Behavioral aspects of sexual selection in mountain gorillas. In *Sexual selection and reproductive competition in primates: New perspectives and directions,* ed. C. B. Jones, 477–501. Norman, OK: American Society of Primatologists.

———. 2007. Gorillas. In *Primates in perspective,* eds. C. J. Campbell, A. Fuentes, K. C. MacKinnon, M. Panger, and S. K. Bearder, 305–320. New York: Oxford University Press.

Robbins, M. M., and A. McNeilage. 2003. Home range and frugivory patterns of mountain gorillas in Bwindi Impenetrable National Park, Uganda. *Int J Primatol 24:* 467–491.

Robbins, M. M., J. B. Nkurunungi, and A. McNeilage. 2006. Variability of the feeding ecology of eastern gorillas. In *Feeding ecology in apes and other primates: Ecological, physical, and behavioral aspects,* eds. G. Hohmann, M. M. Robbins, and C. Boesch, 25–47. Cambridge: Cambridge University Press.

Robbins, M. M., and A. M. Robbins. 2004. Simulation of the population dynamics and social structure of the Virunga mountain gorillas. *Amer J Primatol 63:* 201–223.

Robbins, M. M., A. M. Robbins, N. Gerald-Steklis, and H. D. Steklis. 2005. Long-term dominance relationships in female mountain gorillas: Strength, stability, and determinants of rank. *Behaviour 142:* 779–809.

Robbins, M. M., P. Sicotte, and K. J. Stewart, eds. 2001. *Mountain gorillas: Three decades of research at Karisoke.* Cambridge: Cambridge University Press.

Robbins, M. M., et al. 2004. Social structure and life-history patterns in western gorillas (*Gorilla gorilla gorilla*). *Amer J Primatol 64:* 145–159.

Robinson, J. G., and K. H. Redford. 1989. Body size, diet, and population variation in neotropical forest mammal species: Predictors of local extinction? *Adv Neotrop Mammal 1989:* 567–594.

Robinson, J. G., K. H. Redford, and E. L. Bennett. 1999. Wildlife harvest in logged tropical forests. *Science 284:* 595–596.

Rodman, P. S. 1973. Population composition and adaptive organisation among orang-utans of the Kutai reserve. In *Comparative ecology and behavior of primates,* eds. R. P. Michael, and J. H. Crook, 171–209. London: Academic Press.

———. 1977. Feeding behaviour of orang-utans of the Kutai Nature Reserve, east Kalimantan. In *Primate ecology: Studies of feeding and ranging behaviour in lemurs, monkeys, and apes,* ed. T. H. Clutton-Brock, 384–413. London: Academic Press.

———. 1984. Foraging and social systems of orangutans and chimpanzees. In *Adaptation for foraging in nonhuman primates: Contributions to an organismal biology of prosimians, monkeys, and apes,* eds. P. S. Rodman, and J. G. H. Cant, 134–160. New York: Columbia University Press.

———. 1988a. Resources and group sizes of primates. In *The ecology of social behavior,* ed. C. N. Slobodchikoff, 83–107. New York: Academic Press.

———. 1988b. Diversity and consistency in ecology and behavior. In *Orang-utan biology,* ed. J. H. Schwartz, 31–51. New York: Oxford University Press.

———. 1999. Whither primatology? The place of primates in contemporary anthropology. *Ann Rev Anthropol 28:* 311–339.

———. 2002. Plants of the apes: Is there a hominoid model for the origins of the hominid diet. In *Human diet: Its origin and evolution,* eds. P. S. Ungar, and M. F. Teaford, 77–109. Westport, CT: Bergin & Garvey.

Rodman, P. S., and J. C. Mitani. 1987. Orangutans: Sexual dimorphism in a solitary species. In *Primate societies,* eds. B. B. Smuts, D. L. Cheney, R. M. Seyfarth, R. W. Wrangham, and T. T. Struhsaker, 146–154. Chicago: University of Chicago Press.

Rodseth, L., and R. Wrangham. 2004. Human kinship: A continuation of politics by other means? In *Kinship and behavior in primates,* eds. B. Chapais, and C. M. Berman, 389–419. New York: Oxford University Press.

Rodseth, L., R. W. Wrangham, A. M. Harrigan, and B. B. Smuts. 1991. The human community as a primate society. *Curr Anthrop 32:* 221–241.

Rogers, D. 2005. Congress clears new spending bill; Iraq war costs are covered in $82 billion measure; filibuster fight is signaled. *The Wall Street Journal,* New York, May 11, 2005.

Rogers, M. E., F. Maisels, E. A. Williamson, M. Fernandez, and C. E. G. Tutin. 1990. Gorilla diet in the Lopé Reserve, Gabon: A nutritional analysis. *Oecologia 84:* 326–339.

Rogers, M. E., and E. A. Williamson. 1987. Density of herbaceous plants eaten by gorillas in Gabon: Some preliminary data. *Biotropica 19:* 278–281.

Rogers, M. E., et al. 2004. Western gorilla diet: A synthesis from six sites. *Amer J Primatol 64:* 173–192.

Ron, T. 2005. The Majombe forest in Cabinda: Conservation efforts, 2000–2004. *Gorilla J No. 30, June 2005:* 18–21.

Root, T. L., et al. 2003. Fingerprints of global warming on wild animals and plants. *Nature 421:* 57–60.

Ross, C. 1992. Basal metabolic rate, body weight, and diet in primate: An evaluation of the evidence. *Folia Primatol 58:* 7–23.

Ross, C., and K. E. Jones. 1999. Socioecology and the evolution of primate reproductive rates. In *Comparative primate socioecology,* ed. P. C. Lee, 73–110. Cambridge: Cambridge University Press.

Rothman, J. M., P. J. Van Soest, and A. N. Pell. 2006. Decaying wood is a sodium source for mountain gorillas. *Biology Letters DOI: 10.1098/rsbl.2006.0480.*

Rouse, G. W., S. K. Goffredi, and R. C. Vrijenhoek. 2004. *Osedax:* Bone-eating marine worms with dwarf males. *Science 305:* 668–671.

Rowe, N. 1996. *The pictorial guide to the living primates.* East Hampton, NY: Pogonias Press.

Rubenstein, D. I. 1986. Ecology and sociality in horses and zebras. In *Ecological aspects of social evolution,* eds. D. I. Rubenstein, and R. W. Wrangham, 282–302. Princeton, NJ: Princeton University Press.

———. 1998. Behavioral ecology and conservation policy: On balancing science, applications, and advocacy. In *Behavioral ecology and conservation biology,* ed. T. Caro, 527–553. Oxford: Oxford University Press.

Rubenstein, D. I., and M. Hack. 2004. Natural and sexual selection and the evolution of multi-level societies: Insights from zebras with comparisons to primates. In *Sexual selection in primates: New and comparative perspectives,* eds. P. M. Kappeler, and C. P. van Schaik, 266–279. Cambridge: Cambridge University Press.

Rudran, R. 1973. Adult male replacement in one-male troops of purple-faced langurs (*Presbytis senex senex*) and its effect on population structure. *Folia Primatol 19:* 166–192.

Rudran, R., and E. Fernandez-Duque. 2003. Demographic changes over thirty years in a red howler population in Venezuela. *Int J Primatol 24:* 925–947.

Ruvolo, M. 1997. Molecular phylogeny of the hominoids: Inferences from multiple independent DNA sequence data sets. *Mol Biol Evol 14:* 248–256.

Ruvolo, M., et al. 1994. Gene trees and hominoid phylogeny. *Proc Nat Acad Sci 91:* 8900–8904.

Rylands, A. B., ed. 1993. *Marmosets and tamarins: Systematics, behaviour, and ecology.* Oxford: Oxford University Press.

Sadleir, R. M. F. S. 1969. *The ecology of reproduction in wild and domestic mammals.* London: Methuen.

Sakura, O. 1994. Factors affecting party size and composition of chimpanzees (*Pan troglodytes verus*) at Bossou, Guinea. *Int J Primatol 15:* 167–183.

Sarmiento, E. E. 2003. Distribution, taxonomy, genetics, ecology, and causal links of gorilla survival: The need to develop practical knowledge for gorilla conservation. In *Gorilla biology: A multidisciplinary perspective,* eds. A. B. Taylor, and M. L. Goldsmith, 432–471. Cambridge: Cambridge University Press.

Sarmiento, E. E., T. M. Butynski, and J. Kalina. 1996. Gorillas of Bwindi-Impenetrable forest and the Virunga Volcanoes: Taxonomic implications of morphological and ecological differences. *Amer J Primatol 40:* 1–21.

SAS Institute Inc. 2002. *JMP 5.0.1.2.* Cary, NC: SAS Institute.

Sayer, J. A., and B. M. Campbell. 2004. *The science of sustainable development.* Cambridge: Cambridge University Press.

Sayer, J. A., C. S. Harcourt, and N. M. Collins. 1992. *The conservation atlas of tropical forests. Africa.* London: Macmillan.

Schaefer, H. M., and V. Schmidt. 2002. Vertical stratification and caloric content of the standing fruit crop in a tropical lowland forest. *Biotropica 34:* 244–253.

Schaller, G. B. 1963. *The mountain gorilla: Ecology and behavior.* Chicago: University of Chicago Press.

Schradin, C., and G. Anzenberger. 2002. Why do New World monkey fathers have enhanced prolactin levels. *Evol Anthropol 11 Suppl 1:* 122–125.

Schülke, O. 2005. Evolution of pair-living in *Phaner furcifer. Int J Primatol 26:* 903–919.

Schultz, A. H. 1938. The relative weight of the testes in primates. *Ant Rec 72:* 387–394.

Schürmann, C. L., and J. A. R. A. M. van Hooff. 1986. Reproductive strategies of the orang-utan: New data and a reconsideration of the existing socioecological models. *Int J Primatol 7:* 265–287.

Scott, J., and J. S. Lockard. 2006. Captive female gorilla agonistic relationships with clumped defendable food resources. *Primates 47:* 199–209.

Setchell, J. M., and P. M. Kappeler. 2003. Selection in relation to sex in primates. *Adv St Behav 33:* 87–173.

Seyfarth, R. M. 1976. Social relationships among adult female baboons. *Anim Behav 24:* 917–938.

———. 1977. A model of social grooming among adult female monkeys. *J Theor Biol 65:* 671–698.

———. 1978a. Social relationships among adult male and female baboons. I. Behaviour during sexual consortship. *Behaviour 64:* 204–226.

———. 1978b. Social relationships among adult male and female baboons. II. Behaviour throughout the female reproductive cycle. *Behaviour 64:* 227–247.

Seyfarth, R. M., and D. L. Cheney. 1984. Grooming, alliances, and reciprocal altruism in vervet monkeys. *Nature 308:* 541–543.

———. 2002. What are big brains for? *Proc Nat Acad Sci USA 99:* 4141–4142.

———. 2003. The structure of social knowledge in monkeys. In *Animal social complexity: Intelligence, culture, and individualized societies,* eds. F. B. M. de Waal, and P. L. Tyack, 207–229. Cambridge, MA: Harvard University Press.

Sharma, U. R., and W. W. Shaw. 1993. Role of Nepal Royal Chitwan National Park in meeting the grazing and fodder needs of local people. *Environ Conserv 20:* 139–142.

Shepher, J. 1971. Mate selection among second generation kibbutz adolescents and adults: Incest avoidance and negative imprinting. *Arch Sex Behav 1:* 293–307.

Shi, H., A. Singh, S. Kant, Z. Zhu, and E. Waller. 2005. Integrating habitat status, human population pressure, and protection status into biodiversity conservation priority setting. *Conserv Biol 19:* 1273–1285.

Shields, W. M. 1987. Dispersal and mating systems: Investigating their causal connections. In *Mammalian dispersal patterns: The effects of social structure on population genetics,* eds. B. D. Chepko-Sade, and Z. T. Halpin, 3–24. Chicago: University of Chicago Press.

———. 1993. The natural and unnatural history of inbreeding and outbreeding. In *The natural history of inbreeding and outbreeding: Theoretical and empirical perspectives,* ed. N. W. Thornhill, 143–169. Chicago: University of Chicago Press.

Short, R. V. 1979. Sexual selection and its component parts, somatic and genital selection, as illustrated by man and the great apes. *Adv St Behav 9:* 131–158.

———. 1980. The great apes of Africa. *J Reprod Fertil Suppl 28:* 3–11.

———. 1984. Breast feeding. *Sci Amer 250:* 35–41.

Shultz, S., R. Noë, W. S. McGraw, and R. I. M. Dunbar. 2004. A community-level evaluation of the impact of prey behavioural and ecological characteristics on predator diet composition. *Proc Royal Soc, Biol Sci 271:* 725–732.

Sibly, R. M., D. Barker, M. C. Denham, J. Hone, and M. Pagel. 2005. On the regulation of populations of mammals, birds, fish, and insects. *Science 309:* 607–610.

Sibly, R. M., and R. H. Smith, eds. 1985. *Behavioural ecology: Ecological consequences of adaptive behaviour.* Oxford: Blackwell Scientific.

Sicotte, P. 1993. Inter-group encounters and female transfer in mountain gorillas: Influence of group composition on male behavior. *Amer J Primatol 30:* 21–36.

———. 1994. Effect of male competition on male-female relationships in bi-male groups of mountain gorillas. *Ethology 97:* 47–64.

———. 1995. Interpositions in conflicts between males in bimale groups of mountain gorillas. *Folia Primatol 65:* 14–24.

———. 2000. A case study of mother-son transfer in mountain gorillas. *Primates 41:* 95–103.

———. 2001. Female mate choice in mountain gorillas. In *Mountain gorillas: Three decades of research at Karisoke,* eds. M. M. Robbins, P. Sicotte, and K. J. Stewart, 59–87. Cambridge: Cambridge University Press.

Siegel, S. 1956. *Nonparametric statistics for the behavioral sciences.* Tokyo: McGraw-Hill Kogakusha.

Sievert, J., W. B. Karesh, and V. Sunde. 1991. Reproductive intervals in captive female western lowland gorillas with a comparison to wild mountain gorillas. *Amer J Primatol 24:* 227–234.

Sigg, H., J. Stolba, J. Abegglen, and V. Dasser. 1982. Life history of hamadryas baboons: Physical development, infant mortality, reproductive parameters, and family relationships. *Primates 23:* 473–487.

Silk, J. B. 2002a. Kin selection in primate groups. *Int J Primatol 23:* 849–875.

———. 2002b. The form and function of reconciliation in primates. *Ann Rev Anthropol 31:* 21–44.

———. 2005. Practising Hamilton's rule: Kin selection in primate groups. In *Cooperation in primates and humans: Mechanisms and evolution,* eds. P. M. Kappeler, and C. P. van Schaik, 25–46. Berlin: Springer-Verlag.

Sillén-Tullberg, B., and A. P. Møller. 1993. The relationship between concealed ovulation and mating systems in anthropoid primates—a phylogenetic analysis. *Amer Nat 141:* 1–25.

Simpson, G. G. 1961. *Principles of animal taxonomy.* New York: Columbia University Press.

Simpson, M. J. A. 1973. The social grooming of male chimpanzees: A study of eleven free-living males in the Gombe Stream Reserve, Tanzania. In *Comparative ecology and behaviour of primates,* eds. R. P. Michael, and J. H. Crook, 411–505. London: Academic Press.

Sinclair, A. R. E., S. Mduma, and J. S. Brashares. 2003. Patterns of predation in a diverse predator-prey system. *Nature 425:* 288–290.

Skole, D., and C. Tucker. 1993. Tropical deforestation and habitat fragmentation in the Amazon: Satellite data from 1978 to 1988. *Science 260:* 1905–1910.

Skorupa, J. P. 1986. Responses of rainforest primates to selective logging in Kibale Forest, Uganda: A summary report. In *Primates: The road to self-sustaining populations,* ed. K. Benirschke, 57–70. New York: Springer-Verlag.

Smith, E. A., M. Borgerhoff Mulder, and K. Hill. 2001. Controversies in the evolutionary social sciences: A guide for the perplexed. *Trends Ecol Evol 16:* 128–135.

Smith, R. J., and J. M. Cheverud. 2002. Scaling of sexual dimorphism in body mass: A phylogenetic analysis of Rensch's rule in primates. *Int J Primatol 23:* 1095–1135.

Smith, R. J., and W. L. Jungers. 1997. Body mass in comparative primatology. *J Hum Evol 32:* 523–559.

Smuts, B. B. 1985. *Sex and friendship in baboons.* Hawthorne, NY: Aldine Publishing.

———. 1987. Gender, aggression, and influence. In *Primate societies,* eds. B. B. Smuts, D. L. Cheney, R. M. Seyfarth, R. W. Wrangham, and T. T. Struhsaker, 400–412. Chicago: University of Chicago Press.

Smuts, B. B., D. L. Cheney, R. M. Seyfarth, R. W. Wrangham, and T. T. Struhsaker, eds. 1987. *Primate societies.* Chicago: University of Chicago Press.

Smuts, B. B., and R. W. Smuts. 1993. Male aggression and sexual coercion of females in nonhuman primates and other mammals: Evidence and theoretical implications. *Adv St Behav 22:* 1–63.

Sodhi, N. S., L. P. Koh, B. W. Brook, and P. K. L. Ng. 2004. Southeast Asian biodiversity: An impending disaster. *Trends Ecol Evol 19:* 654–660.

Soltis, J. 2002. Do primate females gain nonprocreative benefits by mating with multiple males? Theoretical and empirical considerations. *Evol Anthropol 11:* 187–197.

———. 2004. Mating systems. In *Macaque societies: A model for the study of social organization,* eds. B. Thierry, M. Singh, and W. Kaumanns, 135–151. Cambridge: Cambridge University Press.

Sommer, V. 2000. The holy wars about infanticide. Which side are you on? and why? In *Infanticide by males and its implications,* eds. C. P. van Schaik, and C. H. Janson, 9–26. Cambridge: Cambridge University Press.

Sommer, V., and L. S. Rajpurohit. 1989. Male reproductive success in harem troops of Hanuman langurs (*Presbytis entellus*). *Int J Primatol 10:* 293–317.

Stacey, P. B., and W. D. Koenig, eds. 1990. *Cooperative breeding in birds.* Cambridge: Cambridge University Press.

Stallmann, R. R., and J. W. Froehlich. 2000. Primate sexual swellings as coevolved signal systems. *Primates 41:* 1–16.

Stamps, J. A. 1988. Conspecific attraction and aggregation in territorial species. *Amer Nat 131:* 329–347.

Stanford, C. B. 1998a. *Chimpanzee and red colobus: The ecology of predator and prey.* Cambridge, MA: Harvard University Press.

———. 1998b. The social behavior of chimpanzees and bonobos: Empirical evidence and shifting assumptions. *Curr Anthropol 39:* 399–420.

———. 2002. Avoiding predators: Expectations and evidence in primate antipredator behavior. *Int J Primatol 23:* 741–757.

———. 2006. The behavioral ecology of sympatric African apes: Implications for understanding fossil hominoid ecology. *Primates 47:* 91–101.

Stanford, C. B., and J. B. Nkurunungi. 2003. Behavioral ecology of sympatric chimpanzees and gorillas in Bwindi Impenetrable National Park, Uganda: Diet. *Int J Primatol 24:* 901–918.

Steenbeek, R. 1996. What a maleless group can tell us about the constraints on female transfer in Thomas's langurs (*Presbytis thomasi*). *Folia Primatol 67:* 169–191.

Steenbeek, R., E. H. M. Sterck, H. De Vries, and J. A. R. A. M. van Hooff. 2000. Costs and benefits of the one-male, age-graded, and all-male phases in wild Thomas's langur groups. In *Primate males: Causes and consequences of variation in group composition,* ed. P. M. Kappeler, 130–145. Cambridge: Cambridge University Press.

Steenbeek, R., and C. P. van Schaik. 2001. Competition and group size in Thomas's langurs (*Presbytis thomasi*): The folivore paradox revisited. *Behav Ecol Sociobiol 49:* 100–110.

Stenseth, N. C. 1983. Causes and consequences of dispersal in small mammals. In *The ecology of animal movement,* eds. I. R. Swingland, and P. J. Greenwood, 63–101. Oxford: Oxford University Press.

Sterck, E. H. M. 1997. Determinants of female dispersal in Thomas langurs. *Amer J Primatol 42:* 179–198.

———. 1999. Variation in langur social organization in relation to the socioecological model, human habitat alteration, and phylogenetic constraints. *Primates 40:* 199–213.

———. 2002. Predator sensitive foraging in Thomas langurs. In *Eat or be eaten: Predator sensitive foraging among primates,* ed. L. E. Miller, 74–91. Cambridge: Cambridge University Press.

Sterck, E. H. M., and J. A. R. A. M. van Hooff. 2000. The number of males in langur groups: Monopolizability of females or demographic processes? In *Primate males: Causes and consequences of variation in group composition,* ed. P. M. Kappeler, 120–129. Cambridge: Cambridge University Press.

Sterck, E. H. M., and A. H. Korstjens. 2000. Female dispersal and infanticide avoidance in primates. In *Infanticide by males and its implications,* eds. C. P. van Schaik, and C. H. Janson, 293–321. Cambridge: Cambridge University Press.

Sterck, E. H. M., and R. Steenbeek. 1997. Female dominance relationships and food competition in the sympatric Thomas langur and long-tailed macaque. *Behaviour 134:* 749–774.

Sterck, E. H. M., D. P. Watts, and C. P. van Schaik. 1997. The evolution of female social relationships in nonhuman primates. *Behav Ecol Sociobiol 41:* 291–309.

Sterck, E. H. M., E. P. Willems, J. A. R. A. M. van Hooff, and S. A. Wich. 2005. Female dispersal, inbreeding avoidance, and mate choice in Thomas langurs (*Presbytis thomasi*). *Behaviour 142:* 845–868.

Stewart, K. J. 1988. Suckling and lactational anoestrus in wild gorillas (*Gorilla gorilla*). *J Reprod Fertil 83:* 627–634.

———. 2001. Social relationships of immature gorillas and silverbacks. In *Mountain gorillas: Three decades of research at Karisoke,* eds. M. M. Robbins, P. Sicotte, and K. J. Stewart, 183–213. Cambridge: Cambridge University Press.

Stewart, K. J., and A. H. Harcourt. 1987. Gorillas: Variation in female relationships. In *Primate societies,* eds. B. B. Smuts, D. L. Cheney, R. M. Seyfarth, R. W. Wrangham, and T. T. Struhsaker, 155–164. Chicago: University of Chicago Press.

———. 1994. Gorillas' vocal behaviour during rest periods: Signals of impending departure. *Behaviour 130:* 29–40.

Stewart, K., A. H. Harcourt, and D. P. Watts. 1988. Determinants of fertility in wild gorillas and other primates. In *Natural human fertility: Social and biological determinants,* eds. P. Diggory, M. Potts, and S. Teper, 22–38. London: Macmillan.

Stewart, K. J., P. Sicotte, and M. M. Robbins. 2001. Mountain gorillas of the Virungas: A short history. In *Mountain gorillas: Three decades of research at Karisoke,* eds. M. M. Robbins, P. Sicotte, and K. J. Stewart, 1–26. Cambridge: Cambridge University Press.

Stockley, P. 2002. Sperm competition risk and male genital anatomy: Comparative evidence for reduced duration of female sexual receptivity in primates with penile spines. *Evol Ecol 16:* 123–137.

Stokes, E. J. 2004. Within-group social relationships among females and adult males in wild western lowland gorillas (*Gorilla gorilla gorilla*). *Amer J Primatol 64:* 233–246.

Stokes, E. J., R. J. Parnell, and C. Olejniczak. 2003. Female dispersal and reproductive success in wild western lowland gorillas (*Gorilla gorilla gorilla*). *Behav Ecol Sociobiol 54:* 329–339.

Strier, K. B. 1999. Why is female kin bonding so rare? Comparative sociality of neotropical primates. In *Comparative primate socioecology,* ed. P. C. Lee, 300–319. Cambridge: Cambridge University Press.

———. 2007. *Primate behavioral ecology.* 3rd. ed. Boston, MA: Allyn and Bacon.

Struhsaker, T. T. 1967. Ecology of vervet monkeys (*Cercopithecus aethiops*) in the Masai-Amboseli Game Reserve, Kenya. *Ecology 48:* 891–904.

———. 2000. The effects of predation and habitat quality on the socioecology of African monkeys: Lessons from the islands of Bioko and Zanzibar. In *Old World monkeys,* eds. P. F. Whitehead, and C. J. Jolly, 393–430. New York: Cambridge University Press.

Stumpf, R. 2005. Does promiscuity preclude choice? Female sexual strategies and mate preferences in chimpanzees of the Taï National Park, Cote d'Ivoire. *Behav Ecol Sociobiol 57:* 511–524.

———. 2007. Chimpanzees and bonobos. In *Primates in perspective,* eds. C. J. Campbell, A. Fuentes, K. C. MacKinnon, M. Panger, and S. K. Bearder, 321–344. New York: Oxford University Press.

Sugardjito, J., I. J. A. te Boekhorst, and J. A. R. A. M. van Hooff. 1987. Ecological constraints on the grouping of wild orang-utans (*Pongo pygmaeus*) in the Gunung Leuser National Park, Sumatra, Indonesia. *Int J Primatol 8:* 17–41.

Sugiyama, Y. 1984. Some aspects of infanticide and intermale competition among langurs, *Presbytis entellus,* at Dharwar, India. *Primates 25:* 423–432.

Sugiyama, Y., and J. Koman. 1979. Social structure and dynamics of wild chimpanzees at Bossou, Guinea. *Primates 20:* 323–339.

Sugiyama, Y., and H. Ohsawa. 1982. Population dynamics of Japanese monkeys with special reference to the effect of artificial feeding. *Folia Primatol 39:* 238–263.

Sussman, R. W., J. M. Cheverud, and T. Q. Bartlett. 1995. Infant killing as an evolutionary strategy: Reality or myth? *Evol Anthropol 3:* 149–151.

Sussman, R. W., and P. A. Garber. 2007. Cooperation and competition in primate social interactions. In *Primates in perspective,* eds. C. J. Campbell, A. Fuentes, K. C. MacKinnon, M. Panger, and S. K. Bearder, 636–651. New York: Oxford University Press.

Sutherland, W. J. 1996. *From individual behaviour to population ecology.* Oxford: Oxford University Press.

———. 1998. The importance of behavioural studies in conservation biology. *Anim Behav 56:* 801–809.

Swedell, L. 2002. Affiliation among females in wild hamadryas baboons (*Papio hamadryas hamadryas*). *Int J Primatol 23:* 1205–1226.

Swenson, J. E., F. S. Arne Söderberg, A. Bjärvall, R. Franzén, and P. Wabakken. 1997. Infanticide caused by hunting of male bears. *Nature 386:* 450–451.

Takahata, Y., M. A. Huffman, S. Suzuki, N. Koyama, and J. Yamagiwa. 1999. Why dominants do not consistently attain high mating and reproductive success: A review of longitudinal Japanese macaque studies. *Primates 40:* 143–158.

Taylor, A. B., and M. L. Goldsmith, eds. 2003. *Gorilla biology: A multidisciplinary perspective.* Cambridge: Cambridge University Press.

te Boekhorst, I. J. A., C. L. Schürmann, and J. Sugardjito. 1990. Residential status and seasonal movements of wild orang-utans in the Gunung Leuser Reserve (Sumatra, Indonesia). *Anim Behav 39:* 1098–1109.

Templeton, C. N., E. Greene, and K. Davis. 2005. Allometry of alarm calls: Black-capped chickadees encode information about predator size. *Science 308:* 1934–1937.

Terborgh, J. 1983. *Five New World primates: A study in comparative ecology.* Princeton, NJ: Princeton University Press.

———. 1999. *Requiem for nature.* Washington, DC: Island Press.

Terborgh, J., and C. H. Janson. 1986. The socioecology of primate groups. *Ann Rev Ecol Syst 17:* 111–135.

Terborgh, J., and C. P. van Schaik. 1987. Convergence vs. nonconvergence in primate communities. In *Organization of communities past and present,* eds. J. H. R. Gee, and P. S. Giller, 185–203. Oxford: Blackwell Scientific.

Terborgh, J., et al. 2001. Ecological meltdown in predator-free forest fragments. *Science 294:* 1923–1926.

Thalmann, O., J. Hebler, H. N. Poinar, S. Paabo, and L. Vigilant. 2004. Unreliable mtDNA data due to nuclear insertions: A cautionary tale from analysis of humans and other great apes. *Mol Ecol 13:* 321–335.

Thierry, B. 2004. Social epigenesis. In *Macaque societies: A model for the study of social organization,* eds. B. Thierry, M. Singh, and W. Kaumanns, 267–290. Cambridge: Cambridge University Press.

———. 2007. The macaques: A double-layered social organization. In *Primates in perspective,* eds. C. J. Campbell, A. Fuentes, K. C. MacKinnon, M. Panger, and S. K. Bearder, 224–239. New York: Oxford University Press.

Thomas, C. D., M. Baguette, and O. T. Lewis. 2000. Butterfly movement and conservation in patchy landscapes. In *Behaviour and conservation,* eds. L. M. Gosling, and W. J. Sutherland, 85–104. Cambridge: Cambridge University Press.

Thomas, R., ed. 2005. Po'o-uli on brink of extinction. *World Birdwatch 27:* 9.

Thornhill, N. W., ed. 1993. *The natural history of inbreeding and outbreeding: Theoretical and empirical perspectives.* Chicago: University of Chicago Press.

Thornhill, R., and J. Alcock. 1983. *The evolution of insect mating systems.* Cambridge, MA: Harvard University Press.

Tinbergen, N. 1963. On aims and methods of ethology. *Z Tierpsychol 20:* 410–433.

Treves, A. 2000. Prevention of infanticide: The perspective of infant primates. In *Infanticide by males and its implications,* eds. C. P. van Schaik, and C. H. Janson, 223–238. Cambridge: Cambridge University Press.

Trivers, R. L. 1972. Parental investment and sexual selection. In *Sexual selection and the descent of man,* ed. B. Campbell, 136–179. London: Heinemann.

Trivers, R. L. 1985. *Social evolution.* Menlo Park, CA: Benjamin/Cummings.

Tsukahara, T. 1993. Lions eat chimpanzees: The first evidence of predation by lions on wild chimpanzees. *Amer J Primatol 29:* 1–11.

Tutin, C. E. G. 1980. Reproductive behaviour of wild chimpanzees in the Gombe National Park, Tanzania. *J Reprod Fertil Suppl 28:* 43–57.

———. 1983. Gorillas feeding on termites in Gabon, West Africa. *J Mammal 64:* 330–331.

———. 1994. Reproductive success story: Variability among chimpanzees and comparisons with gorillas. In *Chimpanzee cultures,* eds. R. W. Wrangham, W. C. McGrew, F. B. M. de Waal, P. G. Heltne, and L. A. Marquardt, 181–193. Cambridge, MA: Harvard University Press.

———. 1996. Ranging and social structure of lowland gorillas in the Lopé Reserve, Gabon. In *Great ape societies,* eds. W. C. McGrew, L. F. Marchant, and T. Nishida, 58–70. Cambridge: Cambridge University Press.

———. 2001. Saving the gorillas (*Gorilla g. gorilla*) and chimpanzees (*Pan t. troglodytes*) of the Congo Basin. *Reprod Fert Dev 13:* 469–476.

Tutin, C. E. G., and M. Fernandez. 1984. Nationwide census of gorilla (*Gorilla g. gorilla*) and chimpanzee (*Pan t. troglodytes*) populations in Gabon. *Amer J Primatol 6:* 313–336.

———. 1985. Foods consumed by sympatric populations of *Gorilla g. gorilla* and *Pan t. troglodytes* in Gabon. *Int J Primatol 6:* 27–43.

———. 1990. Responses of wild chimpanzees and gorillas to the arrival of primatologists: Behaviour observed during habituation. In *Primate responses to environmental change,* ed. H. O. Box, 187–197. London: Chapman & Hall.

———. 1993. Composition of the diet of chimpanzees and comparisons with that of sympatric lowland gorillas in the Lopé Reserve, Gabon. *Amer J Primatol 30:* 195–211.

Tutin, C. E. G., M. Fernandez, M. E. Rogers, E. A. Williamson, and W. C. McGrew. 1991b. Foraging profiles of sympatric lowland gorillas and chimpanzees in the Lopé Reserve, Gabon. *Phil Trans R Soc Lond Ser B 334:* 179–186.

Tutin, C. E. G., R. M. Ham, L. J. T. White, and M. J. S. Harrison. 1997. The primate community of the Lopé Reserve, Gabon: Diets, responses to fruit scarcity, and effects on biomass. *Amer J Primatol 42:* 1–24.

Tutin, C. E. G., and P. R. McGinnis. 1981. Chimpanzee reproduction in the wild. In *Reproductive biology of the great apes: Comparative and biomedical perspectives,* ed. C. E. Graham, 239–264. New York: Academic Press.

Tutin, C. E. G., and R. Oslisly. 1995. *Homo, Pan,* and *Gorilla:* Co-existence over 60,000 years at Lopé in central Gabon. *J Hum Evol 28:* 597–602.

Tutin, C. E. G., R. J. Parnell, L. J. T. White, and M. Fernandez. 1995. Nest building by lowland gorillas in the Lopé Reserve, Gabon: Environmental influences and implications for censusing. *Int J Primatol 16:* 53–76.

Tutin, C. E. G., and A. Vedder. 2001. Gorilla conservation and research in central Africa: A diversity of approaches and problems. In *African rain forest ecology and conservation: An interdisciplinary perspective,* eds. W. Weber, L. J. T. White, A. Vedder, and L. Naughton-Treves, 429–448. New Haven, CT: Yale University Press.

Tutin, C. E. G., E. A. Williamson, M. E. Rogers, and M. Fernandez. 1991a. A case study of a plant-animal relationship: *Cola lizae* and lowland gorillas in the Lopé Reserve, Gabon. *J Trop Ecol 7:* 181–199.

Tutin, C., et al. 2005. *Regional Action Plan for the conservation of chimpanzees and gorillas in western equatorial Africa.* Pp. 36. Conservation International, Washington, DC.

Uchida, A. 1996. What we don't know about great ape variation. *Trends Ecol Evol 11:* 163–168.

Uhde, N. L., and V. Sommer. 2002. Antipredatory behavior in gibbons (*Hylobates lar,* Khao Yai/Thailand). In *Eat or be eaten: Predator sensitive foraging among primates,* ed. L. E. Miller, 268–291. Cambridge: Cambridge University Press.

Utami, S. S., B. Goossens, M. W. Bruford, J. R. de Ruiter, and J. A. R. A. M. van Hooff. 2002. Male bimaturism and reproductive success in Sumatran orang-utans. *Behav Ecol 13:* 643–652.

Utami, S. S., S. A. Wich, E. H. M. Sterck, and J. A. R. A. M. van Hooff. 1997. Food competition between wild orangutans in large fig trees. *Int J Primatol 18:* 909–927.

Utami-Atmoko, S., and J. A. R. A. M. van Hooff. 2004. Alternative male reproductive strategies: Male bimaturism in orangutans. In *Sexual selection in primates: New and comparative perspectives,* eds. P. M. Kappeler, and C. van Schaik, 196–207. Cambridge: Cambridge University Press.

van Hooff, J. A. R. A. M., and C. P. van Schaik. 1992. Cooperation in competition: The ecology of primate bonds. In *Coalitions and alliances in humans and other animals,* eds. A. H. Harcourt, and F. de Waal, 357–389. Oxford: Oxford University Press.

van Noordwijk, M. A., and C. P. van Schaik. 2000. Reproductive patterns of eutherian mammals: Adaptations against infanticide. In *Infanticide by males and its implications,* eds. C. P. van Schaik, and C. H. Janson, 322–360. Cambridge: Cambridge University Press.

van Schaik, C. P. 1983. Why are diurnal primates living in groups? *Behaviour 87:* 120–144.

———. 1989. The ecology of social relationships amongst female primates. In *Comparative socioecology,* eds. V. Standen, and R. A. Foley, 195–218. Oxford: Blackwell Scientific.

———. 1996. Social evolution in primates: The role of ecological factors and male behaviour. In *Evolution of social behaviour patterns in primates and man,* eds. W. G. Runciman, J. Maynard Smith, and R. I. M. Dunbar, 9–31. Oxford: Oxford University Press.

———. 1999. The socioecology of fission-fusion sociality in orangutans. *Primates 40:* 69–86.

———. 2000a. Social counterstrategies against infanticide by males in primates and other mammals. In *Primate males: Causes and consequences of variation in group composition,* ed. P. M. Kappeler, 34–52. Cambridge: Cambridge University Press.

———. 2000b. Infanticide by male primates: The sexual selection hypothesis revisited. In *Infanticide by males and its implications,* eds. C. H. Janson, and C. P. van Schaik, 27–60. Cambridge: Cambridge University Press.

———. 2000c. Vulnerability to infanticide by males: Patterns among mammals. In *Infanticide by males and its implications,* eds. C. P. van Schaik, and C. H. Janson, 61–71. Cambridge: Cambridge University Press.

van Schaik, C. P., and R. O. Deaner. 2002. Life history and cognitive evolution in primates. In *Animal social complexity,* eds. F. B. M. de Waal, and P. L. Tyack. Cambridge, MA: Harvard University Press.

van Schaik, C. P., and R. I. M. Dunbar. 1990. The evolution of monogamy in large primates: A new hypothesis and some crucial tests. *Behaviour 115:* 30–62.

van Schaik, C. P., J. K. Hodges, and C. L. Nunn. 2000. Paternity confusion and the ovarian cycles of female primates. In *Infanticide by males and its implications,* eds. C. P. van Schaik, and C. H. Janson, 361–387. Cambridge: Cambridge University Press.

van Schaik, C. P., and C. H. Janson, eds. 2000. *Infanticide by males and its implications.* Cambridge: Cambridge University Press.

van Schaik, C. P., and P. Kappeler. 1994. Life history, activity period, and lemur social systems. In *Lemur social systems and their ecological base,* eds. P. Kappeler, and J. U. Ganzhorn, 241–260. New York: Plenum Press.

van Schaik, C. P., and P. M. Kappeler. 1997. Infanticide risk and the evolution of male-female association in primates. *Proc Roy Soc, Lond B 1997:* 1681–1694.

———. 2005. Cooperation in primates and humans: Closing the gap. In *Cooperation in primates and humans: Mechanisms and evolution,* eds. P. M. Kappeler, and C. P. van Schaik, 3–21. Berlin: Springer-Verlag.

van Schaik, C. P., and T. Mitrasetia. 1990. Changes in the behaviour of wild long-tailed macaques (*Macaca fascicularis*) after encounters with a model python. *Folia Primatol 55:* 104–108.

van Schaik, C. P., K. A. Monk, and J. M. Y. Robertson. 2001. Dramatic decline in orang-utan numbers in the Leuser ecosystem, northern Sumatra. *Oryx 35:* 14–25.

van Schaik, C. P., G. R. Pradhan, and M. A. van Noordwijk. 2004. Mating conflict in primates: Infanticide, sexual harassment, and female sexuality. In *Sexual selection in primates: New and comparative perspectives,* eds. P. M. Kappeler, and C. P. van Schaik, 131–150. Cambridge: Cambridge University Press.

van Schaik, C. P., and J. A. R. A. M. van Hooff. 1996. Toward an understanding of the orangutan's social system. In *Great ape societies,* eds. W. C. McGrew, L. F. Marchant, and T. Nishida, 3–15. Cambridge: Cambridge University Press.

van Schaik, C. P., and M. A. van Noordwijk. 1985. Evolutionary effect of the absence of felids on the social organization of the macaques on the island of Simeulue (*Macaca fascicularis fusca,* Miller 1903). *Folia Primatol 44:* 138–147.

van Schaik, C. P., M. A. van Noordwijk, and C. L. Nunn. 1999. Sex and social evolution in primates. In *Comparative primate socioecology*, ed. P. C. Lee, 204–240. Cambridge: Cambridge University Press.

van Schaik, C. P., et al. 1995. Estimates of orangutan distribution and status in Sumatra. In *The neglected ape*, eds. R. D. Nadler, B. M. F. Galdikas, L. K. Sheeran, and N. Rosen, 109–116. New York: Plenum Press.

Van Vuren, D. 1998. Mammalian dispersal and reserve design. In *Behavioral ecology and conservation biology*, ed. T. M. Caro, 369–393. New York: Oxford University Press.

Vanleeuwe, H., S. Cajani, and A. Gautier-Hion. 1998. Large mammals at forest clearings in the Odzala National Park, Congo. *Rev Ecol Terre Vie 53:* 171–180.

Vedder, A. L. 1984. Movement patterns of a group of free-ranging mountain gorillas (*Gorilla gorilla beringei*) and their relation to food availability. *Amer J Primatol 7:* 73–88.

Vehrencamp, S. L. 1983. A model for the evolution of despotic versus egalitarian societies. *Anim Behav 31:* 667–682.

Veit, P. G. 1983. Gorilla society. *Nat Hist 91:* 48–59.

Vigilant, L., and B. J. Bradley. 2004. Genetic variation in gorillas. *Amer J Primatol 64:* 161–172.

Vigilant, L., M. Hofreiter, H. Siedel, and C. Boesch. 2001. Paternity and relatedness in wild chimpanzee communities. *Proc Nat Acad Sci USA 98:* 12890–12895.

Vogel, E. R. 2005. Rank differences in energy intake rates in white-faced capuchin monkeys, *Cebus capucinus:* The effects of contest competition. *Behav Ecol Sociobiol 58:* 333–344.

Voltaire. 1770. *Letter to M. le Riche, 6 Feb., 1770.*

Wallace, A. R. 1869. *The Malay Archipelago.* London: Macmillan.

Wallis, J. 1997. A survey of reproductive parameters in the free-ranging chimpanzees of Gombe National Park. *J Reprod Fertil 109:* 297–307.

Walsh, P. D., R. Biek, and L. A. Real. 2005. Wave-like spread of Ebola Zaire. *Pub Lib Sci, Biol 3 (11):* e371, 1–8.

Walsh, P. D., et al. 2003. Catastrophic ape decline in western equatorial Africa. *Nature 422:* 611–614.

Walters, J. R., and R. M. Seyfarth. 1987. Conflict and cooperation. In *Primate societies*, eds. B. B. Smuts, D. L. Cheney, R. M. Seyfarth, R. W. Wrangham, and T. T. Struhsaker, 306–317. Chicago: University of Chicago Press.

Waltham, T. 2005. The Asian tsunami disaster, December 2004. *Geology Today 21:* 22–27.

Walther, G.-R. et al. 2002. Ecological responses to recent climate change. *Nature 416:* 389–395.

Waser, P. M. 1976. *Cercocebus albigena:* Site attachment, avoidance, and intergroup spacing. *Amer Nat 110:* 911–935.

———. 1982. Polyspecific associations: Do they occur by chance? *Anim Behav 30:* 1–8.

Wasser, S. K., and A. K. Starling. 1988. Proximate and ultimate causes of reproductive suppression among female yellow baboons at Mikumi National Park, Tanzania. *Amer J Primatol 16:* 97–121.

Waterman, P. G. 1984. Food acquisition and processing as a function of plant chemistry. In *Food acquisition and processing in primates,* eds. D. J. Chivers, B. A. Wood, and A. Bilsborough, 177–211. New York: Plenum Press.

Waterman, P. J., G. Choo, A. L. Vedder, and D. P. Watts. 1983. Digestability, digestion inhibitors, and nutrients from herbaceous foliage from an African montane flora and its comparison with other tropical flora. *Oecologia 60:* 244–249.

Watson, A., R. Moss, P. Rothery, and R. Parr. 1984. Demographic causes and predictive models of population fluctuations in red grouse. *J Anim Ecol 53:* 639–662.

Watson, L. 2000. Leopard's pursuit of a lone lowland gorilla *Gorilla gorilla gorilla* within the Dzhanga-Sangha Reserve, Central African Republic. *Afr Primates 4:* 74–75.

Watts, D. P. 1984. Composition and variability of mountain gorilla diets in the central Virungas. *Amer J Primatol 7:* 323–356.

———. 1985. Relations between group size and composition and feeding competition in mountain gorilla groups. *Anim Behav 33:* 72–85.

———. 1989a. Infanticide in mountain gorillas: New cases and a reconsideration of the evidence. *Ethology 81:* 1–18.

———. 1989b. Ant eating behavior of mountain gorillas. *Primates 30:* 121–125.

———. 1990a. Ecology of gorillas and its relation to female transfer in mountain gorillas. *Int J Primatol 11:* 21–44.

———. 1990b. Mountain gorilla life histories, reproductive competition, and sociosexual behavior and some implications for captive husbandry. *Zoo Biol 9:* 185–200.

———. 1991a. Strategies of habitat use by mountain gorillas. *Folia Primatol 56:* 1–16.

———. 1991b. Mountain gorilla reproduction and sexual behavior. *Amer J Primatol 24:* 211–225.

———. 1991c. Harassment of immigrant female mountain gorillas by resident females. *Ethology 89:* 135–153.

———. 1992. Social relationships of immigrant and resident female mountain gorillas. 1: Male-female relationships. *Amer J Primatol 28:* 159–181.

———. 1994a. Social relationships of immigrant and resident female mountain gorillas, II: Relatedness, residence, and relationships between females. *Amer J Primatol 32:* 13–30.

———. 1994b. Agonistic relationships between female mountain gorillas (*Gorilla gorilla beringei*). *Behav Ecol Sociobiol 34:* 347–358.

———. 1994c. The influence of male mating tactics on habitat use in mountain gorillas (*Gorilla gorilla beringei*). *Primates 35:* 35–47.

———. 1995. Post-conflict social events in wild mountain gorillas (Mammalia, Hominoidea) I. Social interactions between opponents. *Ethology 100:* 139–157.

———. 1996. Comparative socio-ecology of gorillas. In *Great ape societies,* eds. W. C. McGrew, L. F. Marchant, and T. Nishida, 16–28. Cambridge: Cambridge University Press.

———. 1997. Agonistic interventions in wild mountain gorilla groups. *Behaviour 134:* 23–57.

———. 1998a. Long-term habitat use by mountain gorillas (*Gorilla gorilla beringei*). 1. Consistency, variation, and home range size and stability. *Int J Primatol 19:* 651–680.

———. 1998b. Seasonality in the ecology and life histories of mountain gorillas (*Gorilla gorilla beringei*). *Int J Primatol 19:* 929–948.

———. 1998c. Coalitionary mate guarding by male chimpanzees at Ngogo, Kibale National Park, Uganda. *Behav Ecol Sociobiol 44:* 43–55.

———. 2000a. Causes and consequences of variation in male mountain gorilla life histories and group membership. In *Primate males: Causes and consequences of variation in group composition,* ed. P. M. Kappeler, 169–179. Cambridge: Cambridge University Press.

———. 2000b. Mountain gorilla habitat use strategies and group movements. In *On the move: How and why animals travel in groups,* eds. S. Boinski, and P. A. Garber, 351–374. Chicago: University of Chicago Press.

———. 2001. Social relationships of female mountain gorillas. In *Mountain gorillas: Three decades of research at Karisoke,* eds. M. M. Robbins, P. Sicotte, and K. J. Stewart, 215–240. Cambridge: Cambridge University Press.

———. 2003. Gorilla social relationships: A comparative overview. In *Gorilla biology: A multidisciplinary perspective,* eds. A. B. Taylor, and M. L. Goldsmith, 302–327. Cambridge: Cambridge University Press.

Watts, D. P., and J. C. Mitani. 2002. Hunting and meat sharing by chimpanzees at Ngogo, Kibale National Park, Uganda. In *Behavioural diversity in chimpanzees and bonobos,* eds. C. Boesch, G. Hohmann, and L. F. Marchant, 244–255. Cambridge: Cambridge University Press.

Watts, D. P., M. N. Muller, S. J. Amsler, G. Mbabazi, and J. C. Mitani. 2006. Lethal intergroup aggression by chimpanzees in Kibale National Park, Uganda. *Amer J Primatol 68:* 161–180.

Watts, D. P., and A. E. Pusey. 1993. Behavior of juvenile and adolescent great apes. In *Juvenile primates: Life history, development, and behavior,* eds. M. E. Pereira, and L. A. Fairbanks, 148–167. New York: Oxford University Press.

Weber, A. W. 1987. Socioecologic factors in the conservation of afromontane forest reserves. In *Primate conservation in the tropical rain forest,* eds. C. W. Marsh, and R. A. Mittermeier, 205–229. New York: Alan R. Liss.

———. 1993. Primate conservation and ecotourism in Africa. In *Perspectives on biodiversity: Case studies of genetic resource conservation and development,* eds. C. S. Potter, J. I. Cohen, and D. Janczewski, 129–150. Washington DC: AAAS Press.

Weber, A. W., and A. L. Vedder. 1983. Population dynamics of the Virunga gorillas: 1959–1978. *Biol Conserv 26:* 341–366.

Weladji, R. B., and M. N. Tchamba. 2003. Conflict between people and protected areas within the Bénoué Wildlife Conservation Area, North Cameroon. *Oryx 37:* 72–79.

West, S. A., I. Pen, and A. S. Griffin. 2002. Cooperation and competition between relatives. *Science 296:* 72–75.

Western, D. 1982. Patterns of depletion in a Kenyan rhino population and the conservation implications. *Biol Conserv 24:* 147–156.

Westoby, M., M. R. Leishman, and J. M. Lord. 1995. On misinterpreting the "phylogenetic correction." *J Ecol 83:* 531–534.

White, F. J. 1988. Party composition and dynamics in *Pan paniscus. Int J Primatol 9:* 179–193.

———. 1989. Ecological correlates of pygmy chimpanzee social structure. In *Comparative socioecology,* eds. V. Standen, and R. A. Foley, 151–164. Oxford: Blackwell Scientific.

———. 1996. Comparative socio-ecology of *Pan paniscus.* In *Great Ape Societies,* eds. W. McGrew, L. F. Marchant, and T. Nishida, 29–41. Cambridge: Cambridge University Press.

White, L., and C. E. G. Tutin. 2001. Why chimpanzees and gorillas respond differently to logging: A cautionary tale from Gabon. In *African rain forest ecology and conservation,* eds. A. W. Weber, A. Vedder, and H. Simons Morland, 449–462. New Haven, CT: Yale University Press.

White, F. J., and R. W. Wrangham. 1988. Feeding competition and patch size in the chimpanzee species, *Pan paniscus* and *Pan troglodytes. Behaviour 105:* 148–164.

Whitten, P. L. 1983. Diet and dominance among vervet monkeys. *Amer J Primatol 5:* 139–159.

Whitten, T., D. Holmes, and K. Mackinnon. 2001. Conservation biology: A displacement behavior for academia? *Conserv Biol 15:* 1–3.

Wich, S. A., and E. H. M. Sterck. 2003. Possible audience effect in Thomas langurs (Primates; *Presbytis thomasi*): An experimental study on male loud calls in response to a tiger model. *Amer J Primatol 60:* 155–159.

Wich, S. A., et al. 2004. Life history of wild Sumatran orangutans (*Pongo abelii*). *J Hum Evol 47:* 385–398.

Wildman, D. E., M. Uddin, G. Liu, L. I. Grossman, and M. Goodman. 2003. Implications of natural selection in shaping 99.4% nonsynonymous DNA identity between humans and chimpanzees: Enlarging genus *Homo. Proc Nat Acad Sci USA 100:* 7181–7188.

Wilkie, D. S., and J. F. Carpenter. 1999. Can nature tourism help finance protected areas in the Congo Basin? *Oryx 33:* 332–338.

———. 2000. Bushmeat hunting in the Congo Basin: An assessment of impacts and options for mitigation. In *The apes: Challenges for the 21st. century,* ed. Brookfield Zoo, 212–226. Chicago: Brookfield Zoo. http://www.brookfieldzoo.org/0.asp?nSection515&PageID5195&nLinkID530.

Wilkie, D. S., J. F. Carpenter, and Q. Zhang. 2001. The under-financing of protected areas in the Congo Basin: So many parks and so little willingness-to-pay. *Biodiv Conserv 10:* 691–709.

Wilkie, D. S., J. G. Sidle, and G. C. Boundzanga. 1992. Mechanized logging, market hunting, and a bank loan in Congo. *Conserv Biol 6:* 570–580.

Williams, J., H. Liu, and A. E. Pusey. 2002. Costs and benefits of grouping for female chimpanzees at Gombe. In *Behavioural diversity in chimpanzees and bonobos,* eds. C. Boesch, G. Hohmann, and L. F. Marchant, 192–203. Cambridge: Cambridge University Press.

Williams, J. M., G. W. Oehlert, J. V. Carlis, and A. E. Pusey. 2004. Why do male chimpanzees defend a group range? *Anim Behav 68:* 523–532.

Williamson, E. A., C.E. G. Tutin, M. E. Rogers, and M. Fernandez. 1990. Composition of the diet of lowland gorillas at Lopé in Gabon. *Amer J Primatol 21:* 265–277.

Wilson, E. O. 1975. *Sociobiology.* Cambridge, MA: Belknap Press.

———. 1976. Academic vigilantism and the political significance of sociobiology. *BioScience. 26:* 187–190.

———. 2000. On the future of conservation biology. *Conserv Biol 14:* 1–3.

———. 2002. *The future of life.* New York: Alfred A. Knopf.

Wilson, E. O., and B. Hölldobler. 2005. Eusociality: Origin and consequences. *Proc Nat Acad Sci USA 102:* 13367–13371.

Wilson, M., M. Daly, and S. Gordon. 1998. The evolved psychological apparatus of decision-making is one source of environmental problems. In *Behavioral ecology and conservation biology,* ed. T. M. Caro, 501–523. New York: Oxford University Press.

Wilson, M. L., M. D. Hauser, and R. W. Wrangham. 2001. Does participation in intergroup conflict depend on numerical assessment, range location, or rank for wild chimpanzees? *Anim Behav 61:* 1203–1216.

Wilson, M. L., and R. W. Wrangham. 2003. Intergroup relations in chimpanzees. *Ann Rev Anthropol 32:* 363–392.

Wilson, M. L., W. R. Wallauer, and A. E. Pusey. 2004. New cases of intergroup violence among chimpanzees in Gombe National Park, Tanzania. *Int J Primatol 25:* 523–549.

Winterhalder, B., and E. A. Smith. 2000. Analyzing adaptive strategies: Human behavioral ecology at twenty-five. *Evol Anthropol 9:* 51–72.

Wolf, A. P. 1970. Childhood association and sexual attraction: A further test of the Westermarch hypothesis. *Amer J Phys Anthropol 72:* 503–515.

Wolff, J. O., and D. W. Macdonald. 2004. Promiscuous females protect their offspring. *Trends Ecol Evol 19:* 127–134.

Wolfheim, J. H. 1983. *Primates of the world: Distribution, abundance, and conservation.* Seattle: University of Washington Press.

Wolters, S., and Zuberbühler. 2003. Mixed-species associations of diana and Campbell's monkeys: The costs and benefits of a forest phenomenon. *Behaviour 140:* 371–385.

Woodford, M. H., T. M. Butynski, and W. B. Karesh. 2002. Habituating the great apes: The disease risks. *Oryx 36:* 153–160.

Woodroffe, R., and J. R. Ginsberg. 1998. Edge effects and the extinction of populations inside protected areas. *Science 280:* 2126–2128.

World Resources Institute. 2004. EarthTrends. The Environmental Information Portal. World Resources Institute, 2004. http://earthtrends.wri.org/.

World Resources Institute. 2005. EarthTrends. The Environmental Information Portal. World Resources Institute, 2005. http://earthtrends.wri.org/.

Wrangham, R. W. 1977. Feeding behaviour of chimpanzees in Gombe National Park, Tanzania. In *Primate ecology: Studies of feeding and ranging behaviour in lemurs, monkeys, and apes,* ed. T. H. Clutton-Brock, 503–538. London: Academic Press.

———. 1979. On the evolution of ape social systems. *Soc Sci Inform 18:* 334–368.

———. 1980. An ecological model of female-bonded primate groups. *Behaviour 75:* 262–300.

———. 1982. Mutualism, kinship, and social evolution. In *Current problems in sociobiology,* ed. King's College Sociobiology Group, 269–289. Cambridge: Cambridge University Press.

———. 1986. Ecology and social relationships in two species of chimpanzee. In *Ecological aspects of social evolution,* eds. D. I. Rubenstein, and R. W. Wrangham, 352–378. Princeton, NJ: Princeton University Press.

———. 2000. Why are male chimpanzees more gregarious than mothers? A scramble competition hypothesis. In *Primate males: Causes and consequences of variation in group composition,* ed. P. M. Kappeler, 248–258. Cambridge: Cambridge University Press.

Wrangham, R. 2002. The cost of sexual attraction: Is there a trade-off in female *Pan* between sex appeal and received coercion? In *Behavioural diversity in chimpanzees and bonobos,* eds. C. Boesch, G. Hohmann, and L. F. Marchant, 204–215. Cambridge: Cambridge University Press.

Wrangham, R. W., C. A. Chapman, A. P. Clark-Arcadi, and G. Isabirye-Basuta. 1996. Social ecology of Kanyawara chimpanzees: Implications for understanding the costs of great ape groups. In *Great ape societies,* eds. W. C. McGrew, L. F. Marchant, and T. Nishida, 45–57. Cambridge: Cambridge University Press.

Wrangham, R. W., A. P. Clark, and G. Isabirye-Basuta. 1992. Female social relationships and social organization of Kibale Forest chimpanzees. In *Topics in primatology Vol.1: Human origins,* eds. T. Nishida, W. C. McGrew, P. Marler, M. Pickford, and F. B. M. de Waal, 81–98. Tokyo: University of Tokyo Press.

Wrangham, R. W., N. L. Conklin-Brittain, and K. D. Hunt. 1998. Dietary response of chimpanzees and cercopithecines to seasonal variation in fruit abundance. I. Antifeedants. *Int J Primatol 19:* 949–970.

Wrangham, R. W., J. L. Gittleman, and C. A. Chapman. 1993. Constraints on group size in primates and carnivores: Population density and day range as assays of exploitation competition. *Behav Ecol Sociobiol 32:* 199–209.

Wrangham, R. W., and D. I. Rubenstein. 1986. Social evolution in birds and mammals. In *Ecological aspects of social evolution,* eds. D. I. Rubenstein, and R. W. Wrangham, 452–470. Princeton, NJ: Princeton University Press.

Wrangham, R. W., and B. Smuts. 1980. Sex differences in the behavioral ecology of chimpanzees in the Gombe National Park, Tanzania. *J Reprod Fertil Suppl 28:* 13–31.

Wright, P. C. 1998. Impact of predation risk on the behaviour of *Propithecus diadema edwardsi* in the rain forest of Madagascar. *Behavior 135:* 483–512.

Yamagiwa, J. 1983. Diachronic changes in two eastern lowland gorilla groups (*Gorilla gorilla graueri*) in the Mt. Kahuzi region, Zaire. *Primates 24:* 174–183.

———. 1986. Activity rhythm and the ranging of a solitary male mountain gorilla (*Gorilla gorilla beringei*). *Primates 27:* 273–282.

———. 1987a. Intra- and inter-group interactions of an all-male group of Virunga mountain gorillas. *Primates 27:* 1–30.

———. 1987b. Male life history and the social structure of wild mountain gorillas (*Gorilla gorilla beringei*). In *Evolution and coadaptation in biotic communities,* eds. S. Kawanao, J. H. Connell, and T. Hidaka, 31–51. Tokyo: University of Tokyo Press.

———. 1997. Mushamuka's story: The largest gorilla group and the longest tenure. *Gorilla J 15:* 7–9.

———. 1999a. Socioecological factors influencing population structure of gorillas and chimpanzees. *Primates 40:* 87–104.

———. 1999b. Slaughter of gorillas in the Kahuzi-Biega Park. *Gorilla J 19:* 4–6.

———. 2000. Factors influencing the formation of ground nests by eastern lowland gorillas in Kahuzi-Biega National Park: Some evolutionary implications of nesting behaviour. *J Hum Evol 40:* 99–109.

———. 2003. Bushmeat poaching and the conservation crisis in Kahuzi-Biega National Park, Democratic Republic of the Congo. In *War and tropical forests: Conservation in areas of armed conflict,* ed. S. V. Price, 115–135. New York: Food Products Press.

Yamagiwa, J., and A. K. Basabose. 2006. Diet and seasonal changes in sympatric gorillas and chimpanzees in Kahuzi-Biega National Park. *Primates 47:* 74–90.

Yamagiwa, J., K. Basabose, K. Kaleme, and T. Yumoto. 2003b. Within-group feeding competition and socioecological factors influencing social organization of gorillas in Kahuzi-Biega National Park, Democratic Republic of Congo. In *Gorilla biology: A multidisciplinary perspective,* eds. A. B. Taylor, and M. L. Goldsmith, 328–357. Cambridge: Cambridge University Press.

Yamagiwa, J., A. K. Basabose, K. Kaleme, and T. Yumoto. 2005. Diet of Grauer's gorillas in the montane forest of Kahuzi, Democratic Republic of Congo. *Int J Primatol 26:* 1345–1373.

Yamagiwa, J., and J. Kahekwa. 2001. Dispersal patterns, group structure, and reproductive parameters of eastern lowland gorillas at Kahuzi in the absence of infanticide. In *Mountain gorillas: Three decades of research at Karisoke,* eds. M. M. Robbins, P. Sicotte, and K. J. Stewart, 89–122. Cambridge: Cambridge University Press.

Yamagiwa, J., and J. Kahekwa. 2004. First observations of infanticides by a silverback in Kahuzi-Beiga. *Gorilla J 29:* 6–9.

Yamagiwa, J., J. Kahekwa, and A. K. Basabose. 2003a. Intra-specific variation in social organization of gorillas: Implications for their social evolution. *Primates 44:* 359–369.

Yamagiwa, J., T. Maruhashi, T. Yumoto, and N. Mwanza. 1996. Dietary and ranging overlap in sympatric gorillas and chimpanzees in Kahuzi-Biega National Park, Zaïre. In *Great ape societies,* eds. W. C. McGrew, L. F. Marchant, and T. Nishida, 82–98. Cambridge: Cambridge University Press.

Yamagiwa, J., and N. Mwanza. 1994. Day-journey length and daily diet of solitary male gorillas in lowland and highland habitats. *Int J Primatol 15:* 207–224.

Yamagiwa, J., N. Mwanza, T. Yumoto, and T. Marushashi. 1991. Ant eating by eastern lowland gorillas. *Primates 32:* 247–253.

Yamagiwa, J., et al. 1993. A census of the eastern lowland gorillas *Gorilla gorilla graueri* in Kahuzi-Biega National Park with reference to mountain gorillas *G. g. beringei* in the Virunga region, Zaire. *Biol Conserv 64:* 83–89.

Yerkes, R. M., and A. W. Yerkes. 1929. *The great apes.* New Haven, CT: Yale University Press.

Young, T. P. 1995. Landscape mosaics created by canopy gaps, forest edges, and bush-land glades. *Selbyana 16:* 127–134.

Yu, J., and F. S. Dobson. 2000. Seven forms of rarity in mammals. *J Biogeog 27:* 131–139.

Zuberbühler, K., and D. Jenny. 2002. Leopard predation and primate evolution. *J Hum Evol 43:* 873–886.

Zuberbühler, K., R. Noë, and R. M. Seyfarth. 1997. Diana monkey long-distance calls: Messages for conspecifics and predators. *Anim Behav 53:* 589–604.

AUTHOR INDEX

SUBJECT INDEX

Note: f = figure; t = table not otherwise included in page numbers; b = boxed text.